MERCURY CONTAMINATION: A Human Tragedy
   Patricia A. D'Itri and Frank M. D'Itri

POLLUTANTS AND HIGH RISK GROUPS
   Edward J. Calabrese

METHODOLOGICAL APPROACHES TO DERIVING ENVIRONMENTAL AND
OCCUPATIONAL HEALTH STANDARDS
   Edward J. Calabrese

NUTRITION AND ENVIRONMENTAL HEALTH—Volume I: The Vitamins
   Edward J. Calabrese

NUTRITION AND ENVIRONMENTAL HEALTH—Volume II: Minerals and Macronutrients
   Edward J. Calabrese

SULFUR IN THE ENVIRONMENT, Parts I and II
   Jerome O. Nriagu, Editor

COPPER IN THE ENVIRONMENT, Parts I and II
   Jerome O. Nriagu, Editor

ZINC IN THE ENVIRONMENT, Parts I and II
   Jerome O. Nriagu, Editor

CADMIUM IN THE ENVIRONMENT, Parts I and II
   Jerome O. Nriagu, Editor

NICKEL IN THE ENVIRONMENT
   Jerome O. Nriagu, Editor

ENERGY UTILIZATION AND ENVIRONMENTAL HEALTH
   Richard A. Wadden, Editor

FOOD, CLIMATE AND MAN
   Margaret R. Biswas and Asit K. Biswas, Editors

CHEMICAL CONCEPTS IN POLLUTANT BEHAVIOR
   Ian J. Tinsley

RESOURCE RECOVERY AND RECYCLING
   A. F. M. Barton

QUANTITATIVE TOXICOLOGY
   V. A. Filov, A. A. Golubev, E. I. Liublina, and N.A. Tolokontsev

ATMOSPHERIC MOTION AND AIR POLLUTION
   Richard A. Dobbins

Ecogenetics

Ecogenetics

Genetic Variation in
Susceptibility to
Environmental Agents

Edward J. Calabrese
Division of Public Health
University of Massachusetts
Amherst, Massachusetts

A Wiley-Interscience Publication
JOHN WILEY & SONS
New York   Chichester   Brisbane   Toronto   Singapore

Copyright © 1984 by John Wiley & Sons, Inc.

All rights reserved. Published simultaneously in Canada.

Reproduction or translation of any part of this work beyond that permitted by Section 107 or 108 of the 1976 United States Copyright Act without the permission of the copyright owner is unlawful. Requests for permission or further information should be addressed to the Permissions Department, John Wiley & Sons, Inc.

Library of Congress Cataloging in Publication Data:

Calabrese, Edward J., 1946–
  Ecogenetics: genetic variation in susceptibility to environmental agents.

  (Environmental science and technology, ISSN 0194-0287)
  "A Wiley-Interscience publication."
  Includes index.
  1. Environmentally induced diseases—Genetic aspects. 2. Ecological genetics. I. Title. II. Series.
RB152.C34 1984   616.9'8   83-16957
ISBN 0-471-89112-6

Printed in the United States of America

10 9 8 7 6 5 4 3 2 1

To My Parents

Series Preface
Environmental Science and Technology

The Environmental Science and Technology Series of Monographs, Textbooks, and Advances is devoted to the study of the quality of the environment and to the technology of its conservation. Environmental science therefore relates to the chemical, physical, and biological changes in the environment through contamination or modification, to the physical nature and biological behavior of air, water, soil, food, and waste as they are affected by man's agricultural, industrial, and social activities, and to the application of science and technology to the control and improvement of environmental quality.

The deterioration of environmental quality, which began when man first collected into villages and utilized fire, has existed as a serious problem under the ever-increasing impacts of exponentially increasing population and of industrializing society. Environmental contamination of air, water, soil, and food has become a threat to the continued existence of many plant and animal communities of the ecosystem and may ultimately threaten the very survival of the human race.

It seems clear that if we are to preserve for future generations some semblance of the biological order of the world of the past and hope to improve on the deteriorating standards of urban public health, environmental science and technology must quickly come to play a dominant role in designing our social and industrial structure for tomorrow. Scientifically rigorous criteria of environmental quality must be developed. Based in part on these criteria, realistic standards must be established and our technological progress must be tailored to meet them. It is obvious that civilization will continue to require increasing amounts of fuel, transportation, industrial chemicals, fertilizers, pesticides, and countless other products; and that it will continue to produce waste products of all descriptions. What is urgently needed is a total systems approach to modern civilization through which the pooled talents of scientists and engineers, in cooperation with social scientists and the medical profession, can be focused on the development of order and equilibrium in the presently disparate segments of the human environment. Most of the skills and tools that are needed are already in existence. We surely have a right to hope a technology that has created such manifold environmental problems is also capable of solving them. It is our hope that this Series in Environmental Sciences and Technology will not only serve to make this challenge more explicit to the established professionals, but that it also will help to stimulate the student toward the career opportunities in this vital area.

Robert L. Metcalf
Werner Stumm

Preface

Ecogenetics principally deals with the effects of preexisting genetically determined variability on the response to environmental agents. The word environmental is defined broadly to include the physical, chemical, biological, atmospheric, and climatic agents. Ecogenetics, therefore, is an all-embracing term, and concepts such as pharmacogenetics are seen as subcomponents of ecogenetics.

This work grew logically from my book entitled *Pollutants and High Risk Groups* (1978), which presented an overview of the various host factors (i.e., age, heredity, diet, preexisting diseases, life-style) affecting susceptibility to environmentally induced disease. The present work represents a considerably expanded and updated assessment on how genetic factors affect susceptibility to environmental agents.

As in all rapidly expanding areas, this will certainly not be the last word on the topic but will serve as a conduit for present and future biomedical researchers and policymakers as they approach this field. It should be emphasized that the primary intention of *Ecogenetics* is to provide an objective and critical evaluation of the scientific literature pertaining to genetic factors and differential susceptibility to environmental agents, with particular emphasis on those agents typically considered pollutants.

Chapter 1 provides an overview of the historical foundations of ecogenetics. Chapters 2–13 include an assessment of specific risk conditions according to organ system or function: Chapter 2—red blood cell factors; Chapter 3—liver metabolism; Chapter 4—serum disorders; Chapter 5—cardiovascular diseases; Chapter 6—respiratory disorders; Chapter 7—renal diseases; Chapter 8—immunological disorders; Chapter 9—dermatological conditions; Chapter 10—ocular diseases; Chapter 11—DNA repair diseases; Chapter 12 includes conditions not neatly fitting in the other chapters; Chapter 13 summarizes the principal findings of the previous chapters and offers some future directions.

Research supported by the U.S. Congress Office of Technology Assessment contributed significantly to the conceptualization of parts of this book, and that support is gratefully acknowledged. The views expressed herein, do not, of course, necessarily represent those of the Office of Technology Assessment.

Amherst, Massachusetts
December 1983

EDWARD J. CALABRESE

Contents

1. **INTRODUCTION: THE HISTORY OF ECOGENETICS**   1

2. **RED BLOOD CELL CONDITIONS**   13

    A. Introduction   13
    B. G-6-PD Deficiency   17
    C. Animal Models for G-6-PD Deficiency   35
    D. Sickle-Cell Anemia and Sickle-Cell Trait   39
    E. Animal Models for Sickle-Cell Anemia and Trait   44
    F. The Thalassemias   48
    G. Animal Models for Thalassemias   54
    H. NADH Dehydrogenase Deficiency (i.e., MetHb Reductase Deficiency)   56
    I. Animal Models for the Study of MetHb Formation   59
    J. Deficiencies of Catalase   73
    K. Animal Models for Acatalasemia   78
    L. Effects of Environmental Oxidants on Human Erythrocytes   84
    M. Superoxide Dismutase (SOD)   91
    N. Porphyrias   99
    O. ALA Dehydratase Deficiency   105
    P. Rhodanese Variants   105
    Q. Hb Variants   105

3. **LIVER METABOLISM**   109

    A. Defects in Conjugation Mechanisms   109
    B. Oxidation Center Defects   125
    C. Alcohol Dehydrogenase Variants   130
    D. Gout   138
    E. Animal Models for the Study of Hyperuricemia   142
    F. Sulfite Oxidase Deficiency   146
    G. Paraoxonase Variants   158

xii   Contents

    H. Ornithine Carbamoyl Transferase Deficiency 159
    I. Carbon Disulfide Metabolism 160
    J. Wilson's Disease 162
    K. Metabolic Predisposition to Cigarette Smoking-Induced Bladder Cancer 164

4. SERUM VARIANTS   167

    A. Albumin Variants 167
    B. Pseudocholinesterase Variants 171
    C. Serum α_1-Antitrypsin Deficiency 176
    D. Immunoglobulin A Deficiency 185

5. CARDIOVASCULAR DISEASES   189

    A. Hyperlipidemia 189
    B. Homocystinuria 196
    C. Animal Models for Cardiovascular Disease 200

6. RESPIRATORY DISEASES   209

    A. AHH Inducibility and Lung Cancer 209
    B. Cystic Fibrosis 222
    C. Emphysema 224

7. RENAL DISORDERS   235

    A. Diabetes with Respect to Renal Nephropathy 235
    B. Animal Models for Diabetes Study 249
    C. Genetic Diseases 256

8. IMMUNOLOGICAL ASSOCIATIONS/DEFICIENCIES   263

    A. HLA Associations 263
    B. Susceptibility to Infectious Diseases 265
    C. Immunodeficiency Diseases 274
    D. Susceptibility to Isocyanates 276

Contents   xiii

9. **SKIN DISORDERS**   279

    A. Nonallergic Contact Dermatitis   279
    B. Allergic Contact Dermatitis   283
    C. Genetic Susceptibility to Ultraviolet-Induced Skin Cancer   286
    D. Noncarcinogenic Skin Irritation by Ultraviolet Radiation   292

10. **OCULAR DISORDERS**   297

    A. Genetic Susceptibility to Glaucoma   297
    B. Leber's Optic Atrophy   299

11. **DNA REPAIR AND CHROMOSOME INSTABILITY DISEASES**   301

    A. Xeroderma Pigmentosum   301
    B. Ataxia Telangiectasia   306
    C. Fanconi's Anemia   310
    D. Bloom's Syndrome and Other Disorders   313
    E. Down's Syndrome   315

12. **OTHER GENETIC FACTORS**   319

    A. Susceptibility to Environmentally Induced Goiter   319
    B. Susceptibility to Frostbite   320

13. **SUMMARY AND CONCLUSIONS**   323

INDEX   333

Ecogenetics

Introduction: The History of Ecogenetics

One of the most challenging problems in modern biomedical science is to gain an understanding of the causes of human diseases. The quest for greater understanding has led to the elucidation of the role of microorganisms and chemical and/or physical agents in the disease process. From this study we have acquired not only a greater understanding of our world but also ideas on how to develop (1) specific preventive actions to avoid such diseases in the future as well as (2) therapeutic measures in treating such diseases.

While a legitimately strong emphasis has been placed on identification of the causes of human disease, much less concern has focused on the role of host factors in the disease process itself. Nevertheless, while it is well established that certain agents such as vinyl chloride, for example, may cause cancer, it is also apparent that it is very selective in whom it "chooses" to cause cancer. Why are some people victims yet others are seemingly spared despite apparently similar exposures?

The fact of differential susceptibility to toxic and carcinogenic agents in the population is recognized in other examples as well. For example, reams of public health statistics reveal that cigarette smoking is a major risk factor for the development of lung cancer. Nevertheless, thousands of people have smoked heavily for decades, lived to ripe old ages, and may have died of some disease unrelated to the smoking.

Why then do some people seem to develop a smoking-related cancer while others are bypassed? There is clearly something going on that involves a lot more than just the fact that a chemical has the potential to cause cancer or some other adverse effect. It seems to be able to cause this effect significantly more quickly and efficiently in some but not other people.

What earmarks John B. for cigarette-smoking-induced lung cancer and what makes his close friend Bill M. seem nearly impervious to such an array of chemical insults as delivered by years of heavy cigarette smoking? The answer must be found in John B. and in Bill M. What makes them respond so differently? These are two people who otherwise are alike in age, sex, height, weight, education, job, and hobbies. Yet their risks of cigarette-smoking-induced lung cancer are miles apart.

The causes of differential susceptibility have long been sought. In fact, during World War I it was speculated that TNT-induced adverse health effects were markedly enhanced by consumption of inadequate diets (Viscount, 1917). In 1938, the famous geneticist J. B. S. Haldane suggested a possible role for genetic constitution in the occurrence of bronchitis among potters and even endorsed the possibility of eliminating the genetically predisposed person from potential unhealthy work environments. More specifically, Haldane (1938) stated:

I should be interested to know whether, to take a single example, the death rate among potters from bronchitis is still eight times that of the general population. If so, it would not be unreasonable if a certain proportion of the funds devoted to pottery research at Stoke-on-Trent were spent on research on potters rather than on pots. But while I am sure that our standards of industrial hygiene are shamefully low, it is important to realize that there is a side to this question which has so far been completely ignored. The majority of potters do not die of bronchitis. It is quite possible that if we really understood the causation of this disease, we should find that only a fraction of potters are of a constitution which renders them liable to it. If so, we could eliminate potters' bronchitis by rejecting entrants into the pottery industry who are congenitally disposed to it. We are already making the attempt to exclude accident prone workers from certain trades. The principle could perhaps be carried a good deal further. There are two sides to most of these questions involving unfavourable environments. Not only could the environment be improved, but susceptible individuals could be excluded.

While the study of differential susceptibility to environmental agents has a long history, this field acquired a decided focus when Vogel (1959) coined the term "pharmacogenetics" to describe the interaction of heredity and environment in the pathogenesis of disease and that hereditary drug toxicities could be related to susceptibility or resistance to other diseases. It must be emphasized that genetic factors are but one of a number of host characteristics that may affect one's susceptibility to environmental agents. In addition to one's heredity, age, diet, the presence of other diseases, as well as life-style and behavior may affect susceptibil-

Introduction: The History of Ecogenetics

Table 1-1. Developmental Processes that Enhance Susceptibility to Environmental Pollutants

High-Risk Groups	Estimated Number of Individuals in United States Affected	Pollutant(s) to Which High-Risk Group Is (May Be) at Increased Risk
Developmental Processes		
Immature enzyme detoxification systems	Embryos, fetuses, and neonates to the age of approximately 2–3 months	Pesticides, polychlorinated biphenyls (PCBs)
Immature immune system	Infants and children do not reach adult levels of IgA until the age of 10–12	Respiratory irritants
Deficient immune system as a function of age	Progressive degeneration after adolescence	Carcinogens, respiratory irritants
Differential absorption of pollutants as a function of age	Infants and young children	Barium, lead, radium, strontium
Retention of pollutants as a function of age	Individuals above the age of 50	Fluoride, heavy metals
Pregnancy	Approximately several million females per year in the United States	Anticholinesterase insecticides, carbon monoxide, lead
Circadian rhythms including phase shifts	All people have certain periods of the day when they are more susceptible to challenge	Hydrocarbon carcinogens and probably most other pollutants
Infant stomach acidity	Infants	Nitrates

ity. Lists of examples of environmental agents that one may be at greater risk to because of these additional host factors are given in Tables 1-1–1-3.

Consequently, it is important to realize that one's genetic makeup, while important, is but one of an array of host factors contributing to the overall adaptive capacity of the individual. In many instances, it is possible for such factors to interact in ways that may enhance or offset the effect of each other. For example, a person with a hereditary blood disorder such as glucose-6-phosphate dehydrogenase (G-6-PD) deficiency may

Table 1-2. Partial List of Nutritional Factors Affecting Susceptibility to Environmental Pollutants

Nutritional Deficiency	Estimated Number of Individuals in United States Affected	Pollutant(s) to Which High-Risk Group Is (May Be) at Increased Risk
Vitamin A	25% of children between 7 and 12 have lower than recommended dietary allowance (RDA), or slightly higher percent among the lower income groups.	Aflatoxin, dichlorodiphenyl trichloroethane (DDT), hydrocarbon carcinogens, PCBs
Vitamin C	10–30% of infants, children, and adults of low-income groups receive less than the RDA.	Arsenic, cadmium, carbon monoxide, chromium, DDT, dieldrin, lead, mercury, nitrites, ozone
Vitamin E	7% of the general population are "physiologically deficient"	Lead, nitrite, nitrogen dioxide, ozone
Calcium	65% of children between the ages of 2 and 3 receive less than the RDA	Cadmium, fluoride, lead, strontium
Iron	98% of children between the ages of 2 and 3 receive less than the RDA	Cadmium, hydrocarbon carcinogens, lead, manganese
Magnesium	Most U.S. males have a partial deficiency	Fluoride
Phosphorus	Deficiency in people with various kidney diseases	Lead
Selenium	Unknown; deficiency thought to be rare	Cadmium, mercury, ozone
Zinc	Unknown; deficiency in association with various diseases, but not thought to be widespread	Cadmium, ethanol, lead, nitrosamines
Riboflavin	30% of women and 10% of men aged 30–60 ingest less than ⅔ of the RDA	Hydrocarbon carcinogens, lead, ozone
Dietary protein	10% of women and 5% of men aged 30–60 ingest less than ⅔ of the RDA for protein	DDT and other insecticides, industrial solvents

heighten an already high susceptibility to erythrocyte oxidant stress by consuming a diet inadequate in vitamin E, while consumption of a diet with supplementary vitamin E may mitigate the enhanced predisposition of such G-6-PD deficient erythrocytes to environmental oxidant stress. Consequently, none of these host factors can be thought of as acting in isolation of each other.

In my book *Pollutants and High Risk Groups* (1978), the complemen-

Introduction: The History of Ecogenetics

Table 1-3. Preexisting Conditions

High-Risk Groups	Estimated Number of Individuals in United States Affected	Pollutant(s) to Which High-Risk Group Is (May Be) at Increased Risk
Kidney disease	Relates to genetic diseases, bacterial and virus infections, and hypertensive disease	Excessive sodium in diet, fluoride, lead, other heavy metals
Liver disease	Relates to genetic diseases and virus infections	Carbon tetrachloride, DDT and other insecticides, PCBs
Asthmatic diseases	4,000,000–10,000,000 of the general population	Respiratory irritants: nitrogen dioxide, ozone, sulfates, sulfur dioxide
Chronic respiratory disease	6,000,000–10,000,000 of the general population	Respiratory irritants: nitrogen dioxide, ozone, sulfates, sulfur dioxide
Heart disease	15,000,000 of the general population	Cadmium; carbon monoxide; fluoride; respiratory irritants: ozone, sulfur dioxide; sodium

tary role of the various host factors in affecting the susceptibility to environmental agents was thoroughly assessed and documented. While this complementary and interwoven relationship between the respective host factors needs to be recognized, each host factor remains a major area of study in and of itself. For example, while my original book *Pollutants and High Risk Groups* devoted about 50 pages to how nutritional status affects pollutant toxicity, my two-volume set *Nutrition and Environmental Health: Volume 1—The Vitamins* (1980), and *Volume 2—The Minerals and Macro-Nutrients* (1981), collectively devoted about 1100 pages to that topic. Consequently, this book, like the two nutritional environmental health volumes, represents a considerably expanded analysis of the role of genetic factors in affecting the outcome of environmentally induced disease. However, as emphasized above, this highly specific information, to be properly viewed and understood in terms of human health, must be seen in the total context of other known risk factors as well as within the vast array of environmental stressors to which we are exposed.

Regardless of the limitations of genetic factors in totally explaining the occurrence of differential susceptibility to toxic agents, there has been a growing recognition that genetic influences are important explanatory factors in this regard. The principal impetus to provide a focus to this emerging field and that probably inspired Vogel's term pharmacogenetics, was the earlier work by Motulsky (1957) who reported that specific types of drug toxicity were caused by genetic disorders affecting drug-metabolizing enzymes. He clearly articulated a number of examples of such heredity-based abnormal reactions to drugs and differentiated them from toxic responses by immunological mechanisms, themselves also probably having a genetic component. The genetic conditions, according to Motulsky (1977), which got pharmacogenetics going in the mid-1950s were the occurrence of a pseudocholinesterase variation as the cause of suxamethonium sensitivity and abnormalities in red cell glutathione metabolism as seen in primaquine sensitivity. Soon after these initial and unrelated events were reported, genetic differences to isoniazid were uncovered.

Motulsky (1977) noted that the concept of pharmacogenetics became popular and was consistent with the concurrent emerging development into the ubiquity of biochemical polymorphisms. By 1962, Kalow, the discoverer of the pseudocholinesterase and suxamethonium interaction, published the first book on pharmacogenetics that dealt largely with the resistance of microorganisms and insects to drugs and insecticides, respectively, human genetic conditions that affect drug toxicity such as atypical cholinesterase, isoniazid metabolism, primaquine sensitivity, sensitivity to methemoglobin formation, as well as racial differences in response to drugs. The next year Meier (1963) published a book entitled *Experimental Pharmacogenetics,* which reviewed the genetic factors in animal models that affect drug toxicity and to what extent such information may be useful in animal model development.

In 1963, Price-Evans published an extensive review article entitled "Pharmacogenetics" in the *American Journal of Medicine* in which there was much restatement of the Kalow (1962) report.

While there was great movement toward the unification of genetics with pharmacology as inherent in the term pharmacogenetics, there was also a strong inclination to link the knowledge of genetic predispositions to chemical agents with the field of industrial toxicology and hygiene based on the concepts described in pharmacogenetics. At the time Kalow's (1962) book was published, three articles were also published that suggested the use of biochemical indicators to predict an industrial worker's susceptibility to chemical agents (Stokinger, 1962; Zavon, 1962; Jensen, 1962). This first round of papers, which associated genetic factors with a predisposition for industrial pollutant induced toxicity, was soon followed by reports from Brieger (1963), Mountain (1963), and Stokinger and Mountain (1963). For the most part, these papers displayed considerable overlap. However, the paper of Stokinger and Mountain (1963) noted the work of several European researchers who had linked G-6-PD deficiency with susceptibility to a variety of industrial chemicals (Szeinburg et al., 1959) and thalassemia with lead toxicity (Roche et al., 1960)

Introduction: The History of Ecogenetics

and not just medicinal agents as in Kalow (1962). Finally, not to be overlooked was the publication in 1960 of what is now a classic in the field. *The Metabolic Basis of Inherited Disease,* now in its fourth edition, provided the biochemical foundation to support the occurrence of genetic factors influencing the metabolism of a wide variety of xenobiotics.

In 1968, Stokinger attempted to bridge the gap between the pharmacogenetic perspective with its focus on drugs and the view from industrial hygiene with his article entitled "Pharmacogenetics in the Detection of the Hypersusceptible Worker." Several years later, Brewer (1971), in an editorial in the *American Journal of Human Genetics,* coined the term "ecogenetics." He asserted that genetic variation not only was relevant to drug action but needed to be considered in response to any type of environmental agent.

Throughout this time there had been growing interaction of the medical–pharmacological community and the industrial hygiene sectors with respect to use of the knowledge of genetic factors that affect toxicity to their respective agents. While a major review on pharmacogenetics in 1973 by Vesell extended the scope of present knowledge, it did little to advance a synthesis of the drug-pollutant areas. However, five years later in his review entitled "Pharmacogenetics," Propping (1978) actually contained a subheading entitled "Extension to Ecogenetics," and listed three examples there. At about that same time, an international conference was held in Germany that was titled "Human Genetic Variation in Response to Medical and Environmental Agents: Pharmacogenetics and Ecogenetics." The papers presented at this conference showed considerable movement on the part of the medical–pharmacological community toward developing the concept of ecogenetics.

In the early 1970s, the principal spokesman for genetic predispositions and pollutant toxicity was Dr. Herbert Stokinger, then Chief Toxicologist at the National Institute for Occupational Safety and Health (NIOSH) and the executive secretary for the American Conference of Governmental Industrial Hygienists (ACGIH). During the past decade, *The Metabolic Basis of Inherited Disease* had gone through two revisions and in 1972 it listed more than 150 human genetic diseases. In addition, considerable information had been assembled which indicated that genetic predispositions to emphysema existed as well as to a variety of specific pollutants such as CS_2, isocyanates, and others. This information so impressed many industrial toxicologists that a number of industries felt that such knowledge could be used in preemployment physical examinations as a screening tool. Adding further credibility to this perspective was a 1975 report by the National Academy of Sciences (NAS), which devoted a special section on the role of inborn (genetic) metabolic errors as predisposing factors in the development of toxicity from occupational and environmental pollutants. In their section on genetic disorders, the NAS report (1975) listed 92 human genetic disorders that were adapted from McKusick (1970). Table 1-4 lists these 92 genetic disorders and

Table 1-4. Disorders in Which a Deficient Activity of a Specific Enzyme Has Been Demonstrated in Man

Condition	Enzyme with Deficient Activity
Acatalasia[a]	Catalase
Acid phosphatase deficiency	Lysosomal acid phosphatase
Adrenal hyperplasia I	21-hydroxylase[b]
Adrenal hyperplasia II	11-β-hydroxylase[b]
Adrenal hyperplasia III	3-β-hydroxysteroid dehydrogenase[b]
Adrenal hyperplasia V	17-hydroxylase[b]
Albinism[a]	Tyrosinase
Aldosterone synthesis, defect in	18-hydroxylase[b]
Alkaptonuria	Homogentisic acid oxidase
Angiokeratoma, diffuse (Fabry)	Ceramidetrihexosidase
Apnea, drug-induced[a]	Pseudocholinesterase
Argininemia	Arginase
Argininosuccinic aciduria	Argininosuccinase
Aspartylglycosaminuria	Specific hydrolase (AADC-ase)
Carnosinemia	Carnosinase
Cholesterol ester deficiency (Norum's disease)	Lecithin cholesterol acetyl transferase (LCAT)
Citrullinemia	Argininosuccinic acid synthetase
Crigler-Najjar syndrome[a]	Glucuronyl transferase
Cystathioninuria[a]	Cystathionase
Formininotransferase deficiency	Formininotransferase
Fructose intolerance	Fructose-1-phosphate aldolase
Fructosuria	Hepatic fructokinase[b]
Fucosidosis	Fucosidase
Galactokinase deficiency	Galactokinase
Galactosemia	Galactose-1-phosphate uridyl transferase
Gangliosidosis, generalized	β-galactosidase
Gaucher's disease	Glucocerebrosidase
G-6-PD deficiency[a] (favism, primaquine sensitivity, nonspherocytic hemolytic anemia)	Glucose-6-phosphate dehydrogenase
Glycogen storage disease I	Glucose-6-phosphatase
Glycogen storage disease II	α-1-4-glucosidase
Glycogen storage disease III	Amylo-1-6-glucosidase
Glycogen storage disease IV	Amylo (1-4 to 1-6) transglucosidase
Glycogen storage disease V	Muscle phosphorylase
Glycogen storage disease VI	Liver phosphorylase[b]
Glycogen storage disease VII	Muscle phosphofructokinase
Glycogen storage disease VIII	Liver phosphorylase kinase
Gout, primary (one form)	Hypoxanthine guanine phosphoribosyl transferase
Hemolytic anemia	Diphosphoglycerate mutase
Hemolytic anemia[a]	Glutathione peroxidase
Hemolytic anemia[a]	Glutathione reductase

Table 1-4 (Continued)

Condition	Enzyme with Deficient Activity
Hemolytic anemia[a]	Hexokinase
Hemolytic anemia	Hexosephosphate isomerase
Hemolytic anemia	Phosphoglycerate kinase
Hemolytic anemia	Pyruvate kinase
Hemolytic anemia	Triosephosphate isomerase
Histidinemia	Histidase
Homocystinuria[a]	Cystathione synthetase
Hydroxyprolinemia	Hydroxyproline oxidase
Hyperammonemia I	Ornithine transcarbamylase
Hyperammonemia II	Carbamyl phosphate synthetase
Hyperglycinemia, ketotic form	Propionate carboxylase[b]
Hyperlysinemia	Lysine-ketoglutarate reductase
Hyperoxaluria I glycolic aciduria	2-oxo-glutarate-glyoxylate carboligase
II glyceric aciduria	D-glyceric dehydrogenase
Hyperprolinemia I	Proline oxidase deficiency
Hyperprolinemia II	δ-1-pyrroline-5-carboxylate dehydrogenase[b]
Hypophosphatasia	Alkaline phosphatase
Isovalericacidemia	Isovaleric acid CoA dehydrogenase
Lactase deficiency, adult, intestinal	Lactase
Lactose intolerance of infancy	Lactase
Leigh's necrotizing encephalomyelopathy	Pyruvate carboxylase
Lesch-Nyhan syndrome	Hypoxanthine-guanine phosphoribosyl transferase
Lipase deficiency, congenital	Lipase (pancreatic)
Lysine intolerance	L-lysine:NAD-oxido reductase
Mannosidosis	α-mannosidase
Maple sugar urine disease	Keto acid decarboxylase
Metachromatic leukodystrophy	Arylsulfatase A (sulfatide sulfatide sulfatase)
Methemoglobinemia[a]	NADH-methemoglobin reductase
Methylmalonicaciduria	Methylmalonyl-CoA carboxymutase
Myeloperoxidase deficiency with disseminated candidiasis	Myeloperoxidase
Niemann-Pick disease	Sphingomyelinase
Oroticaciduria	Orotidylic pyrophosphorylase orotidylic decarboxylase
Phenylketonuria[a]	Phenylalanine hydroxylase
Porphyria, congenital[a] erythropoietic	Uroporphyrinogen III cosynthetase
Pulmonary emphysema[a]	α_1-antitrypsin
Pyridoxine dependent	Glutamic acid decarboxylase[b]
Pyridoxine-responsive anemia	δ-aminolevulinic acid synthetase[b]
Refsum's disease	Phytanic acid α-oxidase

Table 1-4 (Continued)

Condition	Enzyme with Deficient Activity
Sarcosinemia	Sarcosine dehydrogenase[b]
Sucrose intolerance	Sucrose, isomaltase
Sulfite oxidase deficiency[a]	Sulfite oxidase
Tay-Sachs disease	Hexosaminidase A
Testicular feminization	Δ^4-5α-reductase[b]
Thyroid hormonogenesis, defect in	Iodothyrosine dehalogenase (deiodinase)
Trypsinogen deficiency disease	Trypsinogen
Tyrosinemia I[a]	Para-hydroxyphenylpyruvate oxidase
Tyrosinemia II[a]	Tyrosine transaminase
Valinemia	Valine transaminase
Vitamin D resistant rickets	Cholecalciferase[b]
Wolman's disease	Acid lipase
Xanthinuria	Xanthine oxidase
Xanthurenic aciduria	Kynureninase
Xeroderma pigmentosa[a]	Ultraviolet specific endonuclease[b]

Source: Adapted from McKusick (1970) and NAS (1975).
[a] Associated with the exacerbation of toxic effects of at least one pollutant (see text).
[b] In some conditions marked in this way (as well as some that are not listed) deficiency of a particular enzyme is suspected, but has not been proved by direct study of enzyme activity.

indicates which of the 92 are thought to predispose the affected individuals to the toxic effects of pollutants.

By the mid-1970s, the U.S. Environmental Protection Agency (EPA), which has the legislative mandate to establish ambient air and drinking water standards, became more interested in groups at increased risk to pollutant effects, especially those with genetic predispositions. While no EPA standards at present specifically are designed to protect an identifiable genetic subsegment of the population, they sponsored a conference in 1978 on pollutants and high-risk groups that included a major section on genetic factors (Calabrese, 1979). At present, the EPA requires that their contractors writing health effects documents include a section on persons at higher than normal risk to the agent in question. In this way at least, the EPA is continually cognizant of genetic or other host factors enhancing risk.

While the issue of genetic factors affecting susceptibility to environmental agents has been primarily a scientific matter with some application of this knowledge by certain industries, this quickly changed in February of 1980, when a four-part story in the *New York Times* by Severo was published. In his story, Severo pointed out the lack of any consistent policy toward genetic screening in the workplace by the federal government, and the apparent inconsistency and lack of accord within certain industries, as well as effectively criticizing the scientific basis of the biomedical evidence supporting industrial genetic screening programs in the first place. This series of articles and the spate of popular articles that followed prompted the U.S. Congress to hold hearings on the scope and validity of genetic screening of industrial workers for predisposition for job-related diseases. As a result of the congressional hearings, the Congress designated its research arm, the Office of Technology Assessment (OTA), to provide a report to Congress that would set forth how the Congress should consider the issue. I was one of the OTA contractors and I provided them with the scientific assessment of the issue.

Other indications of considerable activity in this area may be readily discerned by the fact that the ACGIH sponsored a nationwide conference in 1981 on "Protection of the Hypersensitive Worker," and in the fall of 1983, the Cold Springs Harbor Laboratory, under the direction of Nobel Laureate James Watson, sponsored a conference on genetic factors enhancing susceptibility to chemical agents.

The field of genetic predisposition to environmentally induced disease, therefore, is extraordinarily dynamic and of considerable importance today. Its arms extend not only to the field of medicine, but also to the field of environmental protection, including regulatory programs affecting air, food, and water.

In the past few years, there has been a strong emphasis placed on legal, social, and moral implications of how the knowledge of genetically based differential susceptibility to environmental agents should be used. I can only applaud the efforts that have been made along those lines. However, what is needed today is a consolidation and critical analysis of where the science of genetic susceptibility has taken us over the decade. There has been extraordinary progress during this time because of the recognition not only of scientists but also by the regulatory agencies, industry, the medical profession, and the lay person. Therefore, this field not only can affect social and economic policy for years to come, but also this knowledge could be used on the individual level to possibly reduce one's risk to environmental disease. It is the challenge of the remainder of this work to present the reader with the needed scientific assessment so that policymakers will be provided with a sounder base from which to develop their goals and directions.

REFERENCES

ACGIH (1982). Protection of the susceptible worker. Cincinnati, Ohio.

Brewer, G. J. (1971). Annotations: Human ecology, an expanding role for the human geneticist. *Am. J. Hum. Genet.* **23**: 92–94.

Brieger, H. (1963). Genetic bases of susceptibility and resistance to toxic agents. *J. Occup. Med.* **5**(11): 511–515.

Calabrese, E. J. (1978). *Pollutants and High Risk Groups: The Biological Basis of Enhanced Susceptibility to Environmental and Occupational Pollutants.* Wiley, New York.

Calabrese, E. J. (ed.). (1979). Conference on Pollutants and High Risk Groups. *Environ. Health Perspect.* **29**: 1–77.

Calabrese, E. J. (1980). *Nutrition and Environmental Health: The Influence of Nutritional Status on Pollutant Induced Toxicity and Carcinogenicity. Volume 1—The Vitamins; Volume 2—The Minerals and Macronutrients.* Wiley, New York.

Haldane, J. B. S. (1938). *Heredity and Politics.* George Allen and Unwin, London, pp. 179–180.

Jensen, W. N. (1962). Hereditary and chemically induced anemia. *Arch. Environ. Health* **5**: 212–216.

Kalow, W. (1962). *Pharmacogenetics: Heredity and the Response to Drugs.* Saunders, Philadelphia.

McKusick, V. A. (1970). Human genetics. *Annu. Rev. Genet.* **4**: 1.

Meier, H. (1963). *Experimental Pharmacogenetics.* Academic Press, New York.

Motulsky, A. (1957). Drug reactions, enzymes, and biochemical genetics. *J. Am. Med. Assoc.* **165**: 835–837.

Motulsky, A. G. (1977). Ecogenetics: Genetic variation in susceptibility to environmental agents. In: *Human Genetics,* S. Armandares and R. Lisker (eds.). Excerpta Medica, Amsterdam, pp. 375–385.

Mountain, J. T. (1963). Detecting hypersusceptibility to toxic substances. *Arch. Environ. Health* **6**: 357.

NAS (1975). Special risk due to inborn error of metabolism. *Principles for Evaluating Chemicals in the Environment.* National Academy of Sciences, Washington, D.C.

Omenn, G. S. (1974). Ecogenetics: Host variability in health effects of environmental agents. *Proc. Int. Symp. Health Effects Environ. Pollution.* Paris. June 24–28, Abst. 170.

Omenn, G. S. (1982). Predictive identification of hypersusceptible individuals. *J. Occup. Med.* **24**(5): 369–374.

Omenn, G. S. and Motulsky, A. G. (1978). Ecogenetics: Genetic variation in susceptibility to environmental agents. In: *Genetic Issues in Public Health and Medicine.* Charles C Thomas, Springfield, Illinois, pp. 85–111.

Price-Evans, D. A. (1963). Pharmacogenetics. *Am. J. Med.* **34**: 639–662.

Propping, P. (1978). Pharmacogenetics. *Rev. Physiol. Biochem. Pharmacol.* **83**: 124–173.

Roche, L., Lejeune, E., Tolot, F., Mouriquand, C., and Baron, M. M. (1960). Lead poisoning and thalassemia. *Arch. Mal. Prof.* **21**: 329.

Severo, R. (1980). Genetic tests by industry raise questions on rights of workers. *New York Times,* Feb. 3–6.

Stokinger, H. E. (1962). New concepts and future trends in toxicology. *Am. Ind. Hyg. Assoc. J.* **23**: 8.

Stokinger, H. E. and Mountain, J. T. (1963). Test for hypersusceptibility to hemolytic chemicals. *Arch. Environ. Health* **6**: 495–502.

Stokinger, H. E., Mountain, J. T., and Scheel, L. D. (1968). Pharmacogenetics in the detection of the hypersusceptible worker. *Ann. N.Y. Acad. Sci.* **151**: 968–976.

Szeinburg, A., Adams, A., Myers, F., Sheba, C., and Ramot, B. (1959). A hematological survey of industrial workers with enzyme-deficient erythrocytes. *Arch. Indust. Health* **20**: 510.

Vesell, E. S. (1973). Advances in pharmacogenetics. *Prog. Med. Genet.* **9**: 291–367.

Viscount, C. (1917). In discussion on the origin, symptoms, pathology, treatment and prophylaxis of toxic jaundice observed in munitions workers. *Proc. R. Soc. Med.* **10**: 6.

Vogel, F. (1959). Moderne probleme de humangenetik. *Ergeb. Inn. Med. Kinderheilk.* **12**: 52–125.

Zavon, M. R. (1962). Modern concepts of diagnosis and treatment in occupational medicine. *Am. Ind. Hyg. Assoc. J.* **23**: 30.

Red Blood Cell Conditions

A. Introduction

1. HEREDITARY BLOOD DISORDERS AND HEMOLYTIC SUSCEPTIBILITY

There is a broad group of genetic diseases that result in either producing or predisposing afflicted individuals to the development of hemolytic anemias. For example, these diseases include conditions in which (1) abnormal hemoglobin (Hb) occurs (sickle cell), (2) inability to manufacture one or another of the peptide globin chains of Hb occurs (thalassemia), and (3) particular enzyme deficiencies of the Embden-Meyerhoff and/or hexose monophosphate shunt metabolic pathways (e.g., G-6-PD) as well as other cellular enzyme deficiencies (e.g., catalase) may occur.

The state of knowledge concerning these diseases is quite variable with considerable information known about sickle-cell disease, G-6-PD deficiency, and thalassemia, but little is known about most of the enzyme deficiencies of the glycolytic pathway. Special emphasis is given here to sickle-cell disease, G-6-PD deficiency, and thalassemia since these are the most frequently occurring hereditary blood diseases and, as previously stated, those about which most is known. However, a brief consideration of the less frequently occurring and less understood enzyme deficiency conditions associated with hemolytic anemias will be provided. At this time, it must be emphasized that

Table 2-1. Enzyme Deficiencies Associated with Hemolysis

Enzyme Deficiency	Frequency	Genetics
A. Embden–Meyerhoff Pathway		
1. Hexokinase (HK)	Rare	Most likely autosomal recessive
2. Glucose phosphate isomerase (GPI)	Rare	Autosomal recessive
3. Phosphofructokinase	Rare	
4. Triosephosphate isomerase (TPI)	Rare	Autosomal recessive
5. 2,3-Diphosphoglyceromutase (2,3-DPGM)	Rare	Thought to be autosomal recessive
6. Phosphoglycerate kinase (PGK)	Rare	X-chromosome linkage
7. Pyruvate kinase (PK)	Most frequent next to G-6-PD	Autosomal recessive
B. Hexose Monophosphate (HMP) Shunt Pathway		
1. G-6-PD	Common	X chromosome
2. 6-Phosphogluconate dehydrogenase (6-PGD)	Rare	Autosomal recessive
C. Nonglycolytic Pathway		
1. GSH reductase[a]	Uncertain	Uncertain
2. GSH peroxidase	Rare	Uncertain
3. GSH synthetase[b] Reaction II	Rare	Thought to be autosomal recessive
4. ATPase	Unknown	Unknown

Source: Valentine, W. N. and Tanaka, K. (1972). In: *Metabolic Basis of Inherited Disease*, McGraw-Hill, New York, pp. 1338–1339.

[a]Levels of GSH reductase are known to be influenced by riboflavin and FAD in the diet (Beutler, 1969).

[b]Individuals with this deficiency are known to be sensitive to "oxidant-type" drugs such as primaquine (Carson et al., 1961).

although it is suggested that individuals with such genetic deficiencies may be hypersusceptible to the action of some hemolytic agents, research in this area is generally lacking. However, because of their theoretical potential for causing hemolysis under particular "oxidant" stress conditions and for the sake of an overall perspective in this area, the following summary of glycolytic enzyme deficiencies is provided in Table 2-1.

In order to facilitate an understanding of subsequent sections concerning several of the hereditary blood disorders, a brief review of how the normal red blood cell as well as the hemoglobin molecule functions will follow.

2. NORMAL RED BLOOD CELL FUNCTIONING

The developing red blood cell has the complete metabolic potential for replication, differentiation, and maintenance. However, upon emergence into the reticulocyte stage, the red blood cell becomes nonnucleated and has only limited capacity for protein and lipid synthesis and the incorporation of iron into Hb. The next stage, or the mature red blood stage, is also nonnucleated and possesses no DNA, RNA, mitochondria, or other intracellular organelles and only relatively nonfunctional remnants of Krebs cycle metabolism. The mature red blood cell can manufacture no additional protein and thus, no new enzymes. The catalytic proteins each have their own biologic half-life and decay at different rates as the cell ages. However, the mature red blood cell will synthesize certain compounds such as glutathione (GSH), nicotinamide adenine dinucleotide (NAD), flavin-adenine dinucleotide (FAD), and adenosine triphosphate (ATP) (Valentine and Tanaka, 1972).

The main function of the immature red blood cell is to produce Hb. The immature red cell becomes "mature" just shortly after entering the circulatory system. The mature red cell functions to transport oxygen and carbon dioxide. To carry out these activities, it is necessary to have functional Hb molecules and structural integrity of the cell. A functional Hb molecule requires an intracellular reducing mechanism to ensure the constant reduction of methemoglobin (MetHb) to Hb. The mechanism involves the regeneration of NADH and nicotinamide adenine dinucleotide phosphate (NADPH). To maintain the structural integrity of the cell, it is necessary to keep the proper electrolyte gradient across the cell membrane, and this depends on the availability of energy or ATP (Valentine and Tanaka, 1972; Friedman, 1971).

The volume and thickness of the red blood cell relies on the control of ion transport across the cell membrane. To maintain the proper volume and osmotic balance in light of the osmotic pressure by nondiffusible cellular constituents including Hb and some glycolytic intermediates, the red cell must carry on active transport of cations. For example, sodium is transported into the cell "passively" along an electrochemical gradient and is subsequently "pushed" out of the cell via the expenditure of cellular energy. If glycolysis is prevented and if high energy phosphate molecules (ATP) are not produced, or if the rate of passive entry exceeds the capacity of active transport to balance the flow, then the vital regulation of cations and thus, membrane stability will be disrupted (Jandl and Cooper, 1972).

The red cell needs a source of energy production for synthesis activity, maintenance of proper ion concentration gradients, and the reduction of MetHb, which is constantly formed. Because the mature red blood cell has a reduced capacity to use oxygen and is unable to use pyruvate, glucose is primarily metabolized via the anaerobic (Embden-Meyer-

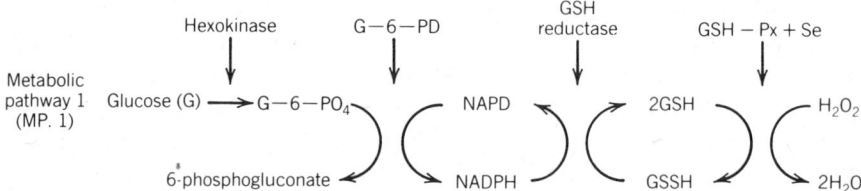

Figure 2-1. Mechanism by which reduced GSH helps maintain membrane stability following "oxidant" stress.

2. HEMOLYTIC AGENTS

Individuals with the G-6-PD deficiency were first recognized as medically important when they exhibited extreme hemolytic sensitivity of erythrocytes on administration of certain antimalarial drugs such as primaquine and pamaquine. Other substances subsequently were definitely shown to cause clinically significant hemolytic anemia in G-6-PD deficients (Table 2-2). An even longer list of substances that may also cause acute hemolysis in G-6-PD deficients exists. However, in contrast to those listed in Table 2-2, it is thought that those in Table 2-3 can primarily be given safely in therapeutic doses. Theoretically, any chemical compound or its metabolite that can accept hydrogen in the defective red blood cell may be suspected as a possible hemolytic agent (Stokinger and Scheel, 1973; Jensen, 1962; Stokinger and Mountain, 1963).

3. INDUSTRIAL EXPOSURES

With respect to industrial practices, it has been two decades since the first suggestion that G-6-PD deficiency may help to explain worker susceptibility to chemically induced anemia (Brieger, 1963; Jensen, 1962; Zavon, 1962).* In 1963, Stokinger and Mountain proposed a partial list of 37 industrial chemicals known to cause hemolysis to which those with a G-6-PD deficiency may be at enhanced risk. They further suggested that screening tests to identify G-6-PD deficients be conducted as part of the preemployment medical examination in order to identify these individuals before job placement. By 1967, Stokinger and Mountain reported that more than 15 industries, research centers, or health-oriented groups were either using the G-6-PD test or had inquired into its use. More specifically, they noted that industries finding the test useful were manufacturers of dyes and dyestuff intermediates, and metals, especially lead

---

*Nutritional status of vitamins C and E may also affect the susceptibility of G-6-PD deficients to oxidant stressor agents (Winterbourne, 1979; Corash et al., 1980; Calabrese, 1978).

Industrial Exposures

Table 2-2. Drugs and Chemicals that Have Clearly Been Shown to Cause Clinically Significant Hemolytic Anemia in G-6-PD Deficiency

Acetanilid	Pentaquine
Methylene blue	Sulfanilamide
Nalidixic acid (Negram)	Sulfacetamide
Naphthalene	Sulfapyridine
Niridazole (Ambilhar)	Sulfamethoxazole (Gantanol)
Nitrofurantoin (Furadantin)	Thiazolesulfone
Phenylhydrazine	Toluidine blue
Primaquine	Trinitrotoluene (TNT)
Pamaquine	

Source: Beutler (1978).

and drugs. In documenting support for their advocacy of G-6-PD deficiency screening, Stokinger and Mountain (1963) cited the report of Szeinburg et al. (1959) who found 25 of 241 male Oriental Jewish industrial workers to have a G-6-PD deficiency (inferred from the GSH stability test). These 25 G-6-PD deficient workers were employed in textile-dyeing plants (9), pharmaceutical factories (2), munitions plants (11), and a rubber-tire factory (3). Of the 25 G-6-PD deficient workers only one

Table 2-3. Drugs that Can Probably Safely Be Given in Normal Therapeutic Doses to G-6-PD Deficient Subjects (without Nonspherocytic Hemolytic Anemia)

Acetaminophen (Paracetamol, Tylenol, Tralgon, Hydroxyacetanilid)	Menapthone
	p-Aminobenzoic acid
Acetophenetidine (phenacetin)	Phenylbutazone
Acetylsalicylic acid (aspirin)	Phenytoin
Aminopyrine (Pyramidone, Amidopyrine)	Probenecid (Benemid)
	Procaine amide hydrochloride (Pronestyl)
Antazoline (Antistine)	
Antipyrine	Pyrimethamine (Daraprim)
Ascorbic acid (vitamin C)	Quinidine
Benzhexol (Artane)	Quinine
Chloramphenicol	Streptomycin
Chlorguanidine (Proguanil, Paludrine)	Sulfacytine
	Sulfadiazine
Chloroquine	Sulfaguanidine
Colchicine	Sulfamerazine
Diphenylhydramine (Benedryl)	Sulfamethoxypyriazine (Kynex)
Isoniazide	Sulfisoxazole (Gantrisin)
L-Dopa	Trimethoprim
Menadione sodium bisulfite (Hykinone)	Tripelennamine (Pyribenzamine)
	Vitamin K

Source: Beutler (1978).

Table 2-4. Thirteen Chemicals to which G-6-PD Deficient Human Erythrocytes are Sensitive

Textile Dyeing Plant
  β-naphthol,[a] aniline,[a] β-oxynaphtholic acid
Pharmaceutical Factory
  Nitrofurantoin, acetylsalicylic acid, menadione sodium bisulfite
Munitions Plant
  Nitroglycerin, TNT, potassium chlorate
Rubber-Tire Factory
  Tetraethyl thiuram disulfide, benzothiazyl disulfite, mercaptobenzothiazole, N-cyclo-hexyl-2-benzothiazole sulfonamide

[a]Zinkham and Childs (1950) and Beutler (1957) had earlier reported G-6-PD deficient erythrocytes were sensitive to these substances.

displayed any hypersusceptibility to exposure (TNT),* which ceased after transferral to another department. This led the authors to conclude that there was no proof of an increased hemolytic hazard in G-6-PD deficient workers in the four types of factories mentioned above. Stokinger and Mountain (1963) were quick to point out that no exposure data were presented, thereby precluding any quantitative risk assessment. While the G-6-PD deficient workers in general did not appear to experience outward effects from occupational exposures at their jobs, Szeinberg et al. (1959) noted that *in vitro* studies revealed that G-6-PD deficient erythrocytes were considerably more sensitive to 13 agents commonly used in the respective four types of plants (Table 2-4). Even though Szeinberg et al. (1959) reported the amount of agent per milliliter of blood, it is not known what this blood level may mean in terms of airborne levels. This would be an important factor for industrial substances shown to be more sensitive to G-6-PD deficient erythrocytes (as measured by the GSH stability test).

Of revelance to the development of occupational exposure standards was a subsequent report by Linch (1974) that G-6-PD deficient workers were more susceptible to chemical (i.e., aromatic nitro and amino compounds) induced cyanosis than normal individuals. As a result of the recognition that G-6-PD deficients were at increased susceptibility to developing cyanosis from the aromatic nitro and amino compounds, DuPont adopted a preemployment screening program to detect such persons. Despite the concern that G-6-PD deficients may be at greater risk to over 30 hemolytic agents used in industry, the documentation of the ACGIH

---

*Note that exposure to TNT has also been associated with hemolytic occurrences in at least six additional G-6-PD deficient workers (see Larizza et al., 1960; Djerassi and Vitany, 1975). In the cases of several workers, hemolysis reoccurred on reexposure to TNT. According to Beutler (1978), the temporal relationship between TNT exposure and the occurrence of severe hemolysis strongly supports the hypothesis.

threshold limit value (TLV) states that for only two substances (i.e., carbon monoxide and naphthalene) the TLV may not be sufficient to protect those with hereditary blood disorders (e.g., those with a G-6-PD deficiency). No evidence was presented by the ACGIH as to why those with hereditary blood disorders may be at potentially increased risk at or below the TLV.

4. ENVIRONMENTAL OXIDANTS

a. Copper

In addition to medical and industrial oxidants as being potential causes of hemolytic anemia in G-6-PD deficients, interest has recently become focused on identifying the effects of environmental oxidants on G-6-PD deficient erythrocytes. The NAS (1977) has speculated that since erythrocytes of the dorset sheep are G-6-PD deficient and are also quite susceptible to copper induced hemolysis, that G-6-PD deficient humans may likewise display an enhanced susceptibility to copper. Subsequent *in vitro* studies by Calabrese et al. (1980) have supported this hypothesis.

b. Ozone

A 1977 hypothesis by Calabrese et al. proposed that G-6-PD deficients may be at enhanced hemolytic risk to ozone toxicity. This hypothesis was derived as follows.

Buckley et al. (1975) have reported that inhaled ozone causes several physical and biochemical changes (e.g., increased lytic sensitivity and decreased GSH levels) affecting the membrane stability of red blood cells of normal human individuals.

Individuals with a G-6-PD enzyme deficiency have been reported to have significantly lower levels of reduced GSH as compared with "normal" individuals. The levels of GSH in whole blood of normal white adults range from 53 to 84 mg % (Zinkham et al., 1958). In contrast, individuals with G-6-PD deficiency have whole blood GSH levels from 38 to 51 mg % (Beutler et al., 1955).

Glutathione is a tripeptide of cysteine, glutamic acid, and glycine that constitutes over 95% of the reduced nonprotein sulfhydryl compounds in the erythrocyte (Beutler et al., 1955). The presence of GSH is necessary to maintain the stability of sulfhydryl-containing enzymes. Reduced GSH is also bound to at least one enzyme, is a cofactor for another, and protects hemoglobin, several enzymes, and coenzymes from oxidation (Tarlov et al., 1962).

Statistical associations based on animal studies have been used to establish a relationship between the rate of spontaneous hemolysis and the degree of GSH oxidation. Felger (1952), using horse blood, indicated that there is no marked increase in the rate of hemolysis until the GSH levels decrease to approximately 40% of its mean initial value.

In addition to these animal studies, human data reveal that a reduction in the level of GSH in whole blood of approximately 34–60% below normal (i.e., in individuals with no G-6-PD deficiency) is frequently associated with the precipitation of acute hemolytic anemia (Zinkham et al., 1958). The basis of the human data is summarized as follows:

1. G-6-PD-deficient individuals usually have only 60–70% of the total GSH of normal individuals when both groups are not under "oxidant" drug stress (Beutler et al., 1955).
2. Acute hemolytic anemia in G-6-PD-deficient individuals treated with primaquine (30 mg) daily is associated with a 14–20% reduction in GSH levels (Tarlov et al., 1962).

It should be made clear that we are not saying that the reduction in the levels of reduced GSH is the cause of hemolysis. Other mechanisms for red blood cell destruction resulting from oxidant stress have been suggested, including the action of hydrogen peroxide (H_2O_2), lipid peroxidation, the toxic effects of GSSG, binding of Heinz bodies to the membrane, and others (Beutler, 1972). Reduced GSH levels are being used here as an indicator of membrane stability and hemolysis onset, and not necessarily the causal agent.

Biochemical research has shown that hemolytic agents produce H_2O_2 in intact red blood cells and that the oxidative effects seen in hemolysis (e.g., loss of GSH) are indications of the presence of H_2O_2 (Cohen and Hochstein, 1964). Further studies have demonstrated that atmospheric ozone causes reduced GSH to be irreversibly oxidized (Menzel, 1971) and that ozone also produces H_2O_2 in the whole blood (Goldstein, 1973). Incubation of erythrocytes with ozone likewise causes hemolytic responses in the red blood cells (Goldstein and Balchum, 1967).

Previously cited studies by Buckley et al. (1975) concerning the effects of ozone on "normal" males have established that a 0.5 ppm exposure for $2\frac{3}{4}$ hr effects a 14% decrease in GSH values. Concomitant with reduced GSH levels, these individuals have demonstrated homeostatic compensatory adaptation by increasing the activity of G-6-PD by 20%. Such an adaptation assists in stabilizing GSH levels. These adaptive "stabilizing" abilities, so evident in normal individuals, are notably diminished in the G-6-PD-deficient individuals.

Thus, if one attempts to determine the percentage of reduction in GSH

Environmental Oxidants

that would have resulted if a G-6-PD-deficient individual had been exposed to the identical amount of ozone, several assumptions are required:

1. The mechanism of action of the drugs used in the glutathione "stability" test affects the GSH levels in an identical fashion as ozone.
2. There is a dose-response relationship between ozone exposure and a hemolytic response.

After exposure to oxidant drugs during the GSH stability test (Beutler, 1957), levels of GSH in normal individuals drop by approximately 20%, at which time the levels of GSH stabilize (Zinkham et al., 1958). However, under identical conditions the GSH level of a G-6-PD-deficient individual does not stabilize and often decreases more than 80% and may even reach 100% depletion (0.0 mg %). Clinical tests have indicated that whole blood from patients with drug-induced hemolytic anemias show GSH levels equal to or less than 20 mg % (Zinkham et al., 1958). The principal reason for this diminished capacity to stabilize GSH levels is the deficiency of the G-6-PD enzyme.

As a result of these data and the above-mentioned assumptions, it is calculated that a 14% decrease in the GSH levels in "normals," as occurred after 0.5 ppm ozone (or 0.4 ppm according to Los Angeles County analytical methods) exposure for $2\frac{3}{4}$ hr is approximately equal to a 56% reduction in G-6-PD-deficient individuals. Consequently, a G-6-PD-deficient individual with a typical GSH level of 40 mg % and exposed to 0.5 ppm ozone for the $2\frac{3}{4}$-hr time period could, based on these calculations and assumptions, show a GSH level of approximately 18 mg %. Levels of GSH equal to or below 20 mg % in the GSH stability test are usually present in G-6-PD-deficient individuals and are associated with the onset of acute hemolysis.

Thus, it is predicted that individuals with a G-6-PD deficiency should be considered at high risk to the hemolytic action of breathable ozone. Furthermore, the model developed here predicts that a G-6-PD-deficient individual may experience an acute hemolytic crisis following less than 3 hr of ozone exposure at 0.5 ppm.

This hypothesis has been difficult to evaluate in an adequate manner since appropriate predictive animal models have not been available (see the next section, which deals with animal models for the high risk group—G-6-PD deficiency). In addition, epidemiological evaluations are usually considerably more expensive and have not been funded by EPA. Nevertheless, recent studies from my laboratory have supported the above hypothesis by *in vitro* experiments that revealed that G-6-PD-deficient erythrocytes were more susceptible to oxidant damage as measured by increases in methemoglobin caused by three possible toxic

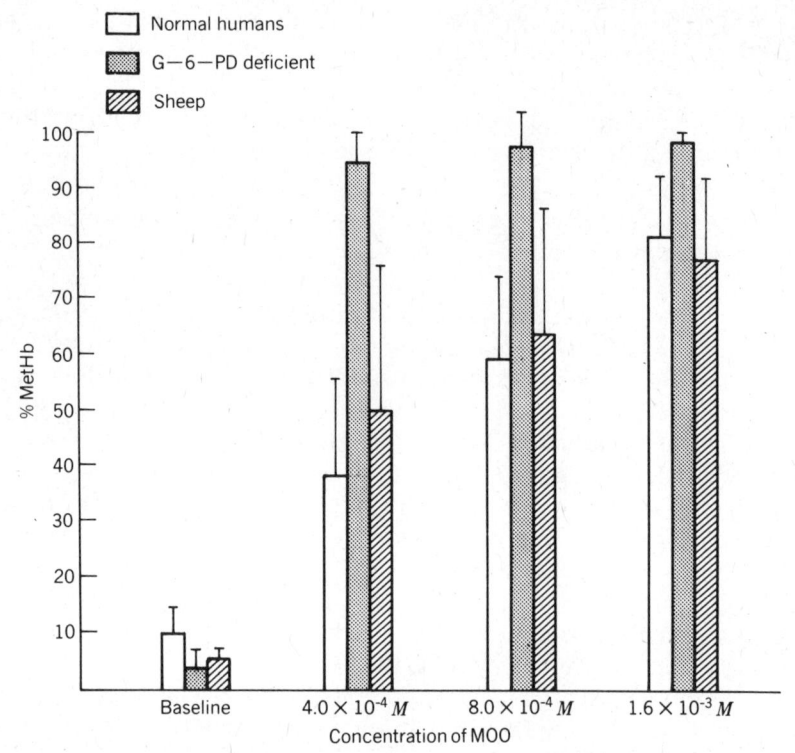

Figure 2-2. The effects of methyl oleate ozonide (MOO) on the erythrocyte percent of MetHb levels of normal and G-6-PD deficient humans and dorset sheep (*in vitro*).

ozone intermediates [i.e., methyl oleate ozonide (Figure 2-2), methyl oleate hydroperoxide, methyl linoleate hydroperoxide] than erythrocytes of normal humans (Calabrese et al., 1982). However, it must be emphasized that it is not definitely known whether any of the above-mentioned "possible toxic ozone intermediates" are actually formed and, if so, how much is formed following ambient ozone exposures and what their biological half-lives are. Nevertheless, such compounds are reasonable candidates for possible intermediates, being consistent with the theoretical foundation for systemic ozone toxicity formulated by Menzel et al. (1975).

c. Chlorite and Related Agents

Greater susceptibility of G-6-PD deficient erythrocytes than normal human red cells to other environmental oxidants (i.e., nitrite, chlorite, and paraquat) was also demonstrated *in vitro* (Calabrese et al., 1980a, 1980b; Moore and Calabrese, 1980). Of particular note in this regard was the striking decrease in reduced GSH levels in G-6-PD-deficient red cells as a result of chlorite incubation. This finding with chlorite was of additional significance since chlorite is an end product of chlorine dioxide (ClO_2) disinfection of potable waters.

The use of chlorine has been almost universally accepted as the method of choice for disinfecting potable water supplies. However, recent studies have demonstrated that the interaction of chlorine with various organic substances in the water results in the formation of trihalomethanes (EPA, 1975; Rook, 1976) that may be carcinogenic (EPA, 1975, 1977; Tardiff, 1977). Therefore, alternative disinfective methods need to be explored. The use of ClO_2 is considered as one method that would reduce the formation of trihalomethanes (EPA, 1977; Stevens et al., 1976). A review of the literature, however, reveals very little published on the human health effects of ClO_2 in potable water supplies. These available reports reveal limited physiological effects from concentrations of ClO_2 as high as 0.7 mg/L (Enger, 1960). However, end products of ClO_2 disinfection of potable waters include chlorate and chlorite (EPA, 1977), and these substances have potential adverse effects to human health (Musil et al., 1964; Richardson, 1937; Ross, 1925).

Two of the primary products resulting from ClO_2 disinfection of surface waters include chlorates (Miltner, 1977) and chlorites (EPA, 1977; Miltner, 1977), with chlorites appearing in concentrations of up to 50% of the ClO_2 demand from pH 4.8 to 9.75, and chlorates increasing from 10 to 30% of the ClO_2 demand at pH increases over the same range (Miltner, 1977). The end products from a 1.5 mg/L ClO_2 dose at pH 7.1 of coagulated, settled, and filtered Ohio river water yielded a final chlorite (ClO_2^-) concentration of 0.72 mg/L and chlorate (ClO_3^-) concentration of 0.41 mg/L after 42 hr of contact (Miltner, 1977). Although the levels of chlorite and chlorate are likely to vary with different water supplies and treatment conditions, it is evident that chlorite and chlorate can exist in significant amounts in ClO_2 dosed water supplies. Consequently, it is important to consider the potential health effects of these chemical species.

The oral administration of chlorates has been shown to produce methemoglobinemia (Richardson, 1937; Jung, 1947), with blood destruction and nephritis in acute poisoning. Richardson (1937) suggested that similar changes might develop in chronic poisoning with manifestations of anemia, uremia, and evidence of nephritis.

Chlorite and hypochlorite are thought to oxidize hemoglobin more rapidly than chlorate (Heubner and Jung, 1941) and may even be involved in the formation of MetHb by chlorate (Heubner, 1949). One investigator reported that the toxicity of chlorite might be similar to chlorate and reported that subjects sensed a "weak furry feeling" after drinking sodium chlorite concentrations of more than 0.3 mg/L (Hopf, 1965), and laboratory technicians drinking a liter of water with a residual of 0.7 mg/L of chlorite experienced nausea and stomach discomfort. According to Musil and Knotek (1965), experiments with laboratory rats showed the sodium chlorite LD$_{50}$ to be 140 mg/kg of live body weight and, furthermore, rats drinking Berounka (Germany) river water dosed with 1.75 mg ClO$_2$/L showed significant differences in both the ratio of liver weight to body weight and hematocrit value. Further, Musil and Knotek (1965) demonstrated chlorite to be a powerful producer of MetHb in rats and recommended that water for consumption contain no ClO$_2^-$ (0.0 mg/L) since it might prove toxic to neonates. It is on the basis of this work that the Norwegian Health Authority recommends a residual chlorite concentration of 0.0 mg/L in potable water supplies.

The formation of MetHb by chlorite is described as an autocatalytic reaction (Koransky, 1952), in that the product of the reaction catalyzes the formation of more product. Similarly, chlorate oxidizes Hb to form MetHb after a short lag phase (Kunzer and Saffer, 1954), and this MetHb then catalyzes the oxidation of more Hb by chlorate (Bodansky, 1951) such that an S-shaped curve typical of an autocatalytic reaction develops. The rate of MetHb formation by chlorate is nearly proportional to the square root of the chlorate concentration (Jung and Kuon, 1951). Addition of MetHb also accelerates the oxidation of Hb by chlorite (Koransky, 1952), indicating a similar mechanism to the autocatalytic oxidation of Hb by chlorate.

The mechanism by which chlorite oxidizes Hb to MetHb has not been fully explained (Kiese, 1974). However, the mechanism of MetHb formation by nitrite has been explained in more detail (Wallace and Caughey, 1975), and chlorite may be similar in this activity because of its other similarities to nitrite in physiologic behavior and atomic structure.

As a result of the public health significance of the potential impact of the product of ClO$_2$ disinfection on persons with a G-6-PD deficiency, controlled exposure studies were undertaken by Lubbers et al. (1981, 1983). They administered chlorite at a concentration of 5 mg/L daily for 12 consecutive weeks with 500 mL of solution given to each volunteer. The investigators reported "no obvious undesirable clinical sequellae by any of the participating subjects. . . ." In several cases, statistically significant trends in certain biochemical or physiological parameters (e.g., MetHb) were associated with treatment; however, none of these trends were judged to have physiological consequences. It should be pointed out that G-6-PD deficiency was defined in their study on the basis of a G-6-PD level of less than 5.0 I.U./g Hb. This is not a particularly low G-6-PD cutoff value with many A-variants being markedly below this value. Mediterranean variant values would be in the range of 1.0 I.U./g Hb. Those with the more extreme deficiency conditions are known to be at greater risk to oxidant stress. Consequently, it is reasonable to assume that there may be considerable variation in the response of G-6-PD deficient individuals to chlorite. It was unfortunate that Lubbers et al. (1981) only gave a cutoff value for G-6-PD activity and not the actual values of the individuals involved in the study, because this makes one treat the G-6-PD deficients as one group when the specific variant of each volunteer needs to be rated for a proper interpretation of the data. Finally, limited epidemiological investigations of the effect of ClO_2 disinfection on residents of Bethesda, Ohio revealed significant deleterious red blood cell effects on only one person of the study and that individual was G-6-PD deficient. While one person's response does not prove that G-6-PD deficients are indeed a real high-risk group to this alternative disinfectant, such data clearly suggest the need for future investigations in this area (Michael et al., 1981).

d. Fungicide

Hemolytic anemia has been reported in a G-6-PD deficient Persian Jew who was exposed to the fungicide zinc ethylene bisdithiocarbamate. However, since the patient also had severe hypocatalasemia and had been given the known hemolytic agent, methylene blue, Beutler (1978) has concluded that the cause-and-effect relationship between fungicide and hemolytic anemia remains to be demonstrated. Further testing by Pinkhas et al. (1963) revealed that the fungicide was able to drastically reduce the levels of GSH in G-6-PD deficients but not normal individuals in the glutathione stability test, thereby adding support to its hemolytic capability in G-6-PD deficients.

e. Lead

Lead exposure has been shown to cause a modest reduction in GSH of lead workers (Stokinger and Mountain, 1967) as well as in *in vitro* experimental conditions (Howard, 1974). Such associations have led McIntire and Angle (1972) to suggest that G-6-PD deficiency (i.e., A-variant) may be a factor enhancing the occurrence of lead induced anemia. In fact, a 1965 report by Ganzoni and Rhomberg (1965) detailed the occurrence of a hemolytic crisis in one G-6-PD deficient patient with lead intoxication. However, no difference in the hemoglobin levels was found among G-6-PD deficients and normal black children who were exposed to lead as a result of their proximity to a battery factory (McIntire and Angle, 1972).

5. CANCER SUSCEPTIBILITY

In 1965, Beaconsfield et al. reported an inverse relationship between cancer mortality and G-6-PD deficiency based on a study of occidental and oriental Jews. More specifically, researchers reported that Ashkenazi Jews (occidental) had a markedly lower incidence of G-6-PD deficiency than oriental Jews (i.e., the French North Africans, the Yemeni, and the Kurdish Jews). However, the age-standardized mortality rates from malignant neoplasms in these two groups were markedly different with the populations having a higher incidence of G-6-PD deficiency displaying a lower cancer incidence. In 1968, Sulis and Spano reported a comparable association in Sardinians while Kessler (1970) offered evidence of a relationship between cancer, G-6-PD activity and diabetes, such that the negative association between cancer and diabetes, especially in males, might be explained partially by G-6-PD activity and the positive association between diabetes and G-6-PD deficiency. These initial associations were further supported by Naik and Anderson (1971) who reported that of 241 American Black cancer patients the number of those with an erythrocyte G-6-PD deficiency was considerably less (4.5% for males and 0.57% for females) than in a control group (9.4% for males and 3.8% for females).

Just why people with a G-6-PD deficiency would be at less risk of developing a cancer was not convincingly explained by the authors of these earlier studies. However, recent *in vitro* studies by Feo and colleagues in Italy have offered an interesting insight as to how this decreased cancer risk may come about. Using fibroblasts they postulated that since G-6-PD deficients exhibit decreased activity of the hexose monophosphate shunt and NADPH/NADP+ ratio and since the mixed function monooxygenase system is affected by G-6-PD activity and the NADPH/NADP+ ratio, that those with a deficiency may not be able to bioactivate precarcinogens as effectively as normal individuals. These investigators evaluated the role of G-6-PD activity as a modulator of the metabolism of several chemical carcinogens by comparing the toxic and transforming effects of dimethylnitrosamine (DMN), BP, and methylnitrosourea (MNU) in normal and G-6-PD deficient cells. They found that deficient fibroblasts exhibit an increased resistance to cell death and growth inhibition induced by DMN and BP, substances whose activation goes through the NADPH-dependent epoxide formation. However, no differences were found in the toxic effect of MNU, a carcinogen not requiring activation.

These findings provide for the first time a biological basis for the epidemiological findings which suggested that G-6-PD deficient individuals may have a decreased cancer risk. Further support for this hypothesis is found in a long-term, peroral treatment with dehydroepiandrosterone (DEA), a potent inhibitor of G-6-PD which reduced the formation of spontaneous mammary cancer in mice (Schwartz, 1979) and reduced aflatoxin B1 and 7,12-dimethylbenzanthracene induced cytotoxicity and transformation in cultured cells. Further research concerning cancer risks of G-6-PD deficient individuals is clearly warranted.

6. FREQUENCY IN THE POPULATION

Numerous surveys of G-6-PD deficiency have been conducted among various groups of people in different geographical locations employing different methods of identification. Beutler (1978) has made a very comprehensive comparison of such surveys amounting to a 20-page table. The frequency of this trait is very high among U.S. black males, being about 16% (Beutler, 1978), and of concern with respect to the U.S. working force. Caucasian population frequencies of this trait are: American, 0.1%; British, 0.1%; Greek, 1–2%; Scandinavians, 1–8%; Indians from India, 0.3%; Mediterranean Jews, 11%; European Jews, 1%. Mongolian population frequencies are: Chinese, 2–5%; Filipinos, 12–13%. There are numerous genetic variants of the G-6-PD deficiency. Of particular importance in this regard is the Mediterranean variant in which G-6-PD activity is only 1–8% of normal compared to the A–variant of the American black males who display G-6-PD activity of about 15–25% of normal. The greater severity of the enzyme deficiency is of clinical concern since the Mediterranean variant is more likely to be considerably more susceptible to oxidant stressor agents and experience a more serious hemolytic crisis (Beutler, 1978).

7. SCREENING TESTS

Numerous screening tests have been designed to identify the G-6-PD deficient erythrocytes. The most simple, reliable, and specific screening procedure for this enzymatic defect is the fluorescent spot test (Beutler, 1966). This procedure has been widely employed (McIntire and Angle, 1972; Yeung et al., 1970; Szeinberg and Peled, 1973; and Dow et al., 1974), and has been shown to be highly reliable in detecting the deficiency. It has also been automated and made available for wide screening (Tan and Whitehead, 1969; Dickson et al., 1973). The tests are specific, and detection of G-6-PD levels lower than about 50% normal is achieved. In a study of 20,810 black Navy and Marine Corps recruits in which 2632 (12.6%) were found to be G-6-PD deficient, total agreement between the automated and manual methods was reported (Uddin et al., 1974). Similar validation of the automated system with the reference manual procedure was also established with complete agreement on 10,000 samples, 12.1% of which were G-6-PD deficient (Dickson et al., 1973).

Despite the remarkable agreement between manual and automated G-6-PD deficiency screening tests, there are possible sources of false positives and negatives. Samples with total hemoglobin concentrations below 7 g/100 mL are likely to give false positive indications of G-6-PD deficiency regardless of the true enzyme activity because of too few cells, thus too little enzyme. If anemia is severe, special modifications in procedure are available to circumvent this potential source of error. A false negative finding for the deficiency is likely to occur in the cases of recent hemolysis or hemorrhage. These situations cause a stimulation of erythropoiesis, thereby causing an enhanced production of young erythrocytes with high G-6-PD activity. This temporarily higher enzyme activity is what results in the false negative conclusion. The automated screening methodology of Tan and Whitehead (1969) allows investigators to identify false positives caused by a deficiency in G-6-PD in the presence of normal G-6-PD. Since the methodology utilizes a measurement of the fluorescence emitted by NADPH generated by both G-6-PD and 6-PGD, this capability to differentiate these respective enzymes is of some importance. While the above methodology detects deficient healthy males (hemizygotes) and homozygous females, it does not very reliably detect heterozygous females.

8. SUMMARY

It has been shown that G-6-PD-deficient individuals are at enhanced hemolytic risk to a number of oxidant drugs and industrial chemicals. Many substances commonly used in industry are known to cause hemolytic changes and have also been speculated to present an enhanced risk to G-6-PD deficients. A limited subset of these substances has been evaluated *in vitro* via the GSH stability test and found to display a greater oxidant stress on G-6-PD-deficient cells. However, the number of specific industrial substances for which reasonably definitive proof exists that G-6-PD deficients are at greater hemolytic risk than normal is limited to certain aromatic amino and nitro compounds (e.g., naphthalene and TNT, respectively). No adequate quantitative risk assessments of the hemolytic actions of these substances on G-6-PD deficients have been published, although it would appear that this information may be available for the aromatic nitro and amino compounds in the paper of Linch (1974) at Dupont. What is emerging is a growing body of *in vitro* evidence strongly implicating the enhanced susceptibility of G-6-PD-deficient erythrocytes to a wide variety of industrial and environmental oxidants. Unfortunately, it has not been possible to relate such *in vitro* exposures to real world exposures. What is quite evident is the nearly total lack of epidemiologic studies designed to assess the impact of industrial and environmental oxidants in G-6-PD deficients. Furthermore, since hemolytic crises may be precipitated in deficients by numerous commonly used medications and bacterial infection, it would be of great interest to assess the interaction of these factors in combination with industrial hemolytic compounds known to present at least an *in vitro* enhanced risk to G-6-PD deficient erythrocytes.

REFERENCES

Albahary, C. (1944). L'origine medullaire des hematies ponctuees et l'origine cytoplasmique des ponctuations basophiles. *Sang* **16**: 341–346.

American Conference of Governmental Industrial Hygienists. Documentation of the TLVs. (1977). Cincinnati, Ohio.

Batolska, A. and Marinova, H. (1970). Modifications du glutathion chez les travnilleurs d'une entreprise metallurgique miniere. *Arch. Mal. Prof. Med. Trav. Secur. Soc.* **31**: 117–222.

Beaconsfield, P., Rainsbury, R. and Kalton, G. (1965). G-6-PD deficiency and the incidence of cancer. *Oncologia* **19**: 11–19.

Beutler, E. (1957). The glutathione instability of drug-sensitive red cells: A new method for the *in vitro* detection of drug sensitivity. *J. Lab. Clin. Med.* **49**: 84.

Beutler, E. (1959). The hemolytic effect of primaquine and related compounds: A review. *Blood* **14**: 103.

Beutler, E. (1966). A series of new screening procedures for pyruvate kinase deficiency, G-6-PD deficiency and glutathione reductase deficiency. *Blood* **28**: 553–562.

Beutler, E. (1972). Glucose-6-phosphate dehydrogenase deficiency. In: *The Metabolic Basis of Inherited Disease*, 3rd ed. J. B. Stanbury, J. B. Wyngaarden, and D. S. Fredrickson (eds.). McGraw-Hill, New York, pp. 1358–1388.

Beutler, E. (1978). *Hemolytic Anemia in Disorders of Red Cell Metabolism*. Plenum Medical, New York, pp. 23–167.

Beutler, E., Dern, R. J., and Alving, A. S. (1955). The hemolytic effect of primaquine. VII. Biochemical studies of drug sensitive erythrocytes to primaquine. *J. Lab. Clin. Med.* **45**: 286–295.

Bodansky, O. (1951). Methemoglobinemia and methemoglobin-producing compounds. *Pharmacol. Rev.* **3**: 144.

Brieger, H. (1963). Genetic bases of susceptibility and resistance to toxic agents. *J. Occup. Med.* **5**: 511–514.

Buckley, R. D., Hackney, J. D., Clark, K., and Posin, C. (1975). Ozone and human blood. *Arch. Environ. Health* **30**: 40–43.

Calabrese, E. J. (1978). *Pollutants and High Risk Groups*. Wiley, New York.

Calabrese, E. J. (1981). Genetic screening for hypersusceptibles in industry. *Med. Hypotheses* **7**: 393–400.

Calabrese, E. J., Kajola, W., and Carnow, B. W. (1977). Ozone: A possible cause of hemolytic anemia in G-6-PD deficient individuals. *J. Toxicol. Environ. Health* **2**: 709–712.

Calabrese, E. J., Moore, G. S., and Ho, S-C. (1980). Low G-6-PD activity in human and sheep red blood cells and susceptibility to copper induced oxidative damage. *Environ. Res.* **21**: 366–372.

Calabrese, E. J., Moore, G. S., and Ho, S-C. (1980a). Low G-6-PD activity and increased sensitivity to paraquat toxicity. *Bull. Environ. Contam. Toxicol.* **24**: 369–373.

Calabrese, E. J., Moore, G., and Ho, S-C. (1980b). Low erythrocyte G-6-PD activity and susceptibility to nitrite-induced methemoglobin formation. *Bull. Environ. Contam. Toxicol.* **26**: 837–840.

Calabrese, E. J., Moore, G. S., and Williams, P. (1982). The effects of three proposed toxic ozone intermediates on G-6-PD deficient humans. Presented at the Society of Toxicology Annual Meeting, Boston, Massachusetts.

Cohen, G. and Hochstein, P. (1964). Generation of hydrogen peroxide in erythrocytes by hemolytic agents. *Biochem.* **3:** 895–900.

Cooper, W. C. (1973). Indicators of susceptibility to industrial chemicals. *J. Occupat. Med.* **15:** 355–359.

Corash, L., Spielberg, S., Bartsocas, C., Boxer, L., Steinherz, R., Sheetz, M., Egan, M., Schlessleman, J., and Schulman, J. D. (1980). Reduced chronic hemolysis during high dose vitamin E administration in Mediterranean type G-6-PD deficiency. *New Eng. J. Med.* **303:** 416–420.

Dickson, L. G., Johnson, L. B., and Johnson, D. R. (1973). Automated fluorometric method for screening for erythrocyte G-6-PD deficiency. *Clin. Chem.* **19:** 301–303.

Djerassi, L. S. and Vitany, L. (1975). Haemolytic episode in G-6-PD deficient workers exposed to TNT. *Br. J. Ind. Med.* **32:** 54–58.

Dow, P. A., Petteway, M. B., and Alperin, J. B. (1974). Simplified method for G-6-PD screening using blood collected on filter paper. *Am. J. Pathol.* **61:** 333–336.

Enger, M. (1960). Treatment of water with chlorine dioxide to improve taste. *G.W.F.* **14:** 340. (Translated for Merc-Library, EPA from original German by Leo Kanner Associates, P.O. Box 5187, Redwood City, California 94043.)

Environmental Protection Agency (1975). Preliminary Assessment of Suspected Carcinogens in Drinking Water. *Report of Congress*, compiled by Office of Toxic Substances, Environmental Protection Agency, Washington, DC.

Felger, G. (1952). Relationship between reduced glutathione content and spontaneous hemolysis in shed blood. *Nature (London)* **170:** 624–625.

Feo, F., Pirisi, L., Pascale, R., Daino, L., Zanetti, S., La Spina, V. and Pani, P. (1982). Modulatory mechanisms of chemical carcinogenesis: the role of the mixed function mono-oxygenase. In: *Recent Trends in Chemical Carcinogenesis*, Vol. 1. P. Pani, F. Feo, and A. Columbano (eds.). ESA, Cagliari.

Feo, F., Piris, L., Pascale, R., Zanetti, S., Daino, L., and La Spina, V. (1982). Modulatory effects of G-6-PD deficiency on the carcinogen activation by "in vitro" growing human fibroblasts. In: *Membranes in Tumor Growth*. T. Galeotti et al. (eds.). Elsevier Biomedical Press, New York, pp. 549–557.

Ganzoni, A. and Rhomberg, F. (1965). Hamolytische. Krise bei mangel on glukose-6-phosphatdehydrogenase und bleiintoxikation. *Acta Haematol.* **34:** 338–346.

Goldstein, B. D. (1973). Hydrogen peroxide in erythrocytes. *Arch. Environ. Health* **26:** 279–280.

Goldstein, B. D. and Balchum, L. J. (1967). Effect of ozone on lipid peroxidation in the red blood cell. *Proc. Soc. Exp. Biol. Med.* **126:** 356–358.

Heubner, W. (1949). Der blutfarbstoff ais katalysator. *Arch. Int. Pharmacodyn. Ther.* **78:** 410.

Huebner, W. and Jung, F. (1941). Zur theorie der chloratvergiftung. *Scheweiz. Med. Wochenschr.* **71:** 247.

Hopf, W. (1965). Experiments with chlorine dioxide treatment of drinking water. GWF **108**(30): 852. (Translation by Leo Kanner Associates.)

Howard, J. K. (1974). Human erythrocyte glutathione reductase and G-6-PD activities in normal subjects and in persons exposed to lead. *Clin. Sci. Mol. Med.* **47:** 515–520.

Jensen, W. N. (1962). Hereditary and chemically induced anemia. *Arch. Environ. Health* **5:** 212–216.

References

Jung, F. (1947). Zur theorie der chlorat vergiftung. III. *Naunyn-Schmiedebergs Arch. Exp. Pathol. Pharmakol.* **204:** 157.

Jung, F. and Kuon, R. (1951). Zum inaktiven hamoglobin des blutes. *Naunyn-Schmiedebergs Arch. Exp. Pathol. Pharmakol.* **214:** 103.

Kessler, I. (1970). A genetic relationship between diabetes and cancer. *Lancet* **i:** 218.

Kiese, M. (1974). *Methemoglobinemia: A Comprehensive Treatise.* CRC Press, Cleveland, Ohio, p. 259.

Koransky, W. (1952). Beitrag zur theorie der chlorat oxydation. *Naunyn-Schmeidebergs Arch. Exp. Pathol. Pharmakol.* **215:** 483.

Kunzer, W. and Saffer, D. (1954). Vergleichende Untersuchung der denaturierung sowie der oxydation von fetalem und erwachsenen-oxy-hamaglobin durch kaluimchlorat. Naunyn-Schmiedebergs *Arch. Exp. Pathol. Pharmakol.* **223:** 501.

Larizza, P., Brunetti, P., and Grignani, F. (1960). Anemie emolitiche enzimopeniche. *Haematologica* **45:** 1–90, 129–212.

Linch, A. L. (1974). Biological monitoring for industrial exposure to cyanogenic aromatic nitro and amine compounds. *Am. Indust. Hyg. Assoc. J.* **35:** 426–432.

Lubbers, J. R., Chauhan, S., and Bianchine, J. R. (1981). Controlled chemical evaluations of chlorine dioxide, chlorite and chlorate in man. *Fundament. Appl. Toxicol.* **1:** 334–338.

Lubbers, J. R., Chauhan, S., Miller, J. K., and Bionchine, J. R. (1983). The effects of chronic administration of chlorite to G-6-PD deficient healthy adult male volunteers. *J. Environ. Pathol. Toxicol.* (in press.)

McIntire, M. S. and Angle, C. R. (1972). Air lead relation to lead in blood of black school children deficient in G-6-PD. *Science* **177:** 520–522.

Menzel, D. B. (1971). Oxidation of biologically active reducing substitutes by ozone. *Arch. Environ. Health* **23:** 149–153.

Menzel, D. B., Slaughter, R. J., Bryant, A. M. and Jauregui, H.O. (1975). Heinz bodies formed in erythrocytes by fatty acid ozonides and ozone. *Arch. Environ. Health* **30:** 296–301.

Michael, G. E., Midey, R. K., Bercz, J. P., Miller, R. G., Grenthouse, D. G., Kraemer, D. F., and Lucas, J. B. (1981). Chlorine dioxide water disinfection: A prospective epidemiology study. *Arch. Environ. Health* **36**(1): 20–27.

Miltner, R. J. (1977). Measurement of chlorine dioxide and related products. Water Supply Research Division. U.S. Environmental Protection Agency, Cincinnati, Ohio.

Moore, G. S. and Calabrese, E. J. (1980). G-6-PD deficiency: A potential high risk group to copper and chlorite ingestion. *J. Environ. Pathol. Toxicol.* **4:** 271–279.

Mountain, J. T. (1963). Detecting hypersusceptibility to toxic substances. An appraisal of simple blood tests. *Arch. Environ. Health* **6:** 357.

Musil, J. and Knotek, Z. (1965). Toxicologic aspects of chlorine dioxide application for the treatment of water containing phenols. *Sb. Vys. Sk. Chem. Technol. Praze Technol. Vody* **8:** 32.

Naik, S. N. and Anderson, D. E. (1971). The association between G-6-PD deficiency and cancer in American Negroes. *Oncology* **25:** 356–364.

National Academy of Sciences. (1977). Drinking Water and Health. Washington, D.C.

Paniker, N. Y. (1975). Red cell enzymes. *Crit. Rev. Clin. Lab. Sci.* (April), p. 469.

Pinkhas, J., Djaldetti, M., Joshua, H., Resnick, C., and deUries, A. (1963). Sulfhemoglobinemia and acute hemolytic anemia with heinz bodies following contact with a fungicide—zinc ethylene bisdithiocarbamate—in a subject with G-6-PD deficiency and hypocatalassemia. *Blood* **21**(4): 484–494.

Piris, L., Pascale, R., Daino, L., Frassetto, S., La Spina, V., Zanetti, S., Gaspa, L., Ledda, G. M., Garceo, R., and Feo, F. (1982). Effect of G-6-PD deficiency on the benz(a)pyrene toxicity for *in vitro* cultured human skin fibroblasts. *Res. Commun. Chem. Path. Pharmacol.* **38**(2): 301–311.

Richardson, A. P. (1937). Toxic potentialities of continued administration of chlorate for blood and tissues. *J. Pharmacol. Exper. Therap.* **59**: 101.

Rook, J. J. (1976). Haloforms in drinking water. *J. Am. Water Works Assoc.* **68**: 615–620.

Ross, V. (1925). Potassium chlorate: Its influence on the blood oxygen binding capacity (hemoglobin concentration), its rate of excretion and quantities found in blood after feeding. *J. Pharmacol. Exp. Ther.* **25**: 47.

Rubino, G. F. et al. (1963). Behavior of glutathione and glutathione-stability test and G-6-PD activity in lead poisoning. *Minerva Med.* **54**: 930.

Schwartz, A. G. (1979). Inhibition of spontaneous breast cancer formation in female C3H (A^{vy}/a) mice by long term treatment with dehydroepiandrosterone. *Cancer Res.* **39**: 1129–1131.

Schwartz, A. G. and Perantoni, A. (1975). Protective effect of dehydroepiandrosterone against aflatoxin B1 and 7,12-dimethylbenzanthracene-induced cytotoxicity and transformation of cultured cells. *Cancer Res.* **35**: 2482–2487.

Stevens, A., Seeger, D., and Slocum, C. J. (1976). Products of chlorine dioxide treatment of organic materials in water. Presented at workshop on ozone/chlorine dioxide oxidation products of organic materials, Nov. 17–19, 1976, Water Supply Research Division, U.S. Environmental Protection Agency, Cincinnati, Ohio.

Stokinger, H. E. and Mountain, J. T. (1963). Tests for hypersusceptibility to hemolytic chemicals. *Arch. Environ. Health* **6**: 495–502.

Stokinger, H. E. and Mountain, J. T. (1967). Progress in detecting the worker hypersusceptible to industrial chemicals. *J. Occup. Med.* **9**: 537–542.

Stokinger, H. E. and Scheel, L. D. (1973). Hypersusceptibility and genetic problems in occupational medicine: A consensus report. *J. Occup. Med.* **15**: 564–573.

Sulis, E., and Spano, G. (1968). Preliminary observations on the incidence of neoplasms and on the enzymatic and proliferative behavior of tumor tissues in persons with G-6-PD deficiency. *Boll. Soc. Ital. Biol. Sper.* **44**: 1246.

Szeinburg, A., Adam, A., Myers, F., Sheba, C., and Ramot, B. (1959). Hematological survey of industrial workers with enzyme-deficient erythrocytes. *Arch. Indust. Health* **20**: 510–516.

Szeinberg, A. and Peled, N. (1973). Detection of G-6-PD deficiency in the newborn using blood specimens dried on filter paper. *Isr. J. Med. Sci.* **9**: 1353–1354.

Tan, I. K. and Whitehead, T. P. (1969). Automated fluorometric determination of G-6-PD and 6-phosphogluconate dehydrogenase activities in red blood cells. *Clin. Chem.* **18**: 440–446.

Tardiff, R. G. (1977). Health effects of organics: Risk and hazard assessment of ingested chloroform. *Am. Water Works Assoc. J.* (December), pp. 658–661.

Tarlov, A. R., Brewer, A. J., Carson, P. E., and Alving, A. S. (1962). Primaquine sensitivity. *Arch. Intern. Med.* **109**: 137–162.

Uddin, D. E., Dickson, L. G., and Brodine, C. E. (1974). G-6-PD deficiency in military recruits. *J. Am. Med. Assoc.* **227**(12): 1408–1409.

Wallace, W. J. and Caughey, W. S. (1975). Mechanism for the autoxidation of hemoglobin by-products, nitrite, and "oxidant" drugs. Peroxide formation by one electron donation to bound dioxygen. *Biochem. Biophys. Res. Commun.* **62**: 561.

Winterbourne, C. C. (1979). Protection by ascorbate against acetylphenylhydrazine-induced heinze body formation in G-6-PD deficient erythrocytes. *Br. J. Haematol.* **41**: 245–252.

Yeung, C. Y., Lai, H. C., and Leung, N. K. (1970). Fluorescent spot test for screening erythrocyte G-6-PD deficiency in newborn babies. *J. Pediatr.* **76**: 931–934.

Zavon, M. R. (1962). Modern concepts of diagnosis and treatment in occupational medicine. *Am. Indus. Hyg. Assoc. J.* **23**: 30.

Zinkham, W. H. and Childs, B. (1958). A defect of glutathione metabolism in erythrocytes from patients with a naphthalene induced hemolytic anemia. *Pediatrics* **22**: 461.

C. Animal Models for G-6-PD Deficiency

Since persons with an erythrocyte G-6-PD deficiency are known to be at increased risk to experiencing hemolytic crises following exposure to a relatively large number of oxidant stressor agents including a variety of medicinal drugs, industrial chemicals, and botanical products, there is considerable interest in finding and/or developing an animal model that would adequately simulate the human G-6-PD deficient individual. While there are numerous human G-6-PD deficient variants, the most appropriate model would most likely represent the human A-variant, which is the most frequent in the American population, or the Mediterranean strain, which is a more severe deficiency.

Surveys of a number of common animals have revealed that there is considerable variation in the amount of erythrocyte G-6-PD activity. For example, erythrocyte G-6-PD activity values range from the nearly negligible values of the yak (Bartels et al., 1963) to those of the normal human, which are between 5–9/I.U., and to those of some mouse strains, which are three times greater than those of normal humans.

Since there are normal and G-6-PD deficient humans, the search for a possible animal model would ideally lead to the discovery of a strain of animal in which there is a characteristically normal enzyme activity and a genetic variant that is deficient in the G-6-PD activity. Once again, it would be ideal if the G-6-PD activity of the normal and deficient erythrocytes were identical to the human condition. For example, the red cell G-6-PD activity of normal humans ranges between 5–9 I.U., while the activity of human A–variants is about 20–30% of normal, being about 1–3 I.U. An animal model with individuals having similar enzyme activity as normal and deficient humans would be excellent in both a relative and absolute sense. However, the search for such a potentially ideal model has been an elusive one.

Short of the ideal one might find a model that seemed appropriate in a relative sense. For instance, if the strain had individuals differing in their G-6-PD activity in a ratio similar to humans of 1:3 or 1:4, there would be a fine relative comparison, but what if the absolute activity was considerably different than the human, say 7 and 21 I.U., respectively? In this case, while 7 I.U. are in the range of normal humans, it would be considered the low or "deficient" animal so as to simulate the human deficient condition. Initially, one would predict that this experimental model would yield false negative toxicological responses on a milligram per kilogram basis. This situation is similar to one that we evaluated in our laboratory at the University of Massachusetts. In this case, two mouse strains C57BR and A/J, differing in red cell G-6-PD activity by a 1:3 ratio, respectively (Hutton, 1971), were compared for their hemolytic sensitivity following exposures to several oxidants such as O_3 and NO_2, in the air, and ClO_2 and, ClO_2^- in the drinking water. The C57BR strain was presumed to simulate G-6-PD deficient individuals while the A/J mice were thought to simulate humans with normal G-6-PD activity levels. This study had at least three inherent problems: (1) the previously mentioned likelihood of false negatives, (2) the concomitant issue of administering larger doses than desired to compensate, and (3) the use of different mouse strains that most likely had other biochemical differences affecting the susceptibility of the red cell to oxidative stress than just G-6-PD activity. One additional problem that emerged in the evaluation of O_3 toxicity was that if very high levels of O_3 ($\geqslant$ 0.5 ppm) were administered, the O_3 would cause extreme lung damage resulting in pulmonary edema and presumably reduce the amount of oxidant stress reaching the erythrocytes. In fact, the extent of red cell changes induced by O_3 were considerably greater at 0.3 ppm O_3 than at 0.9 and 1.5 ppm (Calabrese et al., 1983). This type of dissatisfaction with the relativistic comparison led to the search for an actual G-6-PD deficient model whose enzyme activity would be nearly equal to the human Mediterranean or A–variants. There are a limited number of potential animal models with erythrocyte G-6-PD activity, which falls in the general range of a G-6-PD deficient. However, the only animal given serious consideration for being a possible model for a human G-6-PD deficient is the sheep (Budtz-Olsen et al., 1963; Salvidio et al., 1963; Smith et al., 1965). Other possible models have not proved even initially viable for different reasons. For example, Smith et al. (1976) noted that an inherited G-6-PD deficiency in rats was lost through death by infection (Werth and Verenbarer, 1967) while reports of G-6-PD deficiency in cattle (Naik and Anderson, 1971) and dogs (Naik et al., 1971) were most likely due largely to polymorphism of other enzymes.

A major problem with the use of sheep as a model rests with the fact that sheep have only low G-6-PD activity and there is the lack of a high or normal substrain. This makes it difficult to assess if any enhanced susceptibility that the sheep may have is because of their low erythrocyte G-6-PD activity. There are sheep with high and low erythrocyte GSH levels while having similarly low G-6-PD activity. The high GSH sheep also display a significantly higher level of glutathione peroxidase than the low GSH sheep (Agar and Smith, 1973). Despite this inherent lack of control or normal G-6-PD group, several investigations attempted to compare the responses of sheep erythrocytes to those of human deficients when exposed to various oxidant stressor agents. Maronpot (1972) reported that only two (i.e., acetylphenylhydrazine, pyrimethamine) of six chemical agents known to cause intravascular hemolysis in human G-6-PD deficients caused similar hemolysis while there was a lack of hemolytic response to methylene blue, primaquine, nitrofurantoin, and fava beans. Such ambivalent findings led Maronpot (1972) to conclude

... that although sheep have very low erythrocyte G-6-PD levels, they are not G-6-PD deficient in the same sense that men with similar levels would be. The resistance to hemolysis which these sheep exhibited to substantial doses of potentially hemolytic drugs and to ingestion of fava beans supports this conclusion and suggests that sheep are probably not good animal models for the study of erythrocyte G-6-PD deficiency.

This conclusion of Maronpot (1972) has been recently supported by Williams (1982) who demonstrated that G-6-PD-deficient human erythrocytes were considerably more sensitive than sheep erythrocytes to proposed toxic ozone intermediates (i.e., methyl oleate ozonide, methyl lineolate hydroperoxide, and methyl oleate hydroperoxide).

Despite the lack of naturally occurring animal models to simulate humans with red cell disorders, a relatively recent experimental model has been developed to study human red cell sickling. Since no model's blood behaves sufficiently like the human, it was reasoned that the best alternative to direct testing on humans was to transfuse human cells to an otherwise appropriate model and to study the condition. However, before this could be accomplished it was necessary to block the ability of the model to reject (i.e., phagocytize) the foreign red cells. This was accomplished via the injection of two agents, one (i.e., ethyl palmitate) that blocked the reticuloendothelial system and a nontoxic factor in cobra venom that inactivated the third component of complement resulting in a profound inhibition of intravascular hemolysis. Initial experiments on survival time of the transfused red blood cells with this model (rats) have revealed that the normal and sickle human red blood cells displayed the same relative survival rates as they did in humans. These findings led the authors to conclude that this model may be of considerable value in the evaluation of the pathogenesis and therapy of human sickle-cell disease (Castro et al., 1973; Castro and Cochran, 1978). This transfusion-type animal model, in fact, has been successfully exploited by the original researchers and others to assess the effect of different antisickling agents prior to direct human testing (Castro et al., 1976; Schoomaker et al., 1976). Unfortunately, the rat model, as originally developed, achieved only about 12–14 hr for 50% survival time of the transfused human cells presumably because the human red cell is considerably larger than that of the rat, thereby facilitating its sequestration in the spleen. However, recent studies with the guinea pig, a model with comparable red cell size as man, proved to offer a much more acceptable result since the red cells lived for 2–4 days (50% of survival time) after transfusion (Azikime et al., 1978). While the original studies concerned the use of sickle cells, the heterologous rat model has been used successfully to predict the evaluation of hemolytic susceptibility of G-6-PD deficient human erythrocytes as well (Udomratn et al., 1977).

REFERENCES

Agar, N. W., Gruca, M., and Harley, J. D. (1974). Studies of glucose-6-phosphate dehydrogenase, glutathione reductase and regeneration of reduced glutathione in the red blood cells of various mammalian species. *Aust. Exp. Biol. Med. Sci.* **52**(Pt. 4): 607–614.

Agar, N. S., Roberts, J., Mulley, A., Board, P. G., and Harley, J. D. (1975). The effect of experimental anemia on the levels of glutathione and glycolytic enzymes of the erythrocytes of normal and glutathione-deficient merino sheep. *Aust. J. Biol. Sci.* **28**: 233–238.

Agar, N. S. and Smith, J. E. (1973). Erythrocyte enzymes and glycolytic intermediates of high- and low-glutathione sheep. *Anim. Blood Groups Biochem. Genet.* **4**: 133–140.

Azikime, A. N., Castro, O., Osbaldiston, G. W., and Finch, S. C. (1978). Survival of human 51Cr erythrocytes to guinea pigs. *Br. J. Exp. Pathol.* **59**: 248.

Bartels, H., Hilpert, P., Barkey, K., Belke, K., Riegel, K., Land, E. M., and Metcalfe, J. (1963). Respiratory functions of blood of the yak, llama, camel, Dybowski deer and African elephant. *Am. J. Physiol.* **205**: 331–336.

Budtz-Olsen, O. E., Axten, B., and Haigh, S. (1963). Glucose-6-phosphate dehydrogenase deficiency in erythrocytes of sheep and goats. *Nature* **198**: 1101.

Calabrese, E. J. (1978). Animal model: Mice with low levels of G-6-PD. *Am. J. Pathol.* **91**(2): 409–411.

Calabrese, E. J., Moore, G. J., and Grinberg-Funes, R. (1983). The effects of ozone and vitamin E status on mice strains with different levels of erythrocyte G-6-PD. *J. Environ. Toxicol. Pathol.* (in press).

Castro, O. and Cochran, J. D. (1978). Study of irreversibly sickled cells in an animal model. *J. Nat. Med. Assoc.* **70**(1): 23–26.

Castro, O., Orlin, J., Rosen, M. W., and Finch, S. C. (1973). Survival of human sickle-cell erythrocytes in heterologous species: Responses to variations in oxygen tension. *Proc. Nat. Acad. Sci. U.S.A.* **70**(8): 2356–2359.

Castro, O., Osbaldiston, G. W., Aponte, L., Roth, R., Orlin, J., and Finch, S. (1976). Oxygen-dependent incubation of sickle erythrocytes. *J. Lab. Clin. Med.* **88**: 732–744.

Hutton, J. J. (1971). Genetic regulation of glucose-6-phosphate dehydrogenase activity in the inbred mouse. *Biochem. Genet.* **5**: 315–331.

Kurian, M. and Iyer, G. Y. N. (1977). Oxidant effect of acetylphenylhydrazine: A comparative study with erythrocytes of several animal species. *Can. J. Biochem.* **55**: 597–601.

Lankisch, P. G., Schroeter, R., Lege, L., and Vogt, W. (1973). Reduced glutathione and glutathione reductase—a comparative study of erythrocytes from various species. *Comp. Biochem. Physiol.* **46B**: 639–641.

Maronpot, R. R. (1972). Erythrocyte glucose-6-phosphate deficiency and glutathione deficiency in sheep. *Can. J. Comp. Med.* **36**(1): 55–60.

Naik, S. N. and Anderson, D. E. (1971). Glucose-6-phosphate dehydrogenase and hemoglobin types in cattle. *J. Anim. Sci.* **32**: 132–136.

Naik, S. N., Anderson, D. E., Jardine, J. N., and Clifford, D. H. (1971). Glucose-6-phosphate dehydrogenase deficiency, hepatoglobin and hemoglobin variants in dogs. *Anim. Blood Groups Biochem. Genet.* **2**: 89–91.

Paniker, N. V. and Berther, E. (1972). Glucose-6-phosphate and dehydrogenase NADPH diaphorase in cattle erythrocytes. *J. Anim. Sci.* **34**: 75–76.

Salvidio, E., Pannacivlli, I., and Tizianello, A. (1963). Glucose-6-phospho-gluconate dehydrogenase activities in the red blood cells of several animal species. *Nature* **200**: 372–373.

Schoomaker, E. B., Bremer, G. J., and Oelschlegel, F. J. (1976). Zinc in the treatment of homozygous sickle cell anemia. Studies in an animal model. *Am. J. Hematol.* **1**: 45–57.

Smith, J. E. (1968). Low erythrocyte glucose-6-phosphate dehydrogenase activity and primaquine insensitivity in sheep. *J. Lab. Clin. Med.* **71**: 826–833.

Smith, J., Barnes, J. K., Kaneko, J. J., and Freedland, R. A. (1965). Erythrocytic enzymes of various animal species. *Nature* **205**: 298–299.

Smith, J. and Holdridge, B. (1967). Comparison of erythrocyte glucose-6-phosphate dehydrogenase of man and various ungulates. *Comp. Biochem. Physiol.* **22**: 737–743.

Smith, J. E., McCants, M., and Parks, P. (1968). Comparison of erythrocyte 6-phosphogluconate dehydrogenases of man and various ungulates. *Comp. Biochem. Physiol.* **21**: 137–142.

Smith, J. E., Ryer, K., and Wallace, L. (1976). Glucose-6-phosphate dehydrogenase deficiency in a dog. *Enzyme* **21**: 379–382.

Thompson, R. H., and Rodd, J. R. (1964). Estimation of glucose-6-phosphate dehydrogenase in sheep erythrocytes. *Nature* **201**: 718.

Udomratn, T., Steinberg, M. H., Campbell, G. D., Jr., and Oelschlegel, F. J., Jr. (1977). Effects of ascorbic acid on G-6-PD deficiency erythrocytes: Studies in an animal model. *Blood* **49**: 471–475.

Werth, G. and Vererbarer, G. M. (1967). Glucose-6-phosphate dehydrogenase-Mangel in Erythrocyten von Rathen. *Klin. Wochenschr.* **45**: 265–269.

Williams, P. (1982). The effects of three possible ozone toxic intermediates on the red cells of normal human, G-6-PD deficient individuals and sheep. Master's thesis. University of Massachusetts, Amherst.

World Health Organization Technical Report Series (1967). Standardization of procedures for the study of glucose-6-phosphate dehydrogenase. Report of a WHO Scientific Group. WHO Tech. Rep. Ser. 366.

D. Sickle-Cell Anemia and Sickle-Cell Trait

1. INTRODUCTION

These genetic conditions result from the presence of an abnormal Hb molecule, Hb S, located in the erythrocytes of affected persons. Hb S differs from the normal adult Hb, Hb A, only by the exchange of the amino acid glutamine by valine, at a single location in the Hb chain that consists of 287 amino acids. Hb S is considerably less soluble than Hb A under conditions of low oxygen tension. This decreased solubility may result in the formation of a gel within red blood cells, distorting them and causing sickling to occur. An individual with sickle-cell anemia is homozygous for Hb S while one with sickle-cell trait is heterozygous for Hb S. In general, the homozygous person has nearly 100% Hb S while the heterozygous person has from 20–40% of Hb S. The person with sickle-cell trait will experience sickling only when blood O_2 tension is greatly reduced (i.e., about 10 mmHg from a normal of 95 mmHg) (Stokinger and Scheel, 1973; Cooper, 1973).

While those with sickle-cell anemia are known to have a considerably reduced life span, the health hazards of sickle-cell trait have been generally considered minimal, if not nonexistent, under normal circumstances (Motulsky, 1974). For example, life expectancy, mortality, and hematologic status have been reported as being normal. Situations that have been found to cause sickling problems in those with sickle-cell trait are: severe hypoxia such as is found in unpressurized airplanes, during skin diving, mountain climbing, or possibly during anesthesia.* In general, the overwhelming majority of those with sickle-cell trait in the United States are thought never to have any problems associated with this genetic condition.

---

*In 1956, testing of U.S. military personnel for Hb S was suggested (see Rutter et al., 1956), as a response to reports of spleenic infarction and hematuria in persons with sickle-cell trait (Hb AS) following flight in unpressurized aircraft (Harvey, 1954; Smith and Conley, 1955). Thereafter, the presence of sickle-cell trait was established to be a disqualifying trait for a job as a flight crew member. Since strong evidence of illness or death associated with hemoglobinopathies was lacking as well as available technology suitable for accurate large-scale surveys, testing all military personnel for such traits was not carried out. However, flight personnel were tested. A 1970 report by Jones et al., indicating an association between sickle-cell trait and deaths in such persons undergoing military training at 4000 feet generated considerable renewed interest. These findings and studies by O'Brien et al. (1972) indicating that very stressful physical activities could precipitate a sickle-cell crisis in those with sickle-cell trait prompted the surveying of over 20,000 black Naval and Marine recruits during 1972 (Uddin et al., 1972). Furthermore, the Air Force Academy in 1972 formulated a policy of not accepting persons with sickle-cell trait after the collapse of two black pilot trainees following strenuous exercise. This type of policy was subsequently supported by a 1973 NAS report, which stated that persons with sickle-cell trait should not be pilots or copilots (Holden, 1981). The policy of the Air Force Academy to exclude persons with sickle-cell trait has led to a recent legal challenge (Holden, 1981). While the above examples associated possession of sickle-cell trait with increased risk of death with the presence of low oxygen environments and strenuous exercise, considerable evidence exists that challenges the hypothesis of persons with sickle-cell trait being at increased risk in low oxygen environments. For example, 10 surviving black U.S. pilots from World War II with sickle-cell trait experienced no health problems associated with this condition. In addition, 6.7% of black football players in the National Football League (NFL) have sickle-cell trait. This is a prevalence not significantly different from that found in the black population in the United States, thereby suggesting that there has not been a selective exclusion from professional football due to physical limitations. Furthermore, the Denver entry into the NFL plays in a stadium that is more than 5000 feet above sea level (Murphy, 1973), and no risk associated with playing at this location has been reported by blacks with sickle-cell trait.

2. SUSCEPTIBILITY TO ENVIRONMENTAL POLLUTANTS

Stokinger and Scheel (1973) have suggested that persons with sickle-cell abnormalities may be at increased risk from (1) anemia producers (e.g., benzene, lead, and cadmium), (2) methemoglobin formers (i.e., aromatic amino and nitro compounds), and (3) blood oxygen tension reducers (e.g., carbon monoxide and cyanide). However, they emphasized that they were not aware of any reports describing any such enhanced risk. A 1978 article by Reinhardt reported that as a service Dupont "routinely offers all black employees a test for sickle-cell trait." This type of screening was initiated in 1972 (see Severo, 1980). At Dupont's Chambers Works Plant, those with sickle-cell trait and a hemoglobin value of < 14 g/100 mL are not permitted to work with hemolytic agents. No evidence to support the premise that persons with sickle-cell trait and a low hemoglobin value were at increased risk to hemolytic chemicals such as nitro and amino compounds was presented by Dupont officials.

There have been several clinical reports which suggest that sickle-cell disease (i.e., anemia) may enhance susceptibility to lead intoxication (Seto and Freeman, 1964; Feldman et al., 1973; Anku and Harris, 1974; Erenberg et al., 1974). However, there is insufficient evidence available to adequately evaluate this association. Finally, a recent report by Synder et al. (1981) indicated that osmotically resistant (but not osmotically more susceptible) red cells of patients with sickle-cell anemia were susceptible to the oxidant damage of H_2O_2 *in vitro*.

3. SICKLE ERYTHROCYTES AND OXIDANT STRESS

While there has been much written about the potential enhanced susceptibility of sickled erythrocytes, there is neither documented evidence to support those claims nor has there been adequate theoretical foundations to support hypothetical pollutant associations. However, a recent report by Das and Nair (1980) has added a considerably new series of insights into the possible enhanced oxidant stress of sickled erythrocytes. They reported that sickled erythrocytes have reduced activity levels of glutathione peroxidase (GSH-Px) (i.e., 50% of normal when considered on a per gram Hb basis), and catalase (i.e., 80% of normal on a per gram Hb basis), while having about 40% greater superoxide dismutase (SOD) activity. The authors concluded that an increased SOD activity may lead to a buildup of H_2O_2 while lowered levels of GSH-Px and catalase result in a longer H_2O_2 elimination time. The net result is that sickled cells should display a greater opportunity for H_2O_2 damage. They hypothesize that the diminished ability to remove H_2O_2 may account for the greater indications of oxidant stress in sickled cells, such as fluorescent pigments, Hb inclusion bodies and a raised malonaldehyde content as compared to normal cell content. Of further theoretical importance is that Das and Nair (1980) found that membranes of sickled cells also contained considerably reduced levels of cholesterol. The lowered amount of cholesterol from the membrane may enhance its susceptibility to free radical attack because the unsaturated lipids are protected from free radical attack by intramembrane cholesterol molecules (Demopoulos, 1973).

Finally, Das and Nair (1980) noted an earlier report of Schaeffer et al. (1968) of increased amounts of copper in sickled erythrocytes. They hypothesized that because sickled erythrocytes have unstable Hb and excessive copper, it is reasonable to assume that such cells are more predisposed to the generation of superoxide radicals than normal red cells.

4. FREQUENCY IN POPULATION

The gene for Hb S is most often found in equatorial Africa, parts of India, countries of the Middle East, and those around the Mediterranean. The frequency of sickle-cell anemia is about 0.2–0.5% among American blacks. However, according to Cooper (1973), one would not be likely to find them in any preemployment physical testing since this disease most likely would have revealed itself in overt illness prior to adulthood. Sickle-cell trait is found in about 8% of U.S. blacks or approximately 2 million U.S. blacks.

5. SCREENING TESTS

Several screening methodologies for sickle-cell abnormalities have been developed for mass screenings. The procedure, consisting of cellulose-acetate electrophoresis (CAE) followed by solubility testing, has been favorably regarded because of speed, cost, simplicity, accuracy, and ability to differentiate the various types of Hb (Barnes et al., 1972). CAE is capable of identifying Hb S, F, C, A_2, and A. Quantification of Hb types is easily performed. In order to verify the electrophoresis procedure for Hb S, any blood found to have a Hb S is subsequently evaluated via the solubility test since Hb S displays abnormal solubility. However, Hb C Harlem also gives a positive reaction.

False positive findings from the dithionite-phosphate solubility test may result from polycythemia, macroglobulinemia, or technician error from using too much blood in relation to reagent. False negative solubility tests may result from extremely anemic sickle-cell trait carriers with less than 3.2 g Hb/100 mL. Using packed red cells rather than whole blood eliminates this possibility. CAE followed by a solubility test to confirm the presence of Hb S has been the procedure recommended by the

National Sickle Cell Disease Program and the National Hemoglobinopathy Standardization Laboratory at the Center for Disease Control (Schmidt, 1973, 1973a).

6. SUMMARY

The issue of whether sickle-cell trait carriers are at increased risk from the strains of rigorous training at high altitudes remains controversial and unresolved. The evidence suggests that possession of sickle-cell trait is a contributing factor in the reported cases of injury and/or death but that other as yet undetermined factors must also be present.

No experimental or epidemiological evidence exists to support the hypothesis that persons with sickle-cell trait are at increased risk to the three groups (i.e., anemia producers, MetHb formers, and blood oxygen tension reducers) suggested by Stokinger and Scheel (1973). It must be strongly emphasized that the hypothesis of Stokinger and Scheel (1973) has not been discredited either.

REFERENCES

Anku, V. D. and Harris, J. W. (1974). Peripheral neuropathy and lead poisoning in a child with sickle-cell anemia. *J. Pediatr.* **85:** 337–340.

Ashcroft, M. T., Miall, W. E., and Milner, P. F. (1969). Comparison between characteristics of Jamaican adults with normal hemoglobin and those with sickle cell trait. *Am. J. Epidemiol.* **90:** 236.

Barnes, M. G., Komarmy, L., and Novack, A. H. (1972). A comprehensive screening program for hemoglobinopathies. *J. Am. Med. Assoc.* **219:** 701–705.

Boyle, E. Jr., Thompson, C., and Tyroler, H. A. (1968). Prevalence of sickle cell trait in adults of Charleston County. *Arch. Environ. Health* **17:** 891.

Bullock, W. H. and Jilly, P. N. (1975). Hematology. In: *Textbook of Black Related Diseases.* McGraw-Hill, New York.

Cooper, W. C. (1973). Indicators of susceptibility to industrial chemicals. *J. Occup. Med.* **15:** 355–359.

Das, S. K. and Nair, R. C. (1980). Superoxide dismutase, glutathione peroxidase, catalase and lipid peroxidation of normal and sickled erythrocytes. *Br. J. Haematol.* **44:** 87–92.

Demopoulos, H. B. (1973). Control of free radicals in biologic systems. *Fed. Proc.* **32:** 1903–1908.

Erenberg, G., Rinsler, S. S., and Fish, B. G. (1974). Lead neuropathy and sickle cell disease. *Pediatrics* **54**(4): 438–441.

Feldman, R. G., Haddow, J., Kopito, L., and Schwachman, H. (1973). Altered peripheral nerve conduction velocity: Chronic lead intoxication in children. *Am. J. Dis. Child.* **125:** 39.

Harvey, C. M. (1954). Sickle cell crisis without anemia occurring during air flight. *Mil. Surg.* **115:** 271–275.

Holden, C. (1981). Air Force challenged on sickle trait policy. *Science* **211:** 257.

Jones, S. R., Binder, R. A., and Donowho, E. M., Jr. (1970). Sudden death in sickle cell trait. *N. Engl. J. Med.* **282:** 323–325.

Motulsky, A. G. (1974). Screening for sickle cell hemoglobinopathy and thalassemia. In: *Genetic Polymorphisms and Diseases in Man.* B. Ramot, A. Adam, B. Bonne, R. M. Goodman, and A. Szeinberg (eds.). Academic Press, New York, pp. 215–223.

Murphy, J. R. (1973). Sickle cell hemoglobin (HbAS) in black football players. *J. Am. Med. Assoc.* **225:** 981–982.

O'Brien, R. T. et al. (1972). Splenic infarction and sickle cell trait. *N. Engl. J. Med.* **282:** 323–325.

Reinhardt, C. F. (1978). Chemical hypersusceptibility. *J. Occup. Med.* **20:** 319–322.

Rutter, R., Luttgens, F., Petterson, W. L., Stock, A. E., and Motulsky, A. (1956). Splenic infarction in sicklemia during airplane flight. Pathogenesis, hemoglobin analysis and clinical features of six cases. *Ann. Intern. Med.* **44:** 257–270.

Schaeffer, K., Lofton, J. A., Powell, S. C., Osborne, H. H., and Foster, L. H. (1968). Occurrence of copper in sickling erythrocytes. *Proc. Soc. Exp. Biol. Med.* **128:** 734–737.

Schmidt, R. M. (1973). Laboratory diagnosis of hemoglobinopathies. *J. Am. Med. Assoc.* **224:** 1276–1280.

Schmidt, R. M. (1973a). Standardization in detection of abnormal hemoglobins: Solubility tests for hemoglobin S. *J. Am. Med. Assoc.* **225:** 1225.

Seto, D. S. Y. and Freiman, J. M. (1964). Lead neuropathy in children. *Am. J. Dis. Child.* **107:** 337.

Severo, R. (1980). Genetic tests by industry raise questions on rights of workers. *New York Times,* Feb. 3–6.

Smith, E. W. and Conley, C. L. (1955). Sicklemia and infarction of the spleen during aerial flight. *Bull. Johns. Hopkins Hosp.* **96:** 35–41.

Stokinger, H. E. and Scheel, L. D. (1973). Hypersusceptibility and genetic problems in occupational medicine—A consensus report. *Jour. Occup. Med.* **15:** 564–573.

Synder, L. M., Sauberman, N., Condara, H., Dolan, J., Jacobs, J., Szymanski, I. and Fortier, N. L. (1981). Red cell membrane responses to hydrogen peroxide–sensitivity in hereditary xerocytosis and in other abnormal red cells. *Brit. Jour. Haematol.* **48:** 435–444.

Uddin, D. E., Dickson, L. G., and Brodine, C. E. (1974). Screening of military recruits for hemoglobin variants. *J.A.M.A.* **227:** 1405–1407.

E. Animal Models for Sickle-Cell Anemia and Trait

The sickling of erythrocytes is not a trait unique to humans. In fact, the first reported occurrence of sickling was made by Gulliver in 1840 with deer while the first report of human sickling was not until 70 years later in 1910 by Herrick. In humans, sickle cells result when erythrocytes containing the abnormal Hb S are deoxygenated. This Hb is fairly insoluble in the reduced state and upon deoxygenation it develops intracellular tactoid, which distorts the cell membrane, thereby causing the sickle shape. The deer, however, have their sickle cells induced by strikingly different mechanisms. For example, Undritz (1946; 1947; 1950) induced sickling *in vitro* by passing O_2 through the deer blood. This sickling phenomenon can be reversed by the administration of reducing agents such as $NaHSO_3$ and $Na_2S_2O_5$. Sickling in deer has also been reported to occur when the blood becomes saturated with N_2 or CO presumably because of a replacement of the CO_2 and the subsequent "respiratory alkalosis" (Undritz et al., 1960). These findings have been replicated by other investigators using the same and different strains of deer (Table 2-5).

Sickling of erythrocytes has also been reported in several other types of animals including the mongoose, goat, sheep, genet as well as raccoon, hamster, and squirrel. While increasing the pH of the blood (or oxygenation) enhances sickling in goats, sheep and genets, increasing the salt concentration in the blood of the mongoose, raccoon, hamster, and squirrel enhances sickling in these species (Table 2-5).

Ball et al. (1976) have emphasized that erythrocyte sickling is associated with pathological symptoms only in man. An early report by O'Rouke (1936) revealed anemia in white-tailed deer with sickle cells but later studies have not supported this (Kitchen et al., 1964; Whitten, 1967; Ball et al., 1976). In fact, Ball et al. (1976) state that sickling is a "physiological rather than a pathological condition in deer and genets." They suggest that the absence of clinical symptoms is probably best explained by the fact that the conditions that can induce sickling do not exist in circulating blood. *In vivo* sickling has been induced in deer but only after an infusion of bicarbonate (Whitten, 1967; Parshall et al., 1975). Furthermore, Whitten (1967) has suggested that the relatively small size of deer red cells (i.e., about 26 MCV/CuU as compared to 91 MCV/CuU for humans) may reduce their likelihood of becoming entangled to form a blocking clump if sickling does occur *in vivo*. Ball et al. (1976) noted that red cells of genets and mongooses are also smaller than those of humans by about 50–60%.

These collective findings clearly reveal that the causes of human red blood cell sickling do not resemble those for all examples of presently reported animal species. It is apparent that these examples of sickling in the animal kingdom cannot be presently exploited by toxicologists as models to simulate the human sickling phenomenon.

In 1973 Castro et al. presented a novel idea for creating an animal model approach for studying human sickle cells. They decided that since there were not any adequate sickle-cell models that simulated the human condition, why not place the human red cells with Hb S in the rat model after its phagocytizing capacity had been knocked out by chemical agents.* In this way the responses of normal and sickle cells of humans could be evaluated in different testing schemes. The initial experimentation was encouraging since the relative survival time of sickle cells compared to human normal cells were comparable in the animal model.

While only limited data are available on the model (Castro et al., 1973;

---

*Treatment with ethyl palmitate prevents reticuloendothelial responses while cobra venom knocks out complement.

Table 2-5. The Occurrence of Erythrocyte Sicklings in Animals Other than Man

Deer Blood	*In vitro*: in general, sickling arising in deer as a result of increasing the pH and/or by increasing oxygenation	Gulliver (1840)
Three tropical deer species: *Cervus mexicanus, C. reevesii, C. porcinus*		
Persian deer		
C. elaphus		
Dama dama		Undritz et al. (1946)
C. unicolor		
Pseudoxis hortulorum		
Axis axis		
Elaphurus davidianus		Also Ball et al. (1976)
White-tailed deer		Taylor and Easley (1974)
C. nippon	Administer sodium bicarbonate to produce a transient state of alkalosis and *in vivo* erythrocyte sickling. The percent of sickle cells increased as the blood pH increased. Sickling was enhanced by 100% O_2 ventilation.	Parshall et al. (1975)
Mongoose Blood	*In vitro* caused by increasing the salt concentration of the surrounding medium; not influenced by oxygen tension	Hawkey and Jordan (1967)
Herpestes sanguiners		Ball et al. (1976)
Raccoon, Hamster, and Squirrel Blood	Can have abnormal erythrocyte shapes induced, including sickle-like shapes by storage or treatment with hypertonic solution	Kitchen et al. (1964)
Young Goat Blood	Produced sickling in similar fashion to deer (i.e., dependent on pH—as pH increases so does the sickling)	Holman and Dew (1964)

Table 2-5 (Continued)

Sheep Blood Swaledale lamb North Yorkshire moor Clun Forest	Produced sickling in similar fashion to deer; sickling occurring in whole blood saturated with O_2; oxygenation leads to an increase in pH; sodium metabisulphite reverses the process	Rees Evans (1968)
Genets (2 species)	Sickling induced by heparin 18 hr at 4°C, EDTA, and clotted blood 18 hr at 4°C; oxygenation, 0.9% NaCl with pH 7.4/1 hr. CO_2 treatment reversed sickling as did 2% Na dithionite and 2% Na metabisulphite.	Ball et al. (1976)

Schoomaker et al., 1976; Castro and Cochran, 1978), it appears to offer potential to evaluate the response of human sickle cells in experimental settings. The model, however, requires considerable development before it may offer any practical possibilities for regulatory agencies. One area of principal concern is that the transfused human red cells survive only a very short time with the half-life being generally less than 8 hr for erythrocytes from sickle-cell patients as compared to about 24 hr for normal human blood. This may be because human cells are larger than rat red cells by about 25–33% (Altman and Dittmer, 1961; Schoomaker et al., 1976). The greater size of human cells enhances sequestration in the rat spleen, hence the decreased survival. Perhaps the selection of a model with red cells more closely approaching the size of human red cells (i.e., guinea pigs) would help solve that problem. A report by Castro et al. (1976) has supported this hypothesis.

Finally, further development of such a model not only has potential for assessing sickle-cell responses but also appears to have the distinct potential for serving in the evaluation of other erythrocytic disorders such as G-6-PD deficiency where adequate animal predictive models remain to be found.

REFERENCES

Altman, P. L. and Dittmer, D. S. (1961). Blood and other body fluids. Bethesda, Md. *Fed. Am. Soc. Exp. Biol.*, pp. 110, 116.

Amma, E. L., Sproul, G. D., Wong, S., and Huisman, T. H. J. (1974). Mechanism of sickling in deer erythrocytes. *Ann. N.Y. Acad. Sci.* **241:** 605–613.

Ball, S., Hawkey, C. M., Hime, J. M., Keyner, I. F., and Brombell, M. R. (1976). Red cell sickling in genets. *Comp. Biochem. Physiol.* **54A**: 49–54.
Castro, O., Osbaldiston, G. W., Aponte, L., Roth, R., Orlin, J., Finch, S. C. (1976). Oxygen-dependent circulation of sickle erythrocytes. *J. Lab. Clin. Med.* **88**(5): 732–744.
Castro, O. and Cochran, J. D. (1978). Study of irreversibly sickled cells in an animal model. *J. Nat. Med. Assoc.* **70**(1): 23–26.
Castro, O., Orlm, J., Rosen, M. W., and Finch, S. C. (1973). Survival of human sickle-cell erythrocytes in heterologous species: response to variations in oxygen tension. *Proc. Nat. Acad. Sci.* **70**(8): 2356–59.
Gulliver, G. (1840). On the blood corpuscles of the mammiferous animals. *Phil. Mag. J. Sci.* 3rd series. **17**: 325.
Hawkey, C. M. and Jordan, P. (1967). Sickle-cell erythrocytes in the mongoose *Herpestes sanguineus*. *Trans. R. Soc. Trop. Med. Hyg.* **61**(2): 180–181.
Herrick, J. B. (1910). Peculiar elongated and sickle-shaped red corpuscles in a case of severe anaemia. *Arch. Intern. Med.* **6**: 517–521.
Holman, H. H. and Dew, S. M. (1964). The blood picture of the goat. II. Changes in erythrocytic shape, size and number associated with age. *Res. Vet. Sci.* **5**: 274–285.
Kitchen, H., Putnam, F., and Taylor, W. (1964). Hemoglobin polymorphism: its relation to sickling of erythrocytes in white-tailed deer. *Science* **144**: 1237.
O'Rouke, E. C. (1936). Sickle cell anaemia in deer. *Proc. Soc. Exp. Biol. Med.* **34**: 738–739.
Parshall, C. J., Jr., Vainisi, S. J., Goldberg, M. F., and Wolf, D. (1975). *In vivo* erythrocyte sickling in the Japanese Sika deer (*Cervus nippon*): methodology. *Am. J. Vet. Res.* **36**(6): 749–752.
Rees Evans, E. T. (1968). Sickling phenomenon in sheep. *Nature* **217**: 74–75.
Schoomaker, E. B., Brewer, G. J., and Oelshlegel, F. J., Jr. (1976). Zinc in the treatment of homozygous sickle cell anemia: studies in an animal model. *Am. J. Hematol.* **1**: 45–57.
Simpson, C. F. and Taylor, W. J. (1974). Morphological and submicroscopic comparison of sickle erythrocytes of humans and deer. *Ann. N.Y. Acad. Sci.* **241**: 614–622.
Taylor, W. J. and Easley, C. W. (1974). Sickling phenomena of deer. *Ann. N.Y. Acad. Sci.* **241**: 594–604.
Undritz, E. (1946). *Arch. Vererbungsf.* (Cited in Undritz, 1960).
Undritz, E. (1947). *Verh. Schweiz. Naturforsch. Ges.* **127**: 102. (Cited in Undritz, 1960).
Undritz, E. (1950). *Rev. Hematol.* **5**: 644. (Cited in Undritz, 1960).
Undritz, E., Betke, K., and Lehmann, H. (1960). Sickling phenomenon in deer. *Nature* **187**: 333–334.
Whitten, C. E. (1967). Innocuous nature of the sickling phenomena in deer. *Br. J. Haematol.* **13**: 650–655.

F. The Thalassemias

1. INTRODUCTION

Thalassemia is an erythroblastic anemia, occurring early in life, with its severe form being associated with an enlarged liver and spleen, a leukemoid hematological picture, and characteristic bone changes. In 1925, Cooley and Lee described the severe form that is found in the homozygous state and is now called thalassemia major. The milder condition that is found in the heterozygous state is called thalassemia minor (Lehmann et al., 1966). The classic Mediterranean form of thalassemia is thought to be caused by a deficiency in β-chain production of Hb A with an increase in Hb A_2 from a normal of 2 to 3% to 3 to 5% and persistent synthesis of fetal Hb past the neonatal stage. The amount of fetal Hb will vary markedly among affected individuals. This condition is referred to as β thalassemia. A disruption in α-chain synthesis may also occur and in this instance the disease is referred to as α thalassemia. Since the α chain is common to all three Hb (A, A_2, and F), all three are somewhat depressed in α-thalassemia heterozygotes. The homozygous state for the α condition is fatal, leading to intrauterine death (Lehmann et al., 1966).

Of particular concern to the present report is the health status of the heterozygote because of its considerably greater prevalence in the population than the homozygote. The health status of the heterozygote is difficult to generalize since there appears to be an extremely broad differential expression of the clinical features of the disease even in the carrier state. However, what does seem predictable is that symptoms of the disease, however mild, may be exacerbated when additional stress is encountered; for example, in the presence of bronchopneumonia or during pregnancy. In a thalassemia heterozygote, auxiliary mechanisms of blood production have already been called into action and may no longer be able to handle the expanded activity needed for the maintenance of a normal Hb level (Lehmann et al., 1966).

Since persons heterozygous for the thalassemic trait have a compromised adaptive capacity to maintain blood production, it has been suggested that they may be at increased risk to blood toxins. The research to date has involved clinical assessment of occupational exposures to benzene and lead on persons heterozygous for β thalassemia although the previous biochemical characterization of the disease was not always made.

2. POLLUTANT EFFECTS

a. Benzene

There are several European reports which have investigated or suggested the hypothesis that persons with thalassemia may be at increased risk to the well-characterized blood toxins, benzene and lead. In 1957, Ferrara and Balbo reported observations of a 60-year-old worker with thalassemia who displayed hematological alterations characteristic of benzene intoxication. It was their medical opinion that his thalassemic condition enhanced the occurrence of benzene toxicity. Two years later, Saita and Moreo (1959) reported three cases of hypochromic anemia along with severe leuconeutropenia that they attributed to occupational exposure to benzene. However, as in the case noted above, these authors also felt that the toxic symptoms of benzene poisoning were enhanced by the presence of thalassemia. Based on their findings, Saita and Moreo (1959) favored the preemployment screening for hereditary blood disorders, although generally compatible with good health, but that could enhance the onset of toxicity to hemological toxins, such as benzene in occupational settings. In contrast, Girard et al. (1967) could not find that being a carrier of β thalassemia affected susceptibility to benzene toxicity in 28 workers. Consequently, these authors did not feel that it was reasonable to exclude such persons from working in occupations where benzene was used. Unfortunately, no monitoring of benzene exposure was reported so that it is difficult to properly compare these studies.

b. Lead

Five European papers have related clinical observations of lead intoxication in persons with thalassemia (Saita and Moreo, 1959; Roche et al., 1960; Gaultier et al., 1962; Guerrin et al., 1964; Pernot et al., 1966). Collectively, 17 individuals have been studied, and it is generally thought that those with this red cell disorder are predisposed to lead-induced hematological effects such as basophilic stippling and anemia. In fact, Girard et al. (1967) who did not favor the removal of thalassemics from occupational exposure to benzene, think it would be appropriate for such exclusion from lead exposures.

c. Antioxidant Enzyme Levels and Potential Risk to Environmental Oxidants

It is well-known that superoxide radicals and hydrogen peroxide as intermediates of oxygen metabolism may adversely affect all cells of respiring organisms. Erythrocytes that are especially susceptible to oxidant stress have a complex integrated biochemical system of defense to deal with such stress, including among others SOD for detoxifying superoxide radicals; the enzymes catalase and GSH-Px are used for the elimination of hydrogen peroxide.

It has been hypothesized that β-thalassemia red cells may be at enhanced oxidant risk. In theory, this enhanced risk is based on the premise that red cells containing free α chains are likely to increase superoxide production as seen by *in vitro* studies (Brunori et al., 1975), with an increase of hydrogen peroxide resulting from the decomposition of superoxide radicals by SOD (Winterbourn et al., 1976). In addition, Rachmilewitz et al. (1976) reported an increased lipid membrane peroxidation in thalassemic red cells. However, these authors also acknowledged that the generally low amount of Hb of thalassemics is not thought to be sufficiently adequate to act as a "sump" for the harmful free oxygen radicals.

Since thalassemic red cells are inherently exposed to more oxidant stress in the form of superoxide radicals, Gerli et al. (1980) assessed whether such cells may have adapted by having inherently higher levels of enzymes that would protect against oxidant stress. They found that the antioxidant enzymes SOD, catalase, and GSH-Px were markedly higher in the erythrocyte of individuals with β-thalassemia minor. Their findings are in agreement with those of Balcerzak and Jensen (1966) and Sagone and Balcerzak (1970) who noted a high level of catalase in persons with β-thalassemia minor and with Beutler et al. (1978) and Pescarmona et al. (1978), who noted high levels of GSH-Px in β-thalassemia minor.

Based on their findings, Gerli et al. (1980; 1981) speculated that the increased levels of antioxidant enzymes in those with β-thalassemia minor may be a defense mechanism against oxidant threat, especially against the generation of oxygen radicals. Finally, normal enzyme values were found in those with β-thalassemia major, but these seem best explained by the presence of normal red cells as a result of multiple transfusions.

3. FREQUENCY IN THE POPULATION

The frequency of the α-thalassemia heterozygote in Black Americans is thought to range between 2 and 7% based upon two infant population cord blood surveys (Minnich et al., 1962; Weatherall, 1963). More limited surveys of those of Greek ancestry were reported to have a 2.4% incidence of α thalassemia (Pearson et al., 1973).

The β-thalassemia heterozygote is found in about 4–5% of the Italian-Americans (Neel and Valentine, 1945) and Greek-Americans (Pearson et al., 1973).

4. SCREENING TESTS

Pearson et al. (1973) have developed an electronic measurement of mean corpuscular volume (MCV) that meets the requirements for a screening test for α- and β-thalassemic heterozygotes. The procedure is rapid, automated, and inexpensive. It yielded no false negatives out of a study population of 300. However, it is possible that false positives may occur for persons with an iron deficiency condition while persons who are so-called silent carriers of α or β thalassemia cannot be detected by this screening test. The frequency of the silent carrier is thought to be uncommon for β thalassemia. More recently, Rowley et al. (1979) utilized the methodology of Pearson et al. (1973) in a β-thalassemia screening and consulting study involving 15,534 persons. They required a determination of Hb A_2 volume if the MCV value was < 77 μ^3. If the Hb A_2 value was $> 3.5\%$, then the individual was assumed to be a carrier of β thalassemia. Although not specifically addressed by the researchers, this additional step should have greatly reduced an already low likelihood of false positives (i.e., those with low iron status). Finally, they did not indicate why they selected an initial MCV of < 77 μ^3 since Pearson et al. (1973) utilized 79 μ^3 as their cutoff point.

5. SUMMARY

In light of the clinical nature of these observations, the limited number of individuals studied and the lack of environmental monitoring, it is not possible to conclude that susceptibility to benzene and lead toxicity is enhanced in persons with thalassemia.* However, these clinical studies suggest the need for epidemiological investigations and/or development of a predictive animal model to test the association between thalassemia and lead and/or benzene toxicity.

REFERENCES

Balcerzak, S. P. and Jensen, W. N. (1966). Erythrocyte catalase activity in thalassaemia. *Scand. J. Haematol.* **3**: 245–256.

Benetti, P., DeFilippis, A., Ghezzo, F., and Perini, G. (1970). Attivita' Enzimatiche ematiche E sinomatologia di-beta-talassemici in corso di esposizione professionale a pesticidi. *Arcisp. S. Anna Ferrara* **23**: 33–43.

Beutler, E. (1978) *Hemolytic Anemia in Disorders of Red Cell Metabolism.* Plenum Medical, New York, pp. 23–167.

Brunori, M., Falcioni, G., Fioretti, E., Giardina, B., and Rotilio, G. (1975). Formation of superoxide in the antioxidation of the isolated α and β chains of human hemoglobin and its involvement in hemichrome precipitation. *Eur. J. Biochem.* **53**: 99–104.

Cooley, T. B. and Lee, P. (1925). A series of cases of spenomegaly in children with anemia and peculiar bone changes. *Trans. Am. Ped. Soc.* **37**: 29.

Ferrara, A., and Balbo, W. (1957). *Riv. Inf. Mal. Prof.* **44**: 713.

Gaultier, M., Fournier, E., and Geruais, P. (1962). Thalassemies mineures et Saturnisme. *Bull. Mem. Soc. Med. Hop. Paris,* **113**: 863.

---

*A 1970 report by Benetti et al. indicated that 31 β-thalassemic pesticide handlers did not show enhanced susceptibility to organophosphorus insecticides compared to 42 normal controls even though their initial red cell acetylcholinesterase activity was lower.

References

Gerli, G. C., Beretta, L., Bianchi, M., Pellagatta, A., and Agostoni, A. (1980). Erythrocyte superoxide dismutase, catalase, and glutathione peroxidase activities in β-thalassemia (major and minor). *Scand. J. Haematol.* **25:** 87–92.

Gerli, G. C., Beretta, L., Bianchi, M., and Agostoni, A. (1981). Erythrocyte superoxide dismutase, catalase and glutathione peroxidase in conditions of augmented stress. *Bull. Europ. Physiopath. Resp.* **17**(suppl.): 201–205.

Girard, R. P., Mallein, M. L., Jouvenceau, A., Tolot, F., Revol, L., and Bourret, J. (1967). Etude de la sensibilite aux toxiques industriels des porteurs du trait thalassemique. *J. Med. Lyon* **48:** 1113–1126.

Guerrin, M., Havez, R., Gerard, A., and Roussel, P. (1964). Hemogiobinopathie et saturnisme. *Lille Med.* **9:** 547–549.

Lehmann, H., Huntsman, R. G., and Ager, J. A. M. (1966). The hemoglobinopathies and thalassemia. In: *The Metabolic Basis of Inherited Disease*. 2nd ed. J. B. Stanbury, J. B. Wyngaarden, and D. S. Fredrickson (eds.). McGraw-Hill, New York, pp. 1100–1136.

Minnich, V., Cordonnier, J. K., Williams, W. J., and Moore, C. V. (1962). Alpha, beta and gamma hemoglobin polypeptide chains during the neonatal period with a description of the fetal form of hemoglobin D (St. Louis). *Blood* **19:** 137.

Mountain, J. T. (1963). Detecting hypersusceptibility to toxic substances. *Arch. Environ. Health* **6:** 63–71.

Neel, J. V. and Valentine, W. N. (1945). The frequency of thalassemia. *Am. J. Med. Sci.* **209:** 568–572.

Pearson, H. A., O'Brien, R. T., and McIntosh, S. (1973). Screening for thalassemia trait by electronic measurement of mean corpuscular volume. *N. Engl. J. Med.* **288**(7): 351–353.

Pernot, C., Larcan, A., Barbier, J. M., Kessler, Y., and Petit, J. (1966). Saturnisme et thalasseme. *Ann. Med. Nancy* **5:** 30–36.

Pescarmona, G. P., Sartori, M. L., Bosia, A., and Arese, P. (1978). Modificazioni dell'attivita di alcuni enzimi durante l'mvecchiamento in eritrociti di invidui normali. G6PD-carenti e β-talassemici. Atti XVI Riunione Gruppo dell'Eritrocita, Pescara, December 2.

Rachmilewitz, E. A., Lubin, B. H., and Shohet, S. B. (1976). Lipid membrane peroxidation in β-thalassemia major. *Blood* **47:** 495–505.

Roche, L., Lejeune, E., Tolot, F., Mouriguand, C., Mlle. Baron, Goineau, M., and Soubier, R. (1960). Lead poisoning and thalassemia. *Arch. Mal. Prof.* **21:** 329–333.

Saita, G., and Moreo, L. (1959). Talassemia ed Emopatie professionali. *Med. Lavoro* **50:** 25–36.

Rowley, P. T., Fisher, L., and Lipkin, M., Jr. (1979). Screening and genetic counseling for β-thalassemia trait in a population unselected for interest: Effects on knowledge and mood. *Am. J. Hum. Genet.* **31:** 718–730.

Saita, G. and Moreo, L. (1959). Thalassemia and occupational blood diseases. I. Thalassemia and chronic benzol poisoning. *Med. Lav.* **50:** 25–37.

Sagone, A. L., Jr., and Balcerzak, S. P. (1970). Activity of iron-containing enzymes in erythrocytes and granulocytes in thalassemia and in iron deficiency. *Am. J. Med. Sci.* **259:** 350–357.

Strambi, E. and Righi, E. (1969). Fitness of microcythemic subjects for occupations involving exposure to ionizing radiations. Preliminary considerations. *Med. Lav.* **56:** 664–671.

Weatherall, D. J. (1963). Abnormal hemoglobins in the neonatal period and their relationship to thalassemia. *Br. J. Haematol.* **9:** 265.

Winterbourn, C. C., McGrath, B. M., and Carrell, R. W. (1976). Reactions involving superoxide and normal and unstable haemoglobins. *Biochem. J.* **155:** 493–502.

G. Animal Models for Thalassemias

Limited information exists on the development and predictive efficacy of animal models with thalassemia. However, several different models exist that display a thalassemia-like disorder comparable to that of humans. According to Bannerman et al. (1974), there are several requirements of a potential animal model for thalassemia. These include the following:

1. A hereditary form of anemia in which the red cells show hypochromia and related morphological abnormalities.
2. Evidence of increased Hb catabolism, which is an important feature of the severe forms of thalassemia in man.
3. Evidence of imbalance of globin chain synthesis.

a. Belgrade Rat

One potential animal model for thalassemia is the so-called Belgrade rat. This was first reported by Sladic-Simic et al. (1969) who noted the occurrence of a profound hereditary hypochromic anemia in the offspring of an irradiated albino rat in Belgrade, Yugoslavia. The erythrocyte morphological abnormalities, including hypochromia, microcytosis, target cells, nucleated red cells, stippling, polychromasia, and persistent reticulocytosis closely resemble those found in human thalassemia.

This disease was found to have a genetic basis of an autosomal recessive nature with no expression of the anemia in the carriers (i.e., the heterozygotes). Human heterozygotes for both beta (Weatherall, 1978) and alpha (Weatherall, 1978) thalassemia can be identified in hematological observation with their conditions being considerably milder than that of the homozygous state (Sladic-Simic et al., 1969; Weatherall, 1978).

Of potential significance was that parenteral iron administration produced a significant but incomplete improvement of the anemia in the homozygous rats with the erythrocyte morphology still displaying hypochromasia and microcytosis because the increase in hemoglobin was associated with an increase in the numbers of red cells. This beneficial, though limited, treatment with iron in the Belgrade thalassemic rats does not occur in afflicted humans.

The findings suggest that the Belgrade model may be effective for simulating the human thalassemia in the homozygous condition. Although differences between the model and human thalassemic conditions exist, there is sufficient similarity to warrant active research designed to further assess the comparability of the model with human disease conditions with respect to normal physiology but also in response to oxidant stressor agents via *in vitro* and *in vivo* methods.

b. Other Models

Other animal models displaying characteristics of thalassemia have been summarized by Bannerman et al. (1974). However, there is a general lack of sufficient similarity with the human condition, thereby resulting in a low likelihood that these animals could be employed as adequate predictive models for the human disease. For example, *flex tailed anemia* is an autosomal recessive mouse blood disorder occurring as "a transient siderocytic anemia, white belly spotting, and deformity of the axial skeleton" (Bannerman et al., 1974). Since this condition occurs during gestation and disappears by two weeks of age, it probably cannot be employed as a model for the human adult condition (Mixter and Hunt, 1933). Furthermore, since the primary effects of thalassemia in man do not occur in the fetus and neonate, this model could not be of use for very young humans as well (Weatherall, 1978).

Mouse microcytic anemia, which is characterized by hypochromia and severe erythrocytosis in adults (Russell et al., 1970), has an etiology that is totally unrelated to human thalassemia. This disorder is due to a defect in the absorption of iron into various cell types and not to an imbalance of α- and β-chain (Bannerman et al., 1972) synthesis as occurs in man (Bannerman et al., 1974). A comparable lack of similarity to the human condition is found in the sex-linked anemia of mice (Grewal, 1962). This model's anemia results from an iron deficiency caused by defective absorption of iron into the intestine (Pinkerton and Bannerman, 1967). Consequently, these examples of mouse anemia would not be appropriate models to study the human disease. In addition, two types of hemolytic anemias in mice have been reported that are characterized by red cell hypochromia and are not caused by any defect in iron absorption. However, the precise nature of the defects remains to be assessed.

This synopsis of the current studies of animal models resembling human thalassemia indicates only a very limited current potential for investigators hoping for sufficient comparability to make toxicological predictions meaningful. The present array of models is essentially manmade since such thalassemia-like anemias have generally resulted in offspring of irradiated parents. While such a procedure offers potential for creating novel models that may have applicability to the human condition, Bannerman et al. (1974) offers the ecologic approach as a possibility. He stated that since heterozygote humans have a selective advantage in areas of endemic malaria, similar types of adaptations may have occurred in other animals living in the same habitat. He noted that two kinds of malaria, *Plasmodium berghei* and *P. vinckei* are endemic in rodents of the Congo, Nigeria, and other parts of Africa. He suggested that a survey of wild rodents in these areas may reveal animal models that have also adapted to malarial environments by evolving hemoglobinopathies or thalassemia similar to man and may therefore offer potential models for study.

REFERENCES

Bannerman, R. M., Edwards, J. A., and Kreimer-Birnbaum, M. (1974). Investigation of potential animal models of thalassemia. *Ann. N.Y. Acad. Sci.* **232**: 306–322.

Bannerman, R. M., Edwards, J. A., Kreimer-Birnbaum, M., McFarland, E., and Russell, E. S. (1972). Hereditary microcytic anaemia in the mouse: studies in iron distribution and metabolism. *Br. J. Haemat.* **23**: 235–245.

Grewal, J. S. (1962). A sex-linked anaemia in the mouse. *Genet. Res. Camb.* **3**: 238–247.

Mixter, G. and Hunt, H. R. (1933). Anemia in the flex-tailed mouse, *Mus musculus. Genetics* **18**: 367–387.

Pinkerton, P. H. and Bannerman, R. M. (1967). *Hereditary defect in iron absorption in mice.* Nature **216**: 482–483.

Popp, R. A., Francis, M. C., and Bradshaw, B. S. (1978). Erythrocyte life span in alpha thalassemic mice. *Birth Defects Orig. Artic. Ser.* **14**(1): 181–185.

Russell, E. S., Nash, D. J., Bernstein, S. E., Kent, E. L., McFarland, E. C., Mathews, S. M., and Norwood, M. S. (1970). Characterization and genetic studies of microcytic anemia in house mouse. *Blood* **35**: 838–850.

Sladic-Simic, D., Martinovich, P. N., Ziukovic, N., Kahn, M., and Ranney, H. M. (1969). A thalassemia-like disorder in Belgrade laboratory rats. *Ann. N.Y. Acad. Sci.* **165**: 93–99.

Sladic-Simic, D., Zivkovic, N., Pavic, D., Marinkovic, D., Martinovic, J., and Martinovitch, P. N. (1966). Hereditary hypochromic anemia in the laboratory rat. *Genetics* **53**: 1079.

Weatherall, D. J. (1978). *The Metabolic Basis of Inherited Disease.* McGraw-Hill, New York, p. 1512.

H. NADH Dehydrogenase Deficiency (i.e., MetHb Reductase Deficiency)

1. INTRODUCTION

Of critical importance in the maintenance of human life is the transportation of oxygen to the various tissues. This function is contingent on the capability of Hb to bind oxygen reversibly, which in turn relies on the ferrous iron of the heme moiety. The binding of oxygen by Hb involves the partial transfer of an electron from the heme iron to oxygen. While the "transferred" electron typically returns to the heme iron, there is a limited escape of such electrons, thereby leaving the Hb in an oxidized state called ferri or MetHb. Only a very small percentage (i.e., about 1%)

of total Hb is present in the ferric state because the capacity of the red cell to reduce the ferric state is several hundred times greater than the spontaneous rate of oxidation. MetHb levels accumulate when (1) the rate of oxidation of heme is so great as to exceed the reducing capacity, (2) a change in the globin moiety stabilizes MetHb making it resistant to metabolic reduction (e.g., Hb M disorder), or (3) there is a marked deficiency in the reducing ability of the red cell. The most important metabolic pathway for the reduction of MetHb involves an NADH dehydrogenase, which accounts for 61% of the normal reduction rate (Schwartz and Jaffe, 1978).

2. CAUSES OF METHEMOGLOBINEMIA

Methemoglobinemia in humans was initially reported in persons who were exposed to certain substances that were capable of increasing the rate of oxidation of Hb. Several comprehensive reviews thoroughly documented the diversity of such substances and their proposed mechanisms of action (Kiese, 1974). However, persistently high levels of MetHb have been clinically diagnosed with no known exposure to chemicals that oxidize Hb. Subsequent research of such individuals has frequently revealed an NADH dehydrogenase deficiency as the cause of the high MetHb levels.

Patients displaying hereditary methemoglobinemia display few discernible adverse effects even when MetHb levels are as high as 25%. This disease does not appear to cause cardiac or pulmonary disease; life expectancy is normal, pregnancies are not adversely affected,* and there is no erythrocyte hemolysis. The physiologic mechanisms that compensate for the hypoxia are poorly understood. In contrast to the almost benign nature of hereditary methemoglobinemia is the poorly tolerated and potentially fatal outcome of chemically (e.g., nitrite, aniline derivatives, etc.) induced methemoglobinemia due to heme hypoxia (Schwartz and Jaffe, 1978).

It has repeatedly been shown that persons homozygous or heterozygous for NADH dehydrogenase deficiency display an increased risk of cyanosis following exposure to medicines such as dapsone, primaquine, and chloroquine, which form MetHb. However, no reports of industrial exposures indicating that those with NADH dehydrogenase deficiency are at increased risk to MetHb forming agents have been published although this has been suggested as a likely possibility (Jaffe, 1966). The lack of such findings is not unexpected since there is no mention that industrial screening for such a condition has ever occurred. Typically, if

---
*A minority of patients with hereditary methemoglobinemia are mentally retarded and display a neurologic impairment. This is thought to result from a deficiency of NADH dehydrogenase in the central nervous system (Schwartz and Jaffe, 1978).

industrial cyanosis occurred as Linch (1974) has reported, only G-6-PD deficiency has been checked. In fact, industries that adopted G-6-PD screening most likely followed the suggestion of Stokinger and Mountain (1963) who did not include a recommendation for NADH dehydrogenase deficiency monitoring.

3. FREQUENCY IN POPULATION

It has been reported that there is a very high occurrence of hereditary NADH dehydrogenase deficiency among Alaskan Eskimos and Indians (Scott and Hoskins, 1958), Navajo Indians (Balsamo et al., 1964), and Puerto Ricans (Hsieh and Jaffe, 1971; Schwartz et al., 1972). Heterozygous carriers of this enzyme deficiency display about 50% of the normal enzyme activity with the frequency of such carriers thought to be about 1% as reported by WHO (1973). According to Omenn and Motulsky (1978), the figure of 1% is derived from the Hardy-Weinberg Law assuming that the homozygous recessive is approximately 1 per 40,000.

4. SCREENING TESTS

The definitive diagnosis of hereditary methemoglobinemia requires the demonstration of deficient NADH dehydrogenase activity in red cell hemolysates using the Hegesh assay (Hegesh et al., 1968), or the Scott assay (Scott, 1960). The Hegesh assay is considered preferable because of its greater specificity, accuracy at low enzyme activity levels, and ease of operation.

5. SUMMARY

There is no direct evidence that persons either homo- or heterozygous for NADH dehydrogenase deficiency are at increased risk to MetHb-forming industrial chemicals. Evidence does exist that they are at risk to medicinal compounds capable of causing MetHb formation. There would appear to be good theoretical likelihood that such persons represent a potential industrial high-risk group.

REFERENCES

Balsamo, P., Hardy, W. R., and Scott, E. M. (1964). Hereditary methemoglobinemia due to diaphorase deficiency in Navajo Indians. *J. Pediatr.* **65:** 928.

Cohen, R. J., Sach, J. R., Wicker, D. J., and Conrad, M. E. (1968). Methemoglobinemia provoked by malarial chemoprophylaxis in Vietnam. *N. Engl. J. Med.* **279:** 1127–1131.

Hegesh, E., Calmanovici, N., and Avron, M. (1968). New method for determining ferrihemoglobin reductase. (NADH-methemoglobin reductase) in erythrocytes. *J. Lab. Clin. Med.* **72:** 339.

Hseih, H. S. and Jaffe, E. R. (1971). Electrophoretic and functional variants of NADH methemoglobin reductase in hereditary methemoglobinemia. *J. Clin. Invest.* **50:** 196.

Jaffe, E. (1966). Hereditary methemoglobinemias associated with abnormalities in metabolism of erythrocytes. *Am. J. Med.* **41:** 786–798.

Kiese, M. (1974). *Methemoglobinemia: A Comprehensive Treatise*. CRC Press, Cleveland, Ohio.

Linch, A. L. (1974). Biological monitoring for industrial exposure to cyanogenic aromatic nitro and amine compounds. *Am. Indust. Hyg. Assoc. J.* **35:** 426–432.

Omenn, G. S. and Motulsky, A. G. (1978). "Eco-genetics": Genetic variation in susceptibility to environmental agents. In: *Genetic Issues in Public Health and Medicine*. Thomas, Springfield, Illinois, pp. 83–111.

Schwartz, J. M. and Jaffe, E. R. (1978). Hereditary methemoglobinemia with deficiency of NADH dehydrogenase. In: *The Metabolic Basis of Inherited Disease*. J. B. Stanbury, J. B. Wyngaarden, and D. S. Fredrickson (eds.). 4th ed. McGraw-Hill, New York, pp. 1452–1464.

Schwartz, J. M., Parass, P. S., Ross, J. M., DiPillo, F., and Rizek, R. (1972). Unstable variant of NADH methemoglobin reductase in Puerto Ricans with hereditary methemoglobinemia. *J. Clin. Invest.* **51:** 1594.

Scott, E. M. (1960). The relation of diaphorase of human erythrocytes to inheritance of methemoglobinemia. *J. Clin. Invest.* **39:** 1176.

Scott, E. M. and Hoskins, D. D. (1958). Hereditary methemoglobinemia in Alaskan Eskimos and Indians. *Blood* **13:** 795.

Stokinger, H. E. and Mountain, J. T. (1963). Tests for hypersusceptibility to hemolytic chemicals. *Arch. Environ. Health* **6:** 495–502.

WHO (1973). Pharmacogenetics. WHO Tech. Rep. Ser. No. 524.

I. Animal Models for the Study of MetHb Formation

1. INTRODUCTION

Numerous drugs and environmental pollutants exist that are mild to strong oxidizing agents capable of converting Hb to MetHb. Among the most well-known MetHb producers are aniline derivatives such as acetanilide and acetophenetidine (Lester, 1943), as well as copper sulfate and nitrobenzene, azobenzene, arsine, hydrazine, and sodium nitrite. Kiese (1974) has documented in a systematic manner the types of compounds causing an increase in ferrihemoglobin and their respective mechanism(s) of action.

In light of the relatively large numbers of oxidant stressor agents developed in the drug and chemical industries, as well as those found in the general community environment, it is of value to consider the potential of such agents to cause oxidative stress including ferrihemoglobin (i.e., MetHb) formation in human red blood cells. Limited investigations have evaluated interspecies susceptibilities with regard to ferrihemoglobin formation in hopes of finding an animal model with similar sensitivities as the human. However, before comparing the relative susceptibility to ferrihemoglobin formation among species, a brief description of the ferrihemoglobin reduction mechanisms is helpful. The principal system involved in ferrihemoglobin reduction in mammalian red cells (accounting for at least 60%) is ferrihemoglobin reductase or diaphorase that requires NADH as a cofactor (Smith, 1980). A second reductive process found in normal human and most mammalian red cells is the so-called dormant NADPH-reductase system that requires NADPH as a cofactor. This system appears to be activated only by the administration of exogenous electron carriers such as methylene blue. According to Smith (1980), there is convincing evidence that this system has no significant physiological role in MetHb reduction. Some of the less significant but physiologically active pathways employed for the reduction of ferrihemoglobin involve direct reduction by reduced GSH which may contribute up to 12% of the total red cell reducing ability (Scott et al., 1965) and ascorbic acid which may account for up to 16% of the reductive capacity of the red cell (Smith, 1980). Consequently, it would appear that interspecies differences principally in ferrihemoglobin reductase but also in reduced GSH and ascorbic acid status may contribute to differential susceptibility to MetHb forming agents.

2. INTERSPECIES COMPARISONS

Smith (1980) asserts that the duration of ferrihemoglobinemia after an acute challenge with sodium nitrite is principally contingent on the level of ferrihemoglobin reductase activity in the red cells of various species. He presented evidence as seen in Table 2-6, which indicates wide species variability in MetHb reductase activity. Table 2-6 reveals that the pig and horse erythrocytes have much lower levels of MetHb reductase activity than human erythrocytes, presumably because these species employ plasma lactate instead of glucose to propel the reduction (see Rivkin and Simon, 1965; Robin and Harley, 1966). In contrast, the rat, mouse, guinea pig, and rabbit displayed considerably higher rates of MetHb reductase activity relative to humans. In addition to the above explanation for the slow ferrihemoglobin reduction in pigs and horses, Kiese (1974) suggested that it may be due, at least in the pig, to relatively slow penetration of glucose through the membrane as well (see Kim and McManus, 1971, 1971a).

The first direct interspecies comparisons concerning the formation of ferrihemoglobin were published in 1893 by Szigeti, who noted a marked difference among various laboratory animal models following administration of aniline derivatives. Twenty years later Heubner (1913) extended these findings by reporting similar interspecies variability to a

Interspecies Comparisons

Table 2-6. Spontaneous MetHb Reductase Activity of Mammalian Erythrocytes[a]

	\multicolumn{5}{c}{Investigators}
	(1)	(2)	(3)	(4)	(5)
Species	\multicolumn{5}{c}{Activity in Species/Activity in Humans}
Pig	0.37	0.37		0.09	
Horse	0.75	0.50		0.64	
Cat		0.50	0.85	1.2	1.0
Cow	0.80	0.75		1.1	
Goat	1.1	0.75			
Dog		0.88	1.4	1.3	1.0
Sheep	1.4	1.0		2.1	
Rat		1.4	1.3	1.9	5.0
Guinea pig		1.2	2.4	1.9	4.5
Rabbit		3.5	3.3	3.8	7.5
Mouse					9.5

Source: Smith (1980).

[a] Data from various investigators using nitrited red cells with glucose as a substrate have been normalized by making a ratio of the activity of the species to the activity in human red cells. The indicated investigators are: (1) Smith and Beutler (1966a), (2) Malz (1962), (3) Kiese and Weis (1943), (4) Robin and Harley (1966), (5) Stolk and Smith (1966), Smith et al. (1967), Bolyai et al. (1972).

variety of ferrihemoglobin producing substances. Although these initial reports were of a rather qualitative nature, it established a general trend that carnivorous animals are relatively susceptible to ferrihemoglobin forming agents whereas herbivores were generally insensitive. This enhanced susceptibility of carnivores to the formation of aniline induced MetHb probably is best explained by observations that these animals tend to hydroxylate aniline predominantly in the ortho position while the herbivorous species tested are relatively insensitive because they hydroxylate a greater proportion of the aniline to the nonferrihemoglobin forming para-isomer of aminophenol (Table 2-7) (Parke, 1960).

The next interspecies investigations of susceptibility to ferrihemoglobin formation were published by Lester (1943). He compared the maximum amount of ferrihemoglobin formed following single doses of acetanilide and acetophenetidine in rats, rabbits, cats, dogs, monkeys, and humans. Their findings revealed very large differences in the sensitivity to forming MetHb. The average amounts of ferrihemoglobin produced are shown in Figure 2-3. The data suggest that each species displays a unique threshold dose for ferrihemoglobin formation. Furthermore, the extent of ferrihemoglobin formation is not directly proportionate to the increase in dose. There is also an apparent upper limit to the amount of ferrihemoglobin that can be formed. Larger doses, al-

Table 2-7. Metabolic o- and p-Hydroxylation of Aniline by Various Animal Species

Species	Sex	Oral Dose (mg/kg)	Ratio Para/Ortho
Gerbil	F	250	15
Guinea pig	F	250	11
Golden hamster	F	250	10
Rabbit	F	160–500	6[a]
Rat	M	250	6
	F	250	2.5
Chicken	F	50	4
Mouse	M	250	3
Ferret	F	250	1
Dog	F	200	0.5
Cat	F	200	0.4

Source: From Parke, D. V., *Biochem. J.* **77**, 493 (1960). With permission.
[a] Piotrowski, using much smaller doses of aniline, 0.1 mg/kg intravenously, found the ratio to be 12.

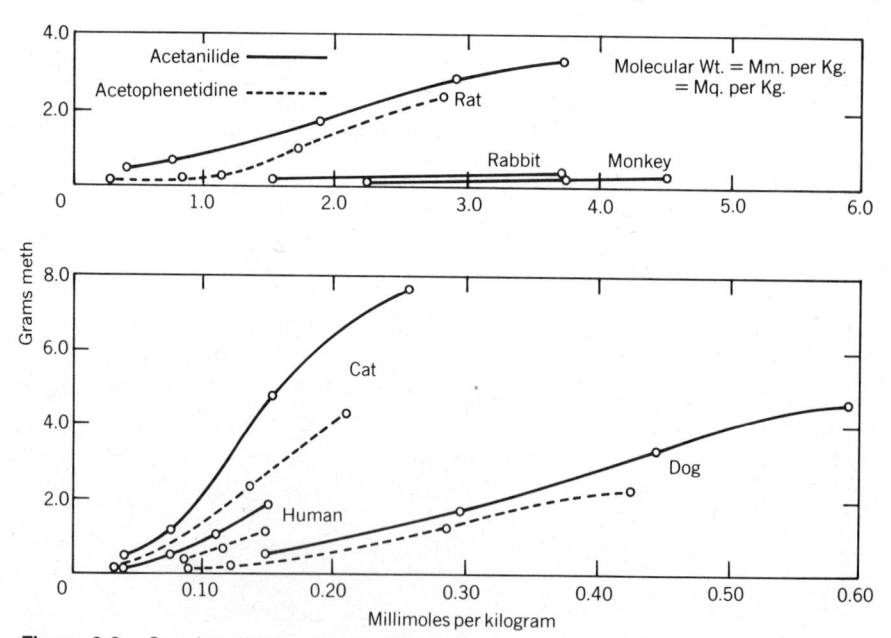

Figure 2-3. Species differences at maximum formation of MetHb for various doses of acetanilide and acetophenetidine. *Source:* Lester (1943).

Interspecies Comparisons

Table 2-8. Comparative Species Sensitivity to Formation of Ferrihemoglobin: Cat Taken as 100

	Acetanilide				Acetophenetidine			
Dose (g)	Cat	Man	Dog	Rat	Cat	Man	Dog	Rat
0.5	100	53	27	6	100	63	32	5
1.0	100	56	30	5	100	62	35	6
1.5	100	56	28	5	100		39	5
2.0	100	60	30	5	100		36	5
2.5	100		30	4	100		32	5
3.0	100		30	4	100			

Source: From Lester, D., *J. Pharmacol. Exp. Ther.* **77**, 154 (1943), with permission.

though not producing higher levels of the ferrihemoglobin, increase the length of time it remains.

With respect to interspecies differences, the rabbit and monkey were essentially nonresponsive. Other investigators have reported a similar lack of responsiveness for monkeys (Smith, 1940) and rabbits (Kruse and McEllroy, 1927; Heubner, 1913; Young and Wilson, 1926) after exposure to acetanilide. The most sensitive animal model was the cat, which was approximately twice as sensitive as man, the second most sensitive responder. Table 2-8 offers a relative comparison of the species differences with respect to ferrihemoglobin formation from exposure of acetanilide and acetophenetidine. The data revealed that humans are slightly more than half as susceptible as the cat, while the dog is half as susceptible as humans and the rat only $\frac{1}{10}$ as sensitive as man.

The causes of the differential susceptibility are difficult to sort out. In the cases of these aniline derivatives one is concerned with the extent to which the substance is converted to the ortho isomer as well as the efficiency by which the species can reduce ferrihemoglobin back to Hb via the action of ferrihemoglobin reductase. Since there are several presumably independent factors affecting the outcome, it is not surprising that the interspecies differences observed cannot be totally explained by either variable.

More recent investigations have supported these earlier studies and have also revealed rather marked interspecies differences in the susceptibility to form ferrihemoglobin (Burger et al., 1966; Smith and Beutler, 1966). These studies revealed that sheep were markedly more sensitive than the pig to both sodium nitrite (Smith and Beutler, 1966) and phenylhydroxylamine (Burger et al., 1966) induced ferrihemoglobin formation with the cow being slightly less sensitive than the sheep.

A direct interspecies comparison of *in vivo* ferrihemoglobin reduction was presented by Kiese and Weger in 1969. After inducing significant levels of ferrihemoglobin by an intravenous injection of 4-dimethyl-

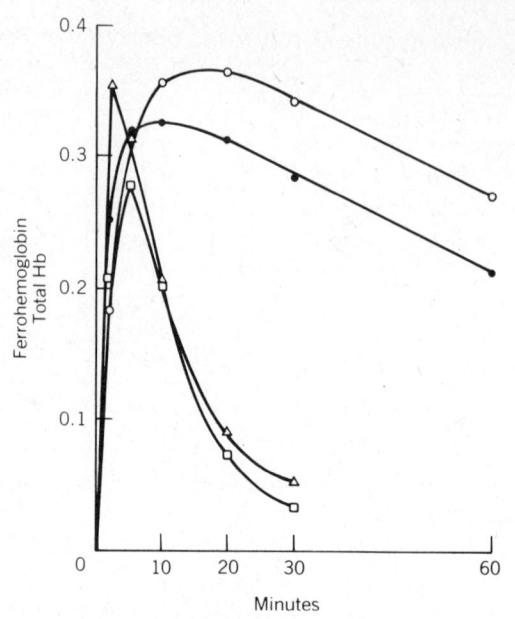

Figure 2-4. Reduction of ferrihemoglobin *in vivo* after formation of ferrihemoglobin by intravenous injection of 4-dimethylaminophenol hydrochloride. ○, dog, 2.0 mg/kg, three experiments; ●, cat, 2.25 mg/kg, seven experiments; □, mouse, 20 mg/kg, three to nine experiments; △, rabbit, 30 mg/kg, four experiments. *Source:* Kiese, M. and Weger, N., *Eur. J. Pharmacol.* **7,** 97 (1979), with permission.

aminophenol the rates of reduction were assessed in cats, dogs, mice, and rabbits. The finding revealed that mice and rabbits very rapidly reduced the ferrihemoglobin, while cats and dogs were quite noticeably less efficient in this regard (Figure 2-4).

A similar comparison of ferrihemoglobin reductase activities in the erythrocytes of a variety of species as assessed in the presence of methylene blue revealed in general that species with high spontaneous rates of reductase activity respond more actively to methylene blue (Smith, 1980; Agar and Harley, 1972). Similar findings have been reported by numerous investigators (Gelinski, 1940; Mathies, 1957; Kiese and Weis, 1943; Malz, 1962; Stolk and Smith, 1966; Smith and Beutler, 1966; Scott, 1967).

3. CAUSES OF INTERSPECIES DIFFERENCES IN FERRIHEMOGLOBIN FORMATION

Susceptibility to ferrihemoglobin formation varies widely among species according to numerous factors. A partial list and explanation of biological factors that may contribute to interspecies differences include the following.

Table 2-9. Rates of Ferrihemoglobin Reduction *In Vitro* in Red Cells of a Variety of Species with Glucose, 250 mg/100 mL, or Lactate, 20 mg/100 mL, as Substrate[a]

	Substrate		
Species	Glucose	Glucose Plus Nile Blue	Lactate
Rabbit	9.0	45	16
Guinea pig	6.5	30	9
Rat	3.5	20	4.4
Dog	3.8	10	4
Human	2.7	28	2.4
Cat	2.3	6	3

Source: From Kiese, M. and Weis, B., *Naunyn Schmiedebergs Arch. Exp. Pathol. Pharmakol.* **202**, 493 (1943), with permission.

[a]Red cells suspended in Ringer-phosphate solution, pH 7.3, at 37° under carbon monoxide. The effect of catalytic dyes was tested with 3×10^{-5} M Nile blue. The figures indicate micrograms of ferrihemoglobin reduced in 1 million red cells in 1 hr.

a. Availability of Substrate

It has been established that ferrihemoglobin is enzymatically reduced in intact red cells when certain substrates are present. Substrates known to facilitate this reaction include glucose (Warburg et al., 1930) and lactate (Wendel, 1930), as well as several other carbohydrates. Kiese (1974) has reported that glucose and other sugars generate NADH via glycolysis with triose phosphate and lactate serving as electron donors for the reduction of NAD. Interspecies (rabbit, guinea pig, rat, dog, human, cat) differences in the rate of ferrihemoglobin formation as influenced by different substrates (i.e., glucose, glucose plus Nile blue, or lactate) consistently demonstrated the rabbit and guinea pig to be the most efficient reducers while the cat was the least efficient followed, respectively, by the human, dog, and rat (Table 2-9) (Kiese and Weis, 1943). Comparable interspecies differences in the rate of ferrihemoglobin reduction were also reported using galactose as a substrate with the rabbit being considerably more effective than the rat and human (Gupta et al., 1972). Finally, it has been previously noted that a reason why pigs are thought to be very slow reducers of ferrihemoglobin is related to the slow penetration of glucose through the cell membrane (Kim and McManus, 1971, 1971a).

b. Differential Gastrointestinal Tract/Bacterial Metabolism

Administration of relatively high levels of inorganic nitrate has not been found to be an impressive producer of ferrihemoglobin *in vivo* in cats

(Pulina, 1942) and dogs (Greene and Hiatt, 1954). However, 0.5 g NaNO$_3$/kg at 24 hr resulted in 20% of the cat Hb being oxidized (Pulina, 1942). Intravenous (2.8 g/kg) or oral (1.4 g/kg) administration in dogs did not produce ferrihemoglobin. In marked contrast, consumption of hay with high nitrate levels is known to be highly toxic to cattle (Breadley et al., 1939). In fact, Breadley et al. (1940) reported that the injection of nitrate (0.5 g/kg or more) resulted in an increase in ferrihemoglobin in the blood of cattle to 80–90% of total Hb, killing these animals, with sheep responding similarly. This interspecies difference with respect to nitrate-induced formation of ferrihemoglobin between cats/dogs and cattle/sheep may be explained in part by the fact that the sheep (Shapiro et al., 1949) and presumably cattle reduced nitrate to nitrite in the rumen.

c. Differential Hydroxylation

It has been demonstrated that the rate and the location of hydroxylation for aniline and its derivatives can affect the formation of ferrihemoglobin metabolites. Dehydroxylation of aminophenols yielding the parent amines is not thought to occur *in vivo* (Kiese, 1974).

d. Influence of Conjugation Rates

Kiese (1974) reported that the extent to which aminophenols are capable of forming ferrihemoglobin is contingent on the rate of conjugation with glucuronic acid and sulfate. Decreasing the rate of these reactions enhances the possibility of increased levels of ferrihemoglobin (Scheff, 1940). The conjugates lack the ability of the free aminophenols to transfer rapidly electrons from ferrihemoglobin to molecular oxyhemoglobin. In fact, Williams (1943) reported a very slow formation of ferrihemoglobin in hemolytes after addition of high concentrations of *o*- or *p*-aminophenylglucuronide. Interestingly, *N*-acetylation prevents the formation of ferrihemoglobin by *p*-aminophenol. These acetylated compounds are prevented from forming ferrihemoglobin unless they themselves or their hydroxylation products are subsequently deacetylated (Kiese, 1974). Furthermore, acetanilide, phenacetin, or prilocaine administration may result in the formation of significant amounts of ferrihemoglobin but only after the release of aniline, *p*-phenetidine, or *O*-toludine following deacetylation. Consequently, the production of ferrihemoglobin by *N*-acylarylamines requires deacetylation before or after *N*- or ring hydroxylation (Kiese, 1974). Similarly, ferrihemoglobin formation in mice by the herbicide propanil, 3,4-dichloropropionanilide was markedly reduced when deacetylation was inhibited by prior treatment with tri-*O*-tolyl phosphate (Singleton and Murphy, 1973). Another con-

Discussion

founding variable making interspecies comparisons more difficult is that of erythrocyte catalase activity that can play a role in preventing the formation of ferrihemoglobin (Aebi, 1970), especially at elevated levels of oxidant exposure.

4. DISCUSSION

In attempting to predict the effects of ferrihemoglobin forming agents on humans *in vivo*, what is the animal model of choice and why? In a practice examination issued by the American Board of Toxicology to help individuals prepare for their formal examination for Board certification in Toxicology, the following question (#100) was posed:

A test material is suspected of producing methemoglobinemia as its major toxicologic effect. The least appropriate species for assessing this possibility is:

A. monkey
B. dog
C. cat
D. rabbit
E. mouse

The correct response according to the answer sheet supplied with the examination was "E. mouse." While an explanation to support the selection of this answer was not provided, strong evidence to support it may be derived from Table 2-6. In Table 2-6 the spontaneous MetHb reductase activity of various mammalian erythrocytes including human, following exposure to nitrite and the substrate glucose were compared by a number of investigators. Of the species evaluated, mice displayed the highest enzyme activity.

Since mice have the greatest ability to reduce ferrihemoglobin following exposure to nitrite in the presence of glucose, they would presumably be very insensitive with respect to exhibiting high ferrihemoglobin levels following nitrite exposure and therefore be an inadequate model for this pollutant.

It must be realized however that the absolutism of the answer given above can be very misleading because sensitivity to ferrihemoglobin producing substances is contingent on many species specific factors and not just the activity of the enzyme, MetHb reductase. The so-called correct answer was presumably based on an evaluation of the effects of but one ferrihemoglobin producing substance, nitrite, and with but one substrate, glucose. Numerous examples have been provided of how susceptibility to ferrihemoglobin producing substances can be modified by other metabolic processes depending on the specific compound and species evaluated. For

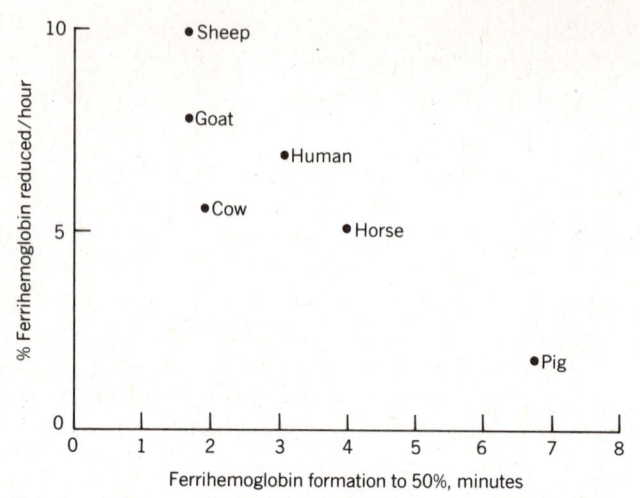

[Figure: scatter plot with y-axis "% Ferrihemoglobin reduced/hour" (0 to 10) and x-axis "Ferrihemoglobin formation to 50%, minutes" (0 to 8). Data points: Sheep (~3, 9.5), Goat (~2, 7.5), Human (~3, 7), Cow (~2.5, 6), Horse (~4, 5), Pig (~6.5, 2).]

Figure 2-5. Relation between rate of ferrihemoglobin formation by nitrite and rate of ferrihemoglobin reduction *in vitro* with red cells of various species. The rate of ferrihemoglobin formation is defined by the time needed for oxidation of 50% of the Hb. Glucose was used as substrate in the reduction experiments. *Source:* Smith, J. E. and Beutler, E., *Am. J. Physiol.* **210**, 347 (1966), with permission.

example, susceptibility of ruminants to nitrate toxicity is caused by the conversion of nitrate to nitrite (Lewis, 1951), which in turn may oxidize the ruminant hemoglobin to ferrihemoglobin (Smith and Beutler, 1966; Betke et al., 1956). These modifying factors clearly demonstrate that an animal model should not be selected for evaluation on the basis of just its ability to reduce ferrihemoglobin but also on its susceptibility for MetHb formation. In fact, Smith and Beutler (1966) hypothesized that it would be logical to assume that the ferrihemoglobin-reducing capacity of a species would be increased if its hemoglobin molecule is inherently susceptible to oxidation. In a comparison of six species, sheep, goats, and cows (ruminants), and humans, horses, and pigs (nonruminants), Smith and Beutler (1966) reported a correlation coefficient of -0.89 ($P \leq 0.05$) between the ability to form and reduce ferrihemoglobin, thereby supporting their hypothesis (Figure 2-5).

The determination of which animal model most accurately predicts the occurrence of ferrihemoglobin formation in normal humans however is an entirely different issue than predicting the most or least sensitive species. If, however, one utilized Table 2-6 to discuss the model(s) responding most similarly to humans it could be concluded that rats, guinea pigs, rabbits, and mice displayed markedly greater reductase activity than man, while pigs and horses were uniformly much lower than humans. The MetHb reductase activity of cats, cows, goats, and dogs fairly closely simulated the responses of humans. Based on these data, it would be tempting to claim that the use of rodents and rabbits would result in false negative findings, while false positives may be expected if

Discussion

the pig and horse were used as models. However, this is too simplistic an approach because ferrihemoglobin levels may be affected by many different factors including the inherent abilities to reduce ferrihemoglobin by the reductase enzyme, the extent of other reducing agents in the blood such as reduced glutathione, cysteine, and ascorbic acid, the inherent sensitivity of the Hb molecule to oxidation, whether the parent compound is the ferrihemoglobin forming compound or a metabolite; if it is a metabolite, then to what extent is it formed, and how long its biological half life is.

To be able to assess whether an animal may be an appropriate model to simulate adequately human responses to potential ferrihemoglobin forming agents is no easy task and the answer may well be different depending on the specific compound under consideration as well as the subgroup of the human population being studied.

The federal EPA has a national maximum drinking water standard limiting the amount of nitrate in drinking water to 10 mg/L as N. This standard is especially designed to protect infants who are considered at increased risk to nitrate toxicity because of several developmental processes unique to the very young. These include the presence of nitrate reducing bacteria living in their stomachs because it has a higher pH than older humans, and that Hb F is more oxidizable than Hb A among other factors. The question then is raised as to what would be the best animal model to use in predicting the *in vivo* effects of nitrate, nitrite, and nitrogen dioxide on normal and high-risk humans. Ideally, one should attempt to compare and contrast the respective models with humans for the known parameters affecting ferrihemoglobin formation. For example, the fact that mice have 9.5 times more erythrocyte MetHb reductase activity than humans would clearly suggest that mice may be considerably less sensitive to nitrite induced ferrihemoglobin formation than humans. In fact, a comparison of mice with six other species (cat, dog, guinea pig, rabbit, sheep) including man with regard to susceptibility to nitrite-induced ferrihemoglobin formation (via three routes of exposure, intravenously, intraperitoneally, and subcutaneously) revealed that mice were the least susceptible to ferrihemoglobin formation closely followed by rabbits (Kiese, 1974). Dogs, cats, and pigs were the most susceptible species. Limited intravenous derived human values suggested a sensitivity close to dogs and cats based on the rate of ferrihemoglobin found per dose of sodium nitrite in milligrams per kilogram (Kiese, 1974). Other studies via peroral administration revealed the human to have a MetHb formed to nitrite dose (mg/kg) ratio of nearly 1.8 while rats display about a 0.7 value, thereby indicating that humans were about twice as sensitive as the rat (Kiese, 1974). In practical terms, this may suggest that if investigators were studying the influence of nitrite or nitrogen dioxide on ferrihemoglobin formation *in vivo* then the rat, but especially the mouse and rabbit, would be expected to yield false negative ferrihemoglobin values while dogs and cats seem to be generally similar to humans. These findings may suggest the use of, say, the dog in inhalation studies with pollutants such as nitrogen dioxide and possibly ozone in an attempt to determine the influence of such pollutants on ferrihemoglobin levels. However, Miller et al. (1978) have reported that 60–90% of the ozone inhaled by dogs becomes extinct (or absorbed) in the nasopharyngeal tract before reaching the lungs, while rabbits closely simulate the human with respect to nasopharyngeal absorption. Thus, the rabbit and dog may be inappropriate models for different reasons.

Interspecies studies of other ferrihemoglobin producing substances have also revealed that mice and rabbits and usually rats as well are much less susceptible than cats and dogs. Human data, when available, are often found in an intermediate position, located between these two divergent groups, although closer to the cats and dogs. This has been found by comparing the findings of various researchers with respect to O-aminophenol hydrochloride, p-methylaminophenol sulfate, p-dimethylaminophenol hydrochloride (Kiese, 1974), nitrobenzene (Kiese, 1974), and pamaquine (Eichholtz, 1927; LeHeux and van Wijngaarden, 1927, 1929). Despite these consistent though limited interspecies differences and their potential application toward attempting to select rationally appropriate animal models, it is quite clear that susceptibility to ferrihemoglobin forming agents remains highly variable depending on the specific pollutant. Yet this inherent variability and initial uncertainty does not preclude the formation of a rational procedure that may permit not only an understanding of the strengths and weaknesses of any selected model including whether they tend to predict either false negatives or positives but also to exploit the variability in responses among the possible animal models when attempting to predict the responses of subgroups within the human population that may markedly differ in their susceptibility to ferrihemoglobin formation for one of a number of possible causes including a deficiency in MetHb reductase activity, Hb molecules being more susceptible to oxidation, G-6-PD deficiency, acatalasemia, microbial reduction of nitrate, differential capacity for acetylation and deacetylation, as well as others. These factors that affect the occurrence of ferrihemoglobin in humans all exist within assessable animal models. A better understanding of these specific human biological processes and those animal models displaying similar capacities offers the possibility of identifying animal models of potential human high-risk groups.

REFERENCES

Aebi, H. (1970). Katalase. In: *Methoden der Enzymatischen Analyse*. Bergmeyer, H. V. (ed.). Verlag Chemie, Weinheim, p. 636.

References

Agar, S. and Harley, J. D. (1972). Erythrocytic methemoglobin reductases of various mammalian species. *Experientia* **28**: 1248.

Bartels, H., Hilpert, P., Barbey, K., Betke, K., Riegel, K., Lang, E. M., and Metcalfe, J. (1963). Respiratory functions of blood of the yak, llama, camel, Dybowski deer, and African elephant. *Am. J. Physiol.* **205**: 331–336.

Betke, J., Kleihaver, E., and Lipps, M. (1956). Vergleichende untersuchungen uber die spontanoxydation von Wabelschnur und Erwachsenenhamoglobin. *Z. Kinderheilkd.* **77**: 549.

Bolyai, J. Z., Smith, R. P., and Gray, C. T. (1972). Ascorbic acid and chemically induced methemoglobinemias. *Toxicol. Appl. Pharmacol.* **21**: 176–185.

Breadley, W. B., Eppson, H. F., and Beath, O. A. (1939). Nitrate as a cause of oat hay poisoning. *J. Am. Vet. Med. Assoc.* **94**: 541.

Breadley, W. B., Eppson, H. F., and Beath, O. A. (1940). Methylene blue as an antidote for poisoning by oat hay and other plants containing nitrates. *J. Am. Vet. Med. Assoc.* **96**: 41.

Burger, A., Stöffler, G., Uehleke, H., and Wagner, J. (1966). Formation of methaemoglobin by phenylhydroxylamine activity of glucose-6-phosphate dehydrogenase in the erythrocytes of different animal species. *Med. Pharmacol. Exp.* **15**: 525–529.

Eichholtz, F. (1927). Beitrag zur Pharmakologie des Plasmochins. Beich. *Arch. Schiffs Trop. Hyg.* **31**: 89.

Gelinski, G. (1940). Rückbildung des Methämoglobins bei vershiedenen Tierarten. *Naunyn Schmiedebergs Arch. Exp. Pathol. Pharmakol.* **195**: 460.

Greene, I. and Hiatt, E. P. (1954). Behavior of the nitrate ion in animal tissues. *Biochem. J.* **22**: 125.

Gupta, J. D., Irvine, S., and Harley, J. D. (1972). Species variation in galactokinase activity of erythrocytes, lens, and liver. *Aust. J. Exp. Biol. Med. Sci.* **50**: 511.

Henschler, D., Hahn, E., Heymann, H., and Wunder, H. (1964). Mechanismus einer Toleranzsteigerung bei wiederholter Einatmung von Lungenödem erzeugenden Gasen. *Naunyn Schmiedebergs Arch. Exp. Pathol. Pharmakol.* **249**: 343.

Henschler, D. and Ludtke, W. (1963). Methämoglobinbildung durch Einatinung niederer Konzentrationen nitroser Gase. *Int. Arch. Gewerb. Path. Hyg.* **20**: 362.

Heubner, W. (1913). *Arch. Exp. Path. Pharmakol.* **72**: 239. (Cited in Lester, 1943.)

Kiese, M. (1974). *Methemoglobinemia: A Comprehensive Treatise.* CRC Press, Cleveland, Ohio.

Kiese, M. and Weger, N. (1969). Formation of ferrihemoglobin with aminophenols in the human for the treatment of cyanide poisoning. *Eur. J. Pharmacol.* **7**: 97.

Kiese, M. and Weis, B. (1943). Die Reduktion des Hämiglobins in den Erythrocyten verschiedener Tiere. *Naunyn Schmiedebergs Arch. Pharmacol.* **202**: 493–501.

Kim, H. D. and McManus, T. J. (1971). Studies on the energy metabolism of pig red cells. I. The limiting role of membrane permeability in glycolysis. *Biochem. Biophys. Acta* **230**: 1.

Kim, H. D. and McManus, T. J. (1971a). Studies on the energy metabolism of pig red cells. II. Lactate formation from free ribose and deoxyribose with maintenance of ATP. *Biochem. Biophys. Acta* **230**: 12.

Kruse and McEllroy (1927). *J. Pharmacol. Exp. Ther.* **31**: 208. (Cited in Lester, 1943.)

LeHeux, J. W. and Lind van Wijngaarden, C. de (1927). Pharmakologische Untersuchungen uber Plasmochin. *Klin. Wochenschr.* **6**: 857.

LeHeux, J. W. and Lind van Wijngaarden, C. de (1929). Uber die pharmakologische Wirtung des Plasmochins. *Naunyn Schmiedebergs Arch. Exp. Pathol. Pharmakol.* **144**: 341.

Lester, D. (1943). Formation of methemoglobin. I. Species differences with acetanilide and acetophenetidine. *J. Pharmacol. Exp. Ther.* **154**: 154–159.

Lewis, D. (1951). The metabolism of nitrate and nitrite in the sheep. I. The reduction of nitrate in the rumen of the sheep. *Biochem. J.* **48**: 175–179.

Malz, E. (1962). Vergleichende Untersuchungen über die Methämoglobin reduktion in kernhaltigen und kernlosen Erythrocyten. *Folia Haematol. (Leipzig)* **78**: 510–515.

Mathies, H. J. (1957). Untersuchungen über die Methämoglobinruckbildung und über die Aldehyddehydrose in kernlosen Erythrocyten. *Wiss. Z. Humboldt Univ.* **6**: 489.

Miller, F. J., Menzel, D. B., and Coffin, D. L. (1978). Similarity between man and laboratory animals in regional pulmonary deposition of ozone. *Environ. Res.* **17**: 84–101.

Parke, D. V. (1960). Studies in detoxication 84. The metabolism of (^{14}C) aniline in the rabbit and other animals. *Biochem. J.* **77**: 493–503.

Parke, D. V. (1968). *The Biochemistry of Foreign Compounds.* Pergamon, New York.

Pflesser, G. (1935). Die Bedentung des Stickstoffmonoxyds bei der Vergiftung durch nitrose Gase. *Naunyn Schmiedebergs Arch. Exp. Pathol Pharmakol.* **179**: 545.

Pulina, B. (1942). Wirkungen von Salpeter und Salpetrigsaaureestern auf das Blut. *Naunyn Schmiedebergs Arch. Exp. Pathol. Pharmakol.* **200**: 324.

Rivkin, S. E. and Simon, E. R. (1965). Comparative carbohydrate catabolism and methemoglobin reduction in pig and human erythrocytes. *J. Cell. Comp. Physiol.* **66**: 49–56.

Robin, H. and Harley, J. D. (1966). Factors influencing response of mammalian species to the methaemoglobin reduction test. *Aust. J. Exp. Biol. Med. Sci.* **44**: 519–526.

Scheff, G. J. (1940). The influence of partial and complete hepatectomy on methemoglobin formation by aniline and *P*-aminophenol. *J. Pharmacol. Exp. Ther.* **70**: 334.

Scott, A. L. (1967). Studies of the peripheral blood reactions to haemolytic agents in different species. *Proc. Eur. Soc. Stud. Drug Toxic.* **8**: 185.

Scott, E. M., Duncan, I. W., and Ekstrand, V. (1965). The reduced pyridine nucleotide dehydrogenases of human erythrocytes. *J. Biol. Chem.* **240**: 481–485.

Shapiro, M. L., Hoflund, S., Clark, R., and Quin, J. L. (1949). Studies on the alimentary tract of the Merino sheep in South Africa. XVI. The nitrate in ruminal ingesta as studied *in vitro*, Onderstepoort. *J. Vet. Sci. Anim. Ind.* **22**: 357.

Singleton, S. D. and Murphy, S. D. (1973). Propanil (3,4-dichloropropionanilide)-induced methemoglobin in mice in relation to acylamidase activity. *Toxicol. Appl. Pharmacol.* **25**: 20.

Smith, J. E. and Beutler, E. (1966). Methemoglobin formation and reduction in man and various animal species. *Am. J. Physiol.* **210**: 347–350.

Smith, P. K. (1940). Changes in blood pigments associated with the prolonged administration of large doses of acetanilid and related compounds. *J. Pharmacol. Exp. Therap.* **70**: 171–178.

Smith, R. P., Alkaitis, A. A., and Shafter, P. R. (1967). Chemically induced methemoglobinemias in the mouse. *Biochem. Pharmacol.* **16**: 317–328.

Smith, R. P. (1980). Toxic responses of the blood. In: *Toxicology: the Basic Science of Poisons.* 2nd ed. J. Doull, C. D. Klaassen, and M. O. Amdur (eds.). MacMillan, New York, pp. 311–331.

Stolk, J. M. and Smith, R. P. (1966). Species differences in methemoglobin reductase activity. *Biochem. Pharmacol.* **15**: 343–351.

Szigeti (1893). *Vjschr. Gericht. Med.* **3**: 9(suppl.). (Cited in Lester, 1943.)

Vehleke, H. (1972). Mechanisms of methemoglobin formation by therapeutic and environmental agents. *Pharmacology and the Future of Man.* Proc. Fifth Int. Congr. Pharmacol. San Francisco, California. **2**: 124–36.

Warburg, O., Kubowitz, F. and Christian, W. (1930). Kohlenhydratverbrennung durch Methamoglobin. Über den Mechanismus einer Methylenblau-Katalyse. *Biochem. Z.* **221**: 494.

Warburg, O., Kubowitz, F., and Christian, W. (1930a). Über die katalytische Wirkung von Methylenblau in lebenden Zellen. *Biochem. Z.* **227**: 245.

Wendel, W. B. (1930). Oxidation of lactate by methemoglobin in erythrocytes with regeneration of hemoglobin. *Proc. Soc. Exp. Biol. Med.* **28**: 401.

Williams, R. T. (1943). The biosynthesis of aminophenyl and sulphonamidoaminophenyl-glucuronides in the rabbit and their action on haemoglobin *in vitro*. *Biochem. J.* **37**: 329.

Young, A. G. and Wilson, J. A. (1926). Toxicological studies of anilin and anilin compounds. III. Toxicological and hematological studies of acetanilid poisoning. *J. Pharmacol. Exp. Therap.* **27**: 133.

J. Deficiencies of Catalase

1. INTRODUCTION

Another disease that can affect the function of red blood cells is acatalasemia, which is a human genetic disorder inherited as an autosomal recessive (Aebi and Suter, 1972). It is characterized as a deficiency in catalase, an enzyme responsible for the breakdown of H_2O_2. Acatalasemia refers to the homozygous condition, while hypocatalasemia denotes the heterozygous carrier. Homozygotes for the trait generally possess little or no catalase activity, while heterozygotes have approximately half the normal level (Takahara et al., 1960; Takahara, 1967). Catalase is normally found in all tissues and is especially high in the liver, kidneys, and erythrocytes.

The acatalasemic condition is most often found without clinical symptoms. The erratic occurrence of symptoms preferentially affect children and usually begin as a small painful oral ulceration in crevices around the neck of teeth which is thought to be the result of H_2O_2-producing bacteria present in the mouth (Aebi and Suter, 1972). The symptoms can appear in mild, moderate, and severe forms.

The physiological role of catalase remains unclear at present. In erythrocytes, it is believed that catalase is important at only high levels of H_2O_2 stress and that GSH-Px is responsible for the elimination of low levels of H_2O_2 in the blood (Cohen and Hochstein, 1963; Jacob et al., 1965). Under physiological conditions, the ability of the erythrocyte to protect Hb from oxidation by H_2O_2 depends on the presence or absence of glucose with GSH-Px critically important in this scheme. Erythrocytes, which are either acatalasemic or treated in such a way as to inactivate catalase, are protected from oxidative hemolysis by H_2O_2 by a compensatory stimulation of the HMP shunt that will result in an increased activity of the GSH regeneration mechanism (Jacob et al., 1965). Acatalasemic erythrocytes under nonstress conditions metabolize glucose through the shunt at three times the normal rate. Under H_2O_2 stress conditions, glucose is metabolized at 12 times the normal rate in the shunt. This provides the underlying mechanism for oxidative protection.

Previously, it was shown that individuals with the G-6-PD deficiency are sensitive to oxidative stress (Beutler, 1972). G-6-PD catalyzes the first reaction in the HMP shunt, and deficiencies in this enzyme result in decreased levels of reduced GSH, a necessary component for GSH-Px to function in breaking down H_2O_2. From the foregoing, it might be predicted that erythrocytes devoid of catalase and also deficient in HMP shunt activity would be extremely vulnerable to oxidant damage by peroxides. When erythrocytes of a G-6-PD deficient individual were inhibited for catalase activity, the presence of H_2O_2 caused a rapid accumulation of irreversibly denatured hemoglobin (Jacob et al., 1965). An individual who had inherited both G-6-PD deficiency and acatalasemia was indeed susceptible to the development of severe hemolytic anemia (Szeinberg et al., 1963). The relationship of G-6-PD deficiency and catalase deficiency is not yet understood. In a screening of 200 American blacks, decreased catalase activity was present only in association with the G-6-PD deficiency (Tarlov and Kellermeyer, 1961). The catalase activity of the G-6-PD deficients was only 60% of normal blacks.

2. RADIATION SENSITIVITY

Taylor et al. (1933) first suggested that catalase might be able to modify the effects of ionizing radiation. They noted that since ionizing radiation is known to affect the production of peroxides such as H_2O_2 in tissue and since catalase is known to detoxify H_2O_2, it is reasonable to hypothesize that catalase may have a modifying effect on the action of radiation on the body. Experiments by Evans (1947) supported the hypothesis of Taylor et al. when they demonstrated that crude extracts of sea urchin sperm containing catalase could prevent death of sperm immersed in irradiated seawater. Barron and Dickman (1949) showed that the addition of catalase (0.3 µg/mL) partially prevented the inactivation of crystalline phosphoglyceraldehyde dehydrogenase by both alpha and beta radiation. They indicated that H_2O_2 contributed about 30% of the total inhibition caused by α radiation and somewhat more for inhibition by β rays.

A Swiss study utilizing human subjects found that red blood cells from acatalasemic people when exposed to irradiation form MetHb about 10–20 times faster than "normals" (Aebi et al., 1962). Furthermore, the effect of gamma irradiation on cancer cells is tremendously reduced by simply adding catalase to the outside of the cell (Thomson, 1963).

3. ROLE OF CATALASE IN CELLULAR DETOXIFICATION

Catalase has been found to be present in both the nucleus and cytoplasm. However, it seems that about 90% of the catalase is apparently present in the cytoplasm (Thompson, 1963; Ludewig and Chanutin, 1950; Creasy, 1960). The location of catalase within the cell is considered important with respect to the metabolism of its substrate (H_2O_2) (Thompson, 1963). Consequently, cytoplasmic catalase may protect against the formation of MetHb and/or changes in cell membranes, while nuclear catalase is thought to offer limited protection to the DNA from H_2O_2. Since only a small percentage of catalase is present in the nucleus, it seems that catalase may not be highly effective in protecting the genetic material from oxidant stress. However, since the volume of the nucleus compared to the entire volume of the cell is often quite small, there may not be the low levels of catalase in the nucleus as compared to levels in the cytoplasm if measured on a per volume basis.

It has been found that catalase actually plays a secondary role to GSH-Px in the detoxification of H_2O_2, and in fact, only when the concentration of H_2O_2 exceeds a certain undefined high level does catalase serve a significant role in H_2O_2 detoxification. This finding may help to explain the widespread conflicting reports concerning the protective effects of catalase on the action of radioactivity. For example, in contrast to previous reports indicating the importance of catalase with regard to detoxifying H_2O_2 following irradiation, a study with hypocatalasemic guinea pigs exposed to various levels of irradiation showed they were not significantly different from the controls with regard to radiation sensitivity and longevity (Rusev et al., 1960).

The dose of radiation that produces the "critical" concentration is probably modified by several factors including natural levels of catalase in the various tissues (levels are specific for each tissue), and serum vitamin E (the absence of vitamin E is associated with the accumulation of peroxides), vitamin C, and selenium (Se) levels.

However, the principal way in which the cell detoxifies H_2O_2 (reduces oxidant stress) is by the action of the enzyme GSH-Px. Thus, in order to maintain continuous functioning of the enzyme, it is necessary to continuously form reduced GSH, which in turn is dependent on other factors such as the supply of glucose, the enzyme G-6-PD to reduce NADP to NADPH, and GSH reductase.

In support of this scheme, human studies (*in vitro*) on the effects of peroxide on normal cells, acatalasemic cells, and cells deficient in G-6-PD, indicated that only cells deficient in the G-6-PD showed marked decreases in reduced GSH levels (Powers, 1963). It should be emphasized that such changes are closely associated with increased cell membrane fragility.

4. FREQUENCY IN THE POPULATION

Catalase deficiency was first recognized in Japan and Korea and has now been detected in western countries (Switzerland, Israel, and Germany).

Observations from the United States suggest the presence of heterozygotes. No gene frequency data exist for the United States, but in Japan the gene frequency is approximately 0.0025 and the Swiss gene frequency is 0.012 (Aebi and Suter, 1972). It should be noted that the frequency of acatalasemia and hypocatalasemia in the United States would probably be different from the calculated frequencies in both Japan and Switzerland.

5. SCREENING TESTS

Large-scale screening for the prevalence of catalase deficiency in man is well documented. For example, in western countries the first acatalasemic individuals were detected during the screening of 73,661 Swiss males reaching the age of 19 in 1961 and 1965 (Aebi et al., 1961; Aebi, 1967). Furthermore, in order to determine the frequency of the acatalasemic gene in the Far East population, Takahara (1967), together with the Atomic Bomb Casualty Commission, investigated the frequency of hypocatalasemia in 82,969 Asian individuals including Japanese and Koreans living in Japan, residents of the Ryukyo Islands and Chinese residing in Taiwan.

For the determination of catalase in the blood, several rapid and accurate techniques are available (Aebi and Suter, 1969, 1972). In future screenings, the automated procedure as devised by Leighton et al. (1968) and Lamy et al. (1967) will probably be the chosen method.

6. SUMMARY

No systematic *in vivo* studies of the responses of humans to toxic substances with either acatalasemia or hypocatalasemia have been published. However, from what limited studies have been conducted and the present understanding of the role of catalase as a cellular antioxidant, it would seem that those heterozygous for an erythrocyte catalase deficiency would not be expected to be at notably increased risk to ambient levels of environmental oxidants.

REFERENCES

Aebi, H. (1967). The investigation of inherited enzyme deficiencies with special reference to acatalasia. *Proc. 3rd. Int. Congr. Hum. Genet.,* Chicago, 1966., J. F. Crow and J. V. Neel (eds.). Johns Hopkins, Baltimore, p. 189.

Aebi, H. and Suter, H. (1969). Catalase. In: *Biochemical Methods in Red Cell Genetics.* J. J. Yunis (ed.). Academic, New York, p. 255.

References

Aebi, H. and Suter, H. (1972). Acatalasemia. In: *The Metabolic Basis of Inherited Disease.* J. B. Stanbury, J. B. Wyngaarden, and D. S. Fredrickson (eds.). McGraw-Hill, New York, pp. 1710–1729.

Aebi, H., Heiniger, J. P., and Suter, H. (1962). Methaemoglobinbildung durch Rontgenbestrahlen in Normalen und Ratalasefreien erythrocyten des menchen. *Experientia* **18:** 129.

Aebi, H., Heiniger, J. P., Butler, R., and Hassig, A. (1961). Two cases of acatalasia in Switzerland. *Experientia.* **17:** 466.

Barron, E. S. G. and Dickman, S. (1949). Studies on the mechanism of action of ionizing radiation. II. Inhibition of sulfhydryl enzymes by alpha, beta and gamma rays. *J. Gen. Physiol.* **32:** 595.

Beutler, E. (1972). G-6-PD deficiency. In: *The Metabolic Basis of Inherited Disease.* J. B. Stanbury, J. B. Wyngaarden, and D. S. Fredrickson (eds.). McGraw-Hill, New York, pp. 1358–1388.

Cohen, G. and Hochstein, P. (1963). Glutathione peroxidase. The primary agent for the elimination of hydrogen peroxide in erythrocytes. *Biochemistry* **2:** 1420.

Creasy, W. A. (1960). The enzymic composition of nuclei isolated from radiosensitive and non-sensitive tissue with special reference to catalase activity. *Biochem. J.* **77:** 5.

Evans, T. C. (1947). Effects of H_2O_2 produced in the medium by radiation on spermatoza of *Arbacia punctulata. Biol. Bull.* **92:** 99.

Jacob, H. S., Ingbar, S. H., and Jandl, J. H. (1965). Oxidative hemolysis and erythrocyte metabolism in hereditary acatalasemia. *J. Clin. Invest.* **44:** 1187.

Lamy, J. N., Namy-Provansal, J., de Russe, J., and Weill, J. D. (1967). Dosage automatique de l'eau oxygene et application a l'etude cinetique de la catalase. *Bull. Soc. Chim. Biol.* **49:** 1167.

Leighton, F., Poole, B., Beaufay, H., Baudhuim, P., Coffey, J. W., Fowler, S., and de Dure, Ch. (1968). The large scale separation of peroxisomes, mitochondria, and lysosomes from the livers of rats injected with triton WR1339. *J. Cell. Biol.* **37:** 482.

Ludewig, S. and Chanutin, A. (1950). Distribution of enzymes in the livers of control and x-irradiated rats. *Arch. Biochem.* **29:** 441.

Powers, E. L., Jr. (1963). Peroxides in radiobiology: A synthesis. *Rad. Res. Sup.* **3:** 270.

Rusev, G., Radev, T., Belikonski, I., and Petkov, B. (1960). Radiochuvstvitelnost na khipokata laznite morski svincheta s ogled na znachenieto na vodorodniya prekis za patogenezata na ostrata lycheva bolest. *Lzvest. Inst. Sravitclna Patol. Domashnite Zhivotni.* **8:** 119–129.

Szeinberg, A., deVries, A., Pinkhas, J., Djaldetti, M., and Ezra, R. (1963). A dual hereditary red blood cell defect in one family: Hypocatalasemia and G-6-PD deficiency. *Acta Genet. Med. Gemellol.* **10:** 2470

Takahara, S. (1967). Acatalasemia. *Asian Med. J.* **10:** 46.

Takahara, S., Hamilton, H. B., Weel, J. V., Hubara, T. Y., Ogura, Y., and Nishimura, E. T. (1960). Hypocatalasemia. A new genetic state. *J. Clin. Invest.* **39:** 610.

Tarlov, A. R. and Kellermeyer, R. W. (1961). The hemolytic effect of primaquine. XI. Decreased catalase activity in primaquine-sensitive erythrocytes. *J. Lab. Clin. Med.* **58:** 204.

Taylor, C. V., Thomas, F. O., and Brown, M. G. (1933). Studies on protozoa. IV. Lethal effects of x-radiation on a sterile culture medium for *Culpidium campylum. Physiol. Zool.* **6:** 467.

Thomson, J. F. (1963). Possible role of catalase in radiation effects on mammals. *Rad. Res. Sup.* **3:** 109.

K. Animal Models for Acatalasemia

There is considerable interspecies variation of erythrocyte catalase activity. This was suggested by several early studies by Battelli and Stern (1905), Itallie (1905), and Krüger and Schuhnecht (1928), and more quantitatively demonstrated by Bondi (1950), Richardson et al. (1953), Paniker and Iyer (1965), and Feinstein (1970). Table 2-10 reveals that humans, monkeys, and cats display the highest erythrocyte catalase activity with the following commonly used laboratory models having the listed approximate percent activity of normal human values: rabbits (65%), albino rats (27%), guinea pigs (25%), and dogs (3%).

It has long been thought that erythrocyte catalase may protect the cell from certain types of oxidative stress by eliminating intracellular H_2O_2 by a catalytic or peroxidative mechanism (Chance, 1952). If the catalase does play a role in protecting the red cell from H_2O_2 stress, then one would predict that those species displaying low erythrocyte catalase activity would be more susceptible than those species exhibiting higher

Table 2-10. Catalase Activity of Erythrocytes

Species	Number of Animals	Catalase Activity (meq H_2O_2/mgHb per 30 min) (mean ± S.D.)
Man	5	15.1 ± 1.4
Monkey (*Macaca radiata*)	3	16.3 ± 0.7
Cat	7	17.8 ± 2.1
Rabbit	14	10.4 ± 0.9
Rat (albino)	4	4.2 ± 0.2
Bat (Pteropus)	2	6.4
Dog	7	0.4 ± 0.4
Cow	6	4.5 ± 0.6
Buffalo	3	5.0 ± 0.14
Goat	6	2.1 ± 1.2
Guinea pig	16	3.8 ± 3.4
Fowl	3	0.56 ± 0.18
Duck	3	0.02 ± 0.01
Pigeon	4	0.1 ± 0.07
Fish		
Ophio cephalus	3	2.5 ± 0.1
Sciana dussumieri	2	1.95
Sillago sihama	1	1.5
Lates calcariper	1	4.8
Etroplus suratensis	1	6.8
Varanus monitor	3	18.6 ± 1.1
Frog (*Rana tigrina*)	35	8.6 ± 2.9

Source: Paniker and Iyer (1965).

Animal Models for Acatalasemia

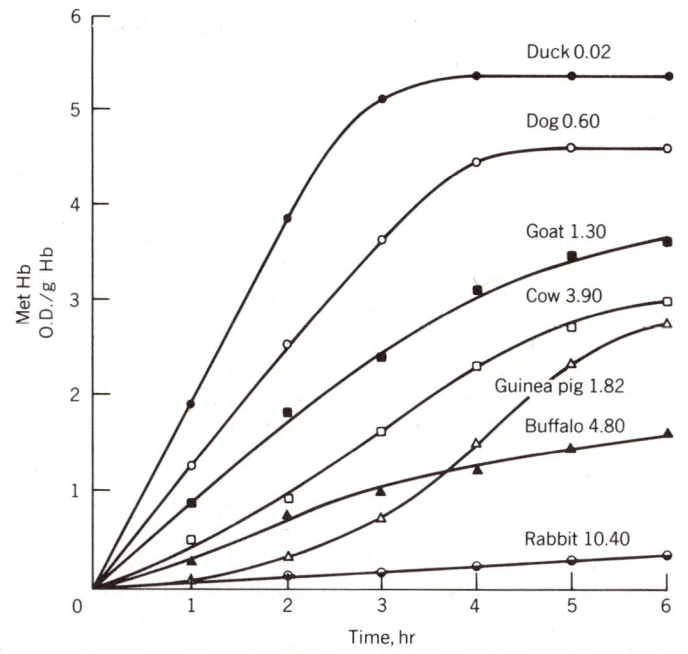

Figure 2-6. Time course of MetHb formation in erythrocytes exposed to hydrogen peroxide. Three milliliters of 5% erythrocyte suspension was placed in the main chamber of 15-mL Warburg flasks and 0.2 mL of 1.75 M H_2O_2 in the center well. The flasks were shaken (90 per minute) at 37° for periods ranging from 1 to 6 hr. Catalase activities of erythrocytes are indicated in the graph. Controls with water in place of H_2O_2 were run and the MetHb values given are suitably corrected.

activity levels. Such an interspecies comparison was performed by Paniker and Iyer (1965) and did, in fact, reveal an inverse relationship between MetHb levels induced by administration of H_2O_2 and red cell catalase activity (Figure 2-6). Not represented in Figure 2-6, although evaluated, were the responses of the human, cat, and monkey red cells since their MetHb formation was particularly low. Additional studies revealed that the administration of catalase enzyme preparation to dog hemolysates offered significant protection from H_2O_2 induced MetHb formation (Table 2-11).

An attempt to define further the role of catalase in protecting erythrocytes from oxidant damage was addressed by Paniker and Iyer (1965). They cited the research of Cohen and Hochstein (1963), which clearly revealed that the metabolism of glucose may protect Hb from H_2O_2 oxidation. Glucose metabolism via the pentose phosphate shunt produces $NADPH_2$, which is a coenzyme for GSH reductase and MetHb reductase, both being significant regulators of MetHb levels. In addition, metabolism of glucose via glycolysis produces $NADH_2$, which also acts as a coenzyme for the above enzymes. If one experimentally removes glucose die from kidney failure caused by red cell casts blocking up the tubules. In this case it would appear that enhanced susceptibility to red cell oxidation may be of some significance.

This type of study has demonstrated that acatalasemic mice are at greater risk than normocatalasemic mice at least to the acute effects of H_2O_2 administration. Whether this low enzyme level enhances their susceptibility to other H_2O_2 producing substances such as O_3 is unknown.

It would appear that future research should directly compare via *in vitro* studies the responses of normal, hypo-, and acatalasemic humans and mice. This type of study, while having its limitations, permits a quick and direct evaluation of the relative toxicological sensitivities between humans and a viable potential animal model.

The recognition that erythrocyte catalase activity may markedly vary among commonly employed animal models may also be utilized in the selection of animal models for toxicological investigations as well as in the interpretation of such studies. This type of knowledge may also allow one to reinterpret previously published findings. For example, several studies have evaluated the influence of high levels of NO_2 on the levels of MetHb formation in the dog (see Kiese, 1974). The question may be asked as to what do the findings of these studies mean for humans? While an accurate answer to the question may not be possible at the present time, several potentially important discriminating variables that may affect the ability of this model to predict human responses are known. Among the biochemical parameters that may affect MetHb formation are the erythrocyte levels of MetHb reductase and catalase. The dog appears to display similar MetHb reductase activity as humans but only about 3% of the normal human catalase activity values. In other words, dogs are more like acatalasemic than normal humans! Does this mean that dogs are not appropriate models to use for predictive toxicity studies for normal humans when MetHb formation is a consideration? That dogs and normal humans display markedly different levels of red cell catalase does not bode especially well for dogs being an accurate predictive model for human responses to MetHb forming agents. However, it must be emphasized that other concurrent biochemical processes affecting MetHb formation and reduction are also operational and may differ between dogs and humans. It could be that these other differences may compensate for the differential catalase activity in terms of MetHb formation and/or reduction. If so, then the dog may be an adequate *functional* model. Most likely, however, such biochemical accommodations may result in highly species specific responses to environmental oxidants. Despite the remaining uncertainties, the limited knowledge available should be utilized, where possible, in a more rational selection of animal models for predictive toxicological studies.

Numerous toxicological studies with O_3 and NO_2 have been performed with various strains of mice. When the strain in one study is A/He while

C57BL/An or NBL/Hf strains are used by other investigators, how are these studies to be compared? Should all mice be expected to respond similarly? Should they respond more similarly than say a mouse strain versus a rat strain? With respect to erythrocyte catalase activity, Hoffman and Recheige (1971) have reported on such activity from 18 inbred strains of mice and found catalase activity to differ by more than 300% (i.e., from 9.99 to 35.50 units). It is evident that the more the investigator knows about the biochemistry of the model, especially those parameters more directly pertaining to the adaptive capabilities of animal vis-à-vis humans, the more relevant one's study may become.

In summary, genetic polymorphisms with respect to catalase activity exist in human erythrocytes. This has resulted in the grouping of humans into three classifications: normal, hypocatalasemic (50% of normal), and acatalasemic (1–2% of normal). Since red cell catalase facilitates the detoxification of exogenous and endogenous H_2O_2, it has been hypothesized that those with a catalase deficiency may be at increased risk to H_2O_2 producing agents such as O_3 or radiation. A survey of numerous animal models has revealed considerable variability in erythrocyte catalase activity. Monkeys and cats have catalase activity most like that of normal humans, while dogs have only $\frac{1}{30}$ the activity of normal humans, thereby resembling the activity of acatalasemic humans. Biochemical geneticists have developed a mouse model exhibiting a trimodal distribution: "normal," hypocatalasemic (40–60%), and acatalasemic (1–2%), similar to human distributional patterns. It is recommended that further research be directed to assess the capability of this model to predict the responses of acatalasemic humans to oxidant stressor agents.

REFERENCES

Adler, H. I. (1963). Catalase, hydrogen peroxide, and ionizing radiation. *Radiat. Res. Suppl.* **3:** 110–129.

Allison, A. C., Rees, W., and Burn, G. P. (1957). Genetically-controlled differences in catalase activity of dog erythrocytes. *Nature* **180:** 649–650.

Battelli, F. and Stern, L. (1905). *Arch. Fisiol.* **2:** 471. (Cited in Richardson et al., 1953.)

Bondi, C. (1950). *Riv. Biol.* **42:** 351. (Cited in Richardson et al., 1953.)

Calabrese, E. J. (1978). *Pollutants and High Risk Groups.* Wiley, New York.

Chance, B. (1952). The state of catalase in the respiring bacterial cell. *Science* **116:** 202–203.

Cohen, G. and Hochstein, P. (1963). The primary agent for the elimination of hydrogen peroxide in erythrocytes. *Biochemistry* **2:** 1420–1428.

Feinstein, R. N. (1970). Acatalasemia in the mouse and other species. *Biochem. Genet.* **4:** 135–155.

Feinstein, R. N., Braun, J. T., and Howard, J. B. (1968). Nature of the heterozygote blood catalase in a hypocatalasemic mouse mutant. *Biochem. Genet.* **1:** 277–285.

Feinstein, R. N., Braun, J. T., and Howard, J. B. (1967). The dog as an example of acatalasemia or hypocatalasemia. *U.S.A.E.C. Argonne National Laboratory*, p. 122.

Feinstein, R. N., Braun, J. T., and Howard, J. B. (1966). Reversal of H_2O_2 toxicity in the acatalasemic mouse by catalase administration: suggested model for possible replacement therapy of inborn errors of metabolism. *J. Lab. Clin. Med.* **68:** 952–957.

Feinstein, R. N., Cotter, G. J., and Hampton, M. M. (1954). Effect on radiation lethality of various agents relevant to the H_2O_2-catalase hypothesis. *Am. J. Physiol.* **177:** 456–460.

Feinstein, R. N., Faulhaber, J. T., and Howard, J. B. (1968). Sensitivity of acatalasemic mice to acute and chronic irradiation and related conditions. *Rad. Res.* **35:** 341–349.

Feinstein, R. N., Howard, J. B., Braun, J. T., and Seaholm, J. E. (1966). Acatalasemic and hypocatalasemic mouse mutants. *Genetics* **53:** 923–933.

Feinstein, R. N., Seaholm, J. E., Howard, J. B., and Russell, W. L. (1964). Acatalasemic mice. *Proc. Nat. Acad. Sci.* **52:** 661–662.

Flynn, R. J., Camden, R. W., Feinstein, R. N., and Grahn, D. (1967). The development of an inbred acatalasemic mouse. *U.S. A.E.C. Argonne National Laboratory*, p. 119.

Foulkes, E. C. and Lemberg, R. (1949–1950). The formation of choleglobin and the role of catalase in the erythrocyte. *Proc. R. Soc. London, Series B.* **136:** 435–438.

Hoffman, H. A. and Recheige, M., Jr. (1971). Erythrocyte catalase in inbred mice. *Enzyme* **12:** 219–225.

Holman, R. A. (1957). A method of destroying a malignant rat tumor *in vivo*. *Nature* **179:** 1033.

Itallie, L. V. (1905). *Akad. Wet. Amsterdam* **14:** 540. (Cited in Richardson et al., 1953.)

Kiese, M. (1974). *Methemoglobinemia: A Comprehensive Treatise*. CRC Press, Cleveland, Ohio.

Krüger, F. V. and Schuhnecht, H. (1928). *Z. Vergl. Physiol.* **8:** 635. (Cited in Richardson et al. 1953.)

Lash, J. W. (1959). Carrier state in human acatalasemia. *Science* **130:** 333–334.

Mills, G. C. and Randall, H. P. (1958). Hemoglobin catabolism. II. The protection of hemoglobin from oxidative breakdown in the intact erythrocyte. *J. Biol. Chem.* **232:** 589–598.

Paniker, N. V. and Iyer, G. Y. N. (1965). Erythrocyte catalase and detoxification of hydrogen peroxide. *Can. J. Biochem.* **43:** 1029–1039.

Radev, T. (1960). Inheritance of hypocatalasaemia in guinea pigs. *J. Genet.* **57:** 169–172.

Richardson, M., Huddleson, I. F., Bethea, R., and Trustdorf, M. (1953). Study of catalase in erythrocytes and bacteria. II. Catalase activity of erythrocytes from different species of normal animals and from normal humans. *Arch. Biochem. Biophys.* **42:** 124–134.

Takahara, S. (1960). International Congress on Hematology 84th, Tokyo, Japan, Sept. 4–10. Abstract p. 286.

Thomson, J. F. (1963). Possible role of catalase in radiation effects on mammals. *Rad. Res. Suppl.* **3:** 93–109.

Warburg, O. K., Gawehn, K., and Geissler, A. W. (1957). Über die wirkung von Wasserstoffperoxyd auf Krebszellen und auf embryonale zellen. *Z. Naturforsch.* **126:** 393–395.

L. Effects of Environmental Oxidants on Human Erythrocytes

1. SELECTION OF PREDICTIVE ANIMAL MODELS

Considerable research efforts have been made to assess the effects of environmental oxidants on erythrocyte metabolism in a variety of animal models and human subjects. How the erythrocyte responds is contingent on many factors including the specific stressor agent, the dose, route, and regimen of exposure as well as the inherent adaptive capacity of the erythrocyte and other factors.

In selecting animal models to study the effects of environmental oxidants on erythrocytes, strong consideration should be given to their ability to predict the response of humans. Unfortunately, there are numerous biological factors that may contribute to interspecies differences depending on the oxidant. For example, Miller et al. (1978) have reported that dogs and rabbits have a different rate of ozone absorption along the nasopharyngeal tract resulting in a different effective dose reaching the deep recesses of the lungs despite an identical external exposure. Other workers have shown that some oxidant drugs require activation in the liver before affecting red cell metabolism and that this metabolism is often species specific (Kiese, 1974). Contributing to the complexity of interspecies comparisons with respect to the responses of erythrocytes to oxidant stressor agents is that there are at least six major enzymes that are designed to help the red cell in dealing with various types of oxidant stress. To what extent and in what manner these different enzymes act and interact to assist the red cell adapt to oxidant stress has been only partially clarified. The absence of any of these enzymes either experimentally induced as with sodium azide inhibiting catalase activity (Paniker and Iyer, 1965) or naturally via genetic processes as in GSH-Px or G-6-PD deficiencies are known to substantially enhance susceptibility to oxidant damage (Calabrese, 1978).

A major difficulty in reviewing the published literature is that most investigators never attempt to assess to what extent their model simulates the human. Regulatory personnel in reviewing and assessing study after study on the effects of ozone or nitrogen dioxide on mice, rats, rabbits, dogs, or monkeys have almost no idea whether these animal models under or over predict human responses.

The EPA has also been interested in developing or uncovering animal models for potential high-risk groups to environmental oxidants. One potential human high-risk group that has been considered are persons with an erythrocyte G-6-PD deficiency. It is well-known that certain breeds of sheep such as the Dorset sheep have an erythrocyte G-6-PD deficiency as well (Calabrese et al., 1980). While sheep experience enhanced hemolytic stress to oxidants such as copper and acetylphenylhydrazine, they respond quite differently than the human deficient by being insensitive to favor beans and primaquine induced hemolysis while human deficients, of course, respond with extreme sensitivity (Maronpot, 1972). These findings clearly indicate that other factors are contributing to the differential susceptibility of human and sheep G-6-PD deficients to such oxidants.

These collective issues strongly suggest the need to provide quantita-

Table 2-12. Interspecies Comparison of Erythrocyte Enzyme Activity Associated with Protecting Against Oxidant Stress. (All values are relative to human values that are given as 1.)

Species	SOD	GSH-Px	Catalase	6-6-PD	GSH Reductase	MetHb Reductase
Humans	1^a	1^a	1^a	1^c	1^c	1^e
Nonhuman primates						
Monkey (unspecified)	0.9^a	0.4^d	1.1^a	0.6^a		
Bonnet monkey		3.7^k				
Pongo				5.9^m		
Aotus				3.2^m		
Sanguinus				3.0^m		
Oedipomidas				3.0^m		
Lagothrix				2.9^m		
Ateles (spider monkey)				2.8^m		
Saimiri				2.8^m		
Cebus				1.0^m		
Macaca (rhesus monkey)		$1.2\ ♀^i$		$\{0.25^i \atop 0.8^m\}$		
Mandrillas				0.8^m		
Papio (baboon)				0.5^m		
Gorilla				0.4^m		
Presbytis				0.4^m		
Theropithecus				0.3^m		
Rat (unspecified)	1.7^a	10.2^a	0.2^a	2.4^g	0.2^g	2.4^e
Sprague-Dawley		$\{4.4\ ♂^i \atop 4.5\ ♂^j\}$		1.4^a		
Rabbit (unspecified)	1.2^a	7.3^a	0.8^a	1^c	0.3^c	4.5^e
Dog (unspecified)	1.5^a	$\{7.8^a \atop 2.2^n\}$	0.02^a	$\{0.7^c \atop 1.1^n\}$	$\{0.1^c \atop 0.4^n\}$	1.1^e
Cat (unspecified)	1.1^a	$\{8.7^a \atop 3.5^n\}$	1.1^a	1.7^n	1.1^n	0.9^e
Guinea pig (unspecified)	1.1^a	8.2^a	0.4^a		1.3^d	2.5^e

Mouse (C57B1/RHo)	1.86^h	27.0^h	0.4^h	9.5^e
A/J				2.6 ♀^{b}
A/HeJ				2.7 ♀^{b}; 2.7 ♂^{b}
AKR/J				1.8 ♀^{b}
BALB/cJ				1.7 ♀^{b}
C3H/HeJ				2.0 ♀^{b}; 2.3 ♂^{b}
C57BL/6J				1.9 ♀^{b}
C58/J				1.8 ♀^{b}
CBA/J				2.1 ♀^{b}
DBA/2J				2.0 ♀^{b}
RF/J				1.8 ♀^{b}
SEL/IReJ				1.9 ♀^{b}; 2.0 ♂^{b}
SJL/J				2.0 ♀^{b}
SWR/J				2.2 ♀^{b}
129/J				2.1 ♀^{b}
C57BR/cdj				1.2 ♀^{b}; 1.2 ♂^{b}
C57L/J				1.1 ♀^{b}; 1.0 ♂^{b}
WBB$_6$F$_1$ +/+				0.7^l
SEL/1Re +/+				0.7^l
WBB$_6$F$_1$-ha/ha				1.1^l
WBB$_6$F$_1$-Nb/Nb				1.3^l
WBB$_6$F$_1$-sph/sph				1.2^l
SEL/1Rr-mk/mk				1.0^l
Cow	1.3^a	8.0^a	0.4^a	0.8^c; 0.1^c
Sheep	1.6^h	5.1^h	0.2^h	0.08^c; 0.2^c
Goat	1.7^a	10.6^a	0.3^a	0.08^c; 0.3^c
Pig	1.0^h	2.5^h	1.1^h	1.7^c; 0.5^c
Horse	1.3^h	{1.8^h / 2.0^h}	0.7^h	{1.7^c / 2.6^n}; 0.2^c / 0.3^n
Llama				2.0^o; 0.9^e
Deer				0.4^o; 1.5^e
Camel				1.7^o; 0.9^e
Elephant				0.55^o; 0.3^e, 0.6^e

Table 2-12 (Continued)

Species	SOD	GSH-Px	Catalase	6-6-PD	GSH Reductase	MetHb Reductase
Yak				0.0^o		
Kangaroo						
gray				4.3^c	0.24^c	
red				0.18^c	0.19^c	

[a] Data taken from Kurian and Iyer (1977). Strains or breeds of animals were not specifically mentioned; SOD assays were performed according to Winterbourn et al.[(1975). *J. Lab. Clin. Med.* **85**: 337–341] with several modifications. The activity was given in terms of the percent inhibition of the reduction of Nitroblue tetrazolium caused by 25 μL of the erythrocyte extract. GSH-Px, given as the micromoles of GSH oxidized per gram of Hb per minute according to the method of Gross et al.[(1967). *Blood* **29**: 481–493]. Catalase was estimated by a standard procedure as given by Sumner, J. B. and Doune, A. L. (1955). In: *Methods in Enzymology*, Vol. II. Academic, New York, page 779.

[b] Data are from Hutton (1971). One unit of G-6-PD activity corresponds to 1 μmole NADPH formed per minute per gram of Hb.

[c] Data from Agar et al. (1974). G-6-PD methodology was the same as given in footnote *b*. Strains or breeds of horses, dogs, sheep, goats, cattle, rabbits, and pigs were not specifically given. GSH reductase activity was given as micromoles of NADPH per minute per gram of Hb. See Agar et al. (1975) for other sheep values.

[d] Data from Lankisch et al. (1973) and expressed as I.U. per gram of Hb.

[e] Data are taken from Smith (1980) who summarized seven published studies; the various investigators used nitrited red cells with glucose as a substrate.

[f] Data were taken from Smith et al. (1965). Ratios were calculated initially with sheep values as the denominator and then with known human values of footnote *c*.

[g] Data taken from Schaich, K. M. and Borg, D. C. (1980). In: *Research Planning Workshop on Health Effects of Oxidants*, Raleigh, North Carolina, Jan. 28–30, 1980.

[h] Data taken from Maral et al. (1977). Strains or breeds not specifically identified; data expressed in SOD units per gram of Hb; one activity unit is defined as the amount of enzyme causing 50% inhibition of the reduction of formazon observed in the blank. A standard curve using pure human erythrocuprein showed that inhibition was linear up to 40%. The various units of SOD per milliliter of blood were obtained by multiplying the percent inhibition by 6.34. GSH-Px units are expressed as micromoles NADPH per minute per gram of Hb. Note that Agar and Smith (1973) found that high and low GSH sheep have markedly different levels of GSH-Px. Catalase activity was expressed as micrograms per gram of hemoglobin.

[i] Data cited from Chow et al. (1975). Enzyme activity units expressed as *n*moles NADPH oxidized or formed per minute per milligram of Hb.

[j] Data taken from Larkin et al. (1978). Sex of rats not given.

[k] Dungsworth et al. (1975).

[l] Hutton and Bernstein (1973).

[m] Barnicott and Cohen (1970).

[n] Harvey and Kaneko (1975).

[o] Bartels et al. (1963).

tive interspecies comparisons of the erythrocyte enzyme activities that play a known important role in affecting susceptibility to oxidant stress. Six enzymes were selected as being essential to acquire information on: SOD, GSH-Px, GSH reductase, G-6-PD, catalase, and MetHb reductase. Surprisingly, no central bank of information concerning the enzyme activities in common laboratory models could be found. The outstanding text of Schalm (1965) entitled *Veterinary Hematology* provided detailed descriptions of the red cells of numerous animal models but it was essentially devoid of enzymatic characterizations.

A literature search revealed that systematic interspecies comparisons of red cell enzymes dealing with oxidant stress have not been published or if so, are not well enough known and/or sufficiently relevant to be cited in whatever literature exists on the subject. Despite this limitation, there is a modest number of articles that have included studies of the specific enzymes of importance here. Table 2-12 reveals that at least very limited data exist on a wide variety of species. Frequently the authors did not mention the strain of mouse, rat, rabbit, dog, or monkey used. In addition, while standard methods were usually cited for the determination of various enzyme activities, it was not uncommon to have "minor modifications" that were not defined thereby reducing further direct comparison. Given these limitations, it has still been possible to achieve a fairly complete set of enzyme values for a number of species of interest including the mouse, rat, rabbit, guinea pig, cat, dog, sheep, cow, goat, pig, horse, monkey, and human. Consideration of Table 2-12 reveals that no species closely resemble the human for all six enzymes. For example, while rabbits displayed a similar G-6-PD activity as humans, they had considerably higher GSH-Px and MetHb reductase activity. Some striking findings were that dogs were fundamentally acatalasemic-displaying less than 5% the activity of normal humans. On the other hand, dogs had subsequently higher SOD and GSH-Px activity than man. Other notable differences included mouse GSH-Px and MetHb reductase activity being 27 and 9.5 times higher than human values, respectively. The findings also revealed considerable variation in enzyme values within the various mouse stains for G-6-PD activity. This indicates the difficulty in generalizing even at the species level. Of particular concern is that little is known about the commonly used mouse and rat strains for most of the enzymes considered. This presents serious limitations in the interpretation of previous and present toxicological studies.

This attempt to catalog the information on these red cell enzyme activities should be considered only the first step in a systematic process that evaluates and compares the enzymes related to oxidant stress in erythrocytes of potential animal models. Completion of such a study will markedly enhance the predictive ability of many toxicological studies relevant to this area as well as providing a more rational basis for the selection of animal models for future studies.

REFERENCES

Agar, N. S., Gruca, M., and Harley, J. D. (1974). Studies of glucose-6-phosphate dehydrogenase, glutathione reductase and regeneration of reduced glutathione in the red blood cells of various mammalian species. *Aust. J. Exp. Biol. Med. Sci.* **52**: 607–614.

Agar, N. S., Roberts, J., Mulley, A., Board, P. G., and Harley, J. D. (1975). The effect of experimental anaemia on the levels of glutathione and glycolytic enzymes of the erythrocytes of normal and glutathione deficient Merino sheep. *Aust. J. Biol. Sci.* **28**: 233–238.

Agar, N. W. and Smith, J. E. (1973). Erythrocyte enzymes and glycolytic intermediates of high and low glutathione sheep. *Anim. Blood Groups Biochem. Genet.* **4**: 133–140.

Barnicott, N. A. and Cohen, P. (1970). Red cell enzymes of primates (Anthropoidea). *Biochem. Genet.* **4**: 41–57.

Bartels, H., Hilpert, P., Barbey, K., Betke, K., Riegel, K., Michael, M., and Metcalf, J. (1963). Respiratory functions of blood of the yak, llama, camel, Dybowski deer, and African elephant. *Am. J. Physiol.* **205**: 331–336.

Calabrese, E. J. (1978). *Pollutants and High Risk Groups: The Biological Basis of Increased Human Susceptibility to Environmental and Occupational Pollutants.* Wiley, New York.

Calabrese, E. J., Moore, G. S., and Ho, S. C. (1980). Low glucose-6-phosphate dehydrogenase (G-6-PD) activity in red blood cells and susceptibility to copper induced oxidative damage. *Environ. Res.* **21**: 366–372.

Chow, C. K., Mustafa, M. G., Cross, E., and Tarkington, B. K. (1975). Effects of ozone exposure on the lungs and the erythrocytes of rats and monkeys: relative biochemical changes. *Environ. Physiol. Biochem.* **5**: 142–146.

Dungworth, D. L., Castleman, W. L., Chow, C. K., Mellick, P. W., Mustafa, M. G., Tarkington, B., and Tyler, W. S. (1975). Effect of ambient levels of ozone on monkeys. *Fed. Proc.* **34**: 1670–1674.

Harvey, J. W. and Kaneko, J. J. (1975). Erythrocyte enzyme activities and glutathione levels of the horse, cat, dog, and man. *Comp. Biochem. Physiol.* **52**: 507–510.

Hutton, J. J. (1971). Genetic regulation of glucose-6-phosphate dehydrogenase activity in the inbred mouse. *Biochem. Genet.* **5**: 315–331.

Hutton, J. J. and Bernstein, S. E. (1973). Metabolic properties of erythrocytes of normal and genetically anemic mice. *Biochem. Genet.* **10**: 297–307.

Kiese, H. (1974). *Methemoglobinemias: A Treatise.* CRC Rubber Co., Cleveland, Ohio.

Kurian, M. and Iyer, G. Y. N. (1977). Oxidant effect of acetylphenylhydrazine; a comparative study with erythrocytes of several animal species. *Can. J. Biochem.* **55**: 597–601.

Lankisch, P. G., Schroeter, R., Lege, L., and Vogt, W. (1973). Reduced glutathione and glutathione reductase—a comparative study of erythrocytes from various species. *Comp. Biochem. Physiol.* **45B**: 639–641.

Larkin, E. C., Kimzey, S. L., and Siler, K. (1978). Response of the rat erythrocyte to ozone exposure. *J. Appl. Physiol.* **45**(6): 893–898.

Maral, J., Puget, K., and Michelson, A. M. (1977). Comparative study of superoxide dismutase, catalase, and glutathione peroxidase levels in erythrocytes of different animals. *Biochem. Biophys. Res. Commun.* **77**(4): 1525–1535.

Maronpot, R. R. (1972). Erythrocyte glucose-6-phosphate dehydrogenase and glutathione deficiency in sheep. *Can. J. Comp. Med.* **36**: 55–60.

Miller, F. J., Menzel, D. B., and Coffin, D. L. (1978). Similarity between man and laboratory animals in regional pulmonary deposition of ozone. *Environ. Res.* **17**: 84–101.

Paniker, N. V. and Iyer, G. Y. N. (1965). Erythrocyte catalase and detoxication of hydrogen peroxide. *Can. J. Biochem.* **43**: 1029–1039.

Smith, R. P. (1980). Toxic responses of the blood. In: *Toxicology*. J. Doull, C. D. Klassen, and M. O. Amdur (eds.). MacMillan, New York, pp. 311–330.

Smith, J., Barnes, J. K., Kaneko, J. J., and Freedland, R. A. (1965). Erythrocytic enzymes of various animal species. *Nature* 205: 298–299.

M. Superoxide Dismutase (SOD)

1. INTRODUCTION

The enzyme SOD is thought to occur in all oxygen-metabolizing cells but is absent in most obligate anaerobes, because its physiological function is to serve as a defense against the potential damaging reactivities of the superoxide radical (O_2^-), which is formed by aerobic metabolic reactions (Oberley and Buettner, 1979). Oxygen is not a very reactive molecule. For it to participate in cellular reactions, it must undergo a stepwise reduction that is typically achieved by the univalent addition of electrons ultimately forming H_2O. The initial step in the activation of oxygen requires energy input and results in the formation of the hydroperoxyl radical ($HO_2 \cdot$), the ion of which is the superoxide anion ($O_2^- \cdot$). Once formed, the superoxide anions can form another oxidant stressor compound, H_2O_2. The enzyme SOD is therefore considered the first line of "antioxidant" defense in aerobically metabolizing cells against "activated" oxygen, since it is capable of dismutating the superoxide radical to $O_2 + H_2O_2$. However, for cells to survive, other enzyme systems such as catalase and GSH-Px must then dissipate the H_2O_2 stress.

This section will give a summary of the research indicating a role of SOD in affecting susceptibility to toxicity of various stressor agents including O_2 at high concentrations, radiation, ozone, nitrogen dioxide, and paraquat. However, prior to these toxicological interactions a review of the effect of age on erythrocyte SOD levels, and its normal range of values as well as the occurrence of genetic variants of SOD will be provided.

2. ONTOGENY OF ERYTHROCYTE SOD ACTIVITY

The developmental changes in human erythrocyte SOD activity have been studied by several researchers. In general, newborn infants display a level of SOD in the range of about 65–80% of adult values (Ueda and Ogata, 1978; Rotilio et al., 1977; Legge et al., 1977). By three months of age, adult levels of erythrocyte SOD are achieved. The levels of SOD display a progressive decrease during old age with values of those aged 70–100 years approximating those of the newborn (Veda and Ogata, 1978). In contrast to the above reports, Michelson et al. (1977) reported no significant differences in SOD values with regard to age. In addition, these studies noted no differences in activity between cord and maternal blood (Yoshioka et al., 1979; Michelson et al., 1977; Kobayashi et al., 1977). However, whether this later lack of difference may also involve a pregnancy effect remains to be assessed.

3. NORMAL RANGES OF ERYTHROCYTE SOD

Regardless of age or sex, there is a considerable range of erythrocyte SOD values. A range of twofold to fourfold in the population within designated age/sex groups has been reported (Legge et al., 1977; Rotilio et al., 1977). While the distribution of SOD values in the population appears to exhibit a unimodal, normal distribution pattern, the question may be raised as to whether those at the lower end of the spectrum may be more susceptible to oxidant stress. While this issue has not been very actively researched, Rotilio et al. (1977) have associated a low SOD value in full-term infants with hyperbilirubinemia. In addition, these researchers also presented data which suggested that a high SOD/GSH-Px ratio was associated with hyperbilirubinemia in infants.

4. SOD LEVELS IN DISEASE STATES

No difference in erythrocyte SOD levels exists between normal individuals and those with rheumatoid arthritis (Scudder et al., 1976) and those with Duchenne muscular dystrophy (Hunter et al., 1981). In contrast, those with Down's syndrome have been repeatedly shown to exhibit elevated levels of erythrocyte SOD (Frischer et al., 1981). These researchers found that those with Down's syndrome displayed 43% greater SOD activity than normal individuals. In addition, GSH-Px was 40% greater in erythrocytes of those with Down's syndrome. These increased values were not accounted for by a decreased mean red cell age.

5. GENETIC VARIANTS OF SOD

Large-scale surveys have revealed an inherited electrophoretic variation of SOD, the red cell isoenzyme, in several human populations. The incidence of the variant allele SOD^2 is quite low in most Caucasian populations. For example, the occurrence of the heterozygote in Britain based on a survey of approximately 11,000 individuals was only 0.62 per 1000. However, several areas of high frequency exist, especially in the Tornadolen region of Sweden (Beckman and Pakarinen, 1973; Welch and Mears, 1972; Carter et al., 1976). If those with this genetic variant may respond differently to superoxide producing agents is unknown.

6. OXYGEN TOXICITY

It is well-known that most mammalian species cannot survive following prolonged exposure to 100% oxygen (1 atmosphere). While death comes as a result of extreme lung damage, the mechanism of this toxicity remains speculative. Perhaps the most creditable theoretical implication is that the oxygen exposure generates the formation of the superoxide anion (O_2^-), which is a very potent free radical. It is hypothesized that since the superoxide anion is an unstable free radical, it is likely that even low concentrations may be harmful to the living cell. It is widely accepted that cells can survive the production of superoxide anion because it is very quickly dissipated by the enzyme SOD, and high concentrations of the radical do not develop. The SOD enzyme catalyzes the dismutation of the superoxide free radical as follows:

$$O_2^- + O_2^- + 2H^+ \xrightarrow{SOD} O_2 + H_2O_2$$

It has been variously speculated that the presence of SOD is protective against the toxic effects of oxygen and that it is required for survival in an oxygen environment. Bacterial studies have supported this proposal in that the ability of these organisms to survive in an oxygen environment is correlated with the level of SOD (Costilow and Keele, 1972; McCord et al., 1971). Subsequent studies also indicated that SOD protected rats from oxygen toxicity. More specifically, after exposure to 85% oxygen for seven days, rats displayed a 50% increase of SOD activity in their lungs and became "tolerant" (i.e., survived more than four days) to 100% oxygen. All control rats died within 72 hr in 100% oxygen. The authors reported that the rate of development of oxygen "tolerance" closely follows the time course for the increase in pulmonary SOD activity. Interestingly, a similar ability of rats to develop a tolerance for oxygen toxicity was not evident in guinea pigs, hamsters, and mice under similar circumstances. These nonadaptive species also did not have as large a change in pulmonary SOD as the rat, thereby supporting the notion that O_2^- contributes to the occurrence of pulmonary toxicity and that SOD provides an *in vivo* defense against this radical (Crapo and Tierney, 1974).

7. OZONE TOXICITY

The occurrence of tolerance in rats to the adverse effects of oxygen has been associated with increased SOD levels in the lung (Crapo and Tierney, 1974). Such findings imply that oxygen toxicity may be related to the production of the superoxide free radical. Additional studies have revealed that increases in SOD appear to reduce O_3 toxicity (Mustafa et al., 1974), while enhancing tolerance to such stress (Reddy et al., 1975). Based on these preliminary findings, Douglas et al. (1977) assessed whether the acquiring of tolerance to O_3 could be associated with increased levels of SOD in lung tissue.

In their testing protocol using mice and rats, no evidence was presented to support the hypothesis that O_3 tolerance was related to lung SOD levels since both controls and O_3 treated groups did not differ in SOD levels. In addition, the mice strains that are tolerant to O_3 were not tolerant to oxygen. The authors concluded that O_3 tolerance is not associated with an increase of SOD and that O_2 and O_3 tolerances must involve different mechanisms.

While SOD levels were not associated with O_3 tolerance in the Douglas et al. (1977) study, it must be emphasized that lower levels of O_3 over a longer period (8 days) have been able to increase SOD levels. Whether such adaptive changes prevent or reduce O_3-induced lung damage is of considerable interest.

Finally, a report by Crapo and Drew (1975) indicated that the levels of SOD are increased in the lung of rats tolerant to NO_2. However, rats that were tolerant to NO_2 were not tolerant to O_2 and vice versa, thereby suggesting that the mechanism of O_2 and NO_2 toxicity is different.

8. RADIATION TOXICITY

Since radiation exposure is thought to generate superoxide radicals in living tissue, it followed that the concentration of SOD may be an important determination of radiation toxicity. A variety of studies have considered various aspects of this issue. For example, the effects of X irradiation on white blood cells were markedly reduced by SOD treatment (Petkav et al., 1978). Another approach has added a drug called diethyldithiocarbamate (DDC), which inactivates SOD in red cells and has enhanced radiation toxicity (Stone et al., 1978). A third approach considered whole animal systems in which intravenous injections of SOD significantly enhanced survival of mice to acute X-ray doses (Petkav et al., 1975, 1978). From a therapeutic perspective, the drug orgotein, which is Cu-Zn SOD, has been successfully tested in a double-blind placebo-controlled study to safely and effectively reduce and/or prevent the side effects of high energy radiation therapy (8400 or 6400 rad) of persons treated for bladder cancer (Menander-Huber et al., 1978).

9. PARAQUAT TOXICITY

The herbicide paraquat (1,1′dimethyl-4′bipyridylium chloride) has been reported to cause fatal intoxication in animals and man, with the mechanism being related to acute lung injury. More specifically, paraquat has been widely used as a suicidal agent with at least 400 suicides being directly related to paraquat (Rhodes and Patterson, 1978).

Paraquat is known to have a low redox potential, and while in the reduced state it can use molecular oxygen as a one electron receptor. Further studies have revealed that its toxicity is markedly enhanced in the presence of greater than ambient concentrations of oxygen. A chemical analogue of paraquat, diquat (1,1'ethylene-2,2'dipyridylium dichloride), was reported to produce superoxide free radicals ($O_2^-\cdot$) when reduced aerobically either by NADPH and GSH reductase or photochemically in the presence of EDTA. Of considerable importance for this chapter is that SOD treatment inhibited the occurrence of superoxide radicals by diquat. These findings led Autor in 1974 to propose that the administration of SOD may prevent acute paraquat toxicity in a variety of animal models.

In a study with male Holtzman rats that were orally treated with paraquat, Autor (1974) supported her hypothesis by showing that SOD administration (i.e., 1 hr prior to paraquat treatment and four additional injections of SOD after paraquat exposure) reduced acute paraquat toxicity regardless of whether normal or 90–95% oxygen was present at atmospheric pressure. However, the greatest protection (i.e., extending lifetime by > twofold) occurred in the group treated with paraquat and maintained in normal air. The author speculated that the amount of SOD needed to protect against paraquat in an oxygen environment, or against the toxic effects of high oxygen concentration, by itself may be considerably larger than employed throughout her study.

Regardless of the differences in response among the various studies, it was concluded (1) that paraquat may produce toxicity via the superoxide radical since SOD reduces paraquat toxicity, and (2) that SOD may be useful in treating paraquat intoxication.

The striking report of Autor (1974) was supported in a subsequent study by Wasserman and Block (1978) who also noted that SOD significantly prolonged survival and reduced histolytic lung disease of male Sprague-Dawley rats, given a single intraperitoneal injection of 50 mg/kg paraquat 1 hr before SOD administration. These authors also speculated, as did Autor (1974), that SOD treatment may have human usefulness in the case of paraquat poisoning. Wasserman and Block (1978) pointed out that several cases of paraquat poisoning had been treated to date and been published (Fairshter et al., 1976; Harley et al., 1977), and in neither instance was there any clear-cut benefit attributable to SOD. However, Wasserman and Block (1978) felt that the doses employed were too low and that the SOD therapy was administered too late in the clinical course.

In contrast to the previous studies, which clearly supported the hypothesis that exogenous administration of SOD may reduce paraquat toxicity, was the report by Rhodes and Patterson (1978) which found that administration of SOD to mice and rats had no beneficial effect on acute paraquat toxicity. These authors earlier had shown that injection of labeled SOD caused only a very small increase in tissue SOD levels relative to endogenous levels. It was for this reason that they thought their study revealed that SOD did not protect against paraquat-induced lung damage. Giri et al. (1982) likewise reported that SOD treatment (intravenous) of four doses at 12-hr intervals (one prior and three after paraquat treatment) did not affect paraquat-induced increased pulmonary vascular permeability and pulmonary edema as well as the mortality rate in mice.

Why the findings of Rhodes and Patterson (1978) and Giri et al. (1981) differed from that of Autor (1974) (i.e., rats) and Wasserman and Block (1978) (i.e., rats) is not obvious. Unfortunately, the Rhodes and Patterson (1978) report did not mention which sex or animal strain they used. Nevertheless, the overall methodologies were generally very comparable. However, Giri et al. (1982) asserted that the reason for such discrepancies may be tied to a species difference since they (Giri et al., 1982) used mice while Autor (1974) and Wasserman and Block (1978) employed rats. Furthermore, the number of animals these studies employed and the percent protection displayed by the SOD treatment was small. They suggested that one of the reasons for the lack of an SOD protective effect may be its rapid elimination from the body with a half-life in mice of 2 hr as well as a lack of penetration to the lungs, although labeled studies by Petkav et al. (1976) seem to preclude this idea.

In the third nonsupportive study, Block (1979), who earlier coauthored the 1978 report with Wasserman, reported that administration of exogenous SOD to vitamin E deficient rats did not offer any protection from acute paraquat toxicity. While the reason for this lack of protection remains to be determined, Block (1979) offered several suggestions:

1. In the absence of vitamin E, it is possible that peroxidative chain reactions may be triggered and sustained by very minute amounts of $O_2\cdot$ escaping detoxification by SOD. Since $O_2\cdot$ generation occurs predominantly within the cell cytoplasm and is enhanced by paraquat, and since most exogenously administered SOD is confined to the extracellular space, it seems likely that some $O_2\cdot$ would avoid dismutation by both exogenous and endogenous SOD.
2. It is possible that transperitoneal transport of SOD or other metalloproteins is impaired in vitamin E deficient animals resulting in low blood or tissue concentrations. Unfortunately, blood and/or tissue levels of SOD were not assayed in the present study.
3. The dosing schedule of SOD used in this study (10 mg/kg IP bid for three days prior to and three days after paraquat) may have been inadequate for modifying the development of acute paraquat toxic-

ity. This possibility is suggested because serum concentrations of SOD peak 1–2 hr after IP administration of SOD and are not detectable 8–12 hr after IP administration of doses comparable to that used in this study. Therefore, more frequent (e.g., every 6 hr) or prolonged administration of SOD might have resulted in a significant improvement in duration and/or percent survival in these rats. Further studies will be required to resolve these possibilities.

In conclusion, there is some evidence to support the idea that exogenous SOD may reduce acute paraquat toxicity in rodents. However, the occurrence of several negative studies in animal model and clinical human studies, while not discrediting the earlier supportive findings, suggests the need to further define the effect of treatment parameters including concentration, duration, and route of exposure on the impact of SOD administration on paraquat toxicity.

10. SUMMARY

SOD has been found to play a critical role in helping living organisms defend against environmental agents and endogenous processes that result in the formation of superoxide radicals. Numerous agents such as copper, paraquat, oxygen, irradiation, as well as a variety of widely used drugs produce superoxide radicals. Evidence is now mounting that SOD may play a principal role in preventing toxicity to such agents. Of considerable importance is that an extremely wide range of SOD values exist in human erythrocytes and it has been found that those infants on the lower end of the spectrum may be at increased risk of developing hyperbilirubinemia. No similar studies of adults have been carried out that would help to establish whether lower SOD levels may be a risk factor in oxidant pollutant toxicity. However, this is a worthy area of further investigation. In addition, at least one genetic variant of SOD exists in the human population. Whether persons with this variant form respond differentially to oxidant stress is unknown.

REFERENCES

Autor, A. P. (1974). Reduction of paraquat toxicity by superoxide dismutase. *Life Sci.* **14:** 1309.

Beckman, G. and Pakarinen, A. (1973). Superoxide dismutase: A population study. *Hum. Hered.* **23:** 346–351.

Block, E. R. (1979). Potentiation of acute paraquat toxicity by vitamin E deficiency. *Lung* **156:** 195–203.

Carter, H. D., Auton, J. A., Welch, S. G., Marshall, W. H., and Fraser, G. R. (1976). Superoxide dismutase variants in Newfoundland—a gene from Scandinavia? *Hum. Hered.* **26:** 4–7.

Costilow, R. H. and Keele, B. B., Jr. (1972). Superoxide dismutase in *Bacillus popilliae*. *J. Bacteriol.* **111:** 628–630.

Crapo, J. D. and Drew, R. T. (1975). Pulmonary biochemical "adaptation" and partial cross tolerance between NO_2 and O_2. *ARRD* **111:** 902–903.

Crapo, J. D. and Tierney, D. F. (1974). Superoxide dismutase and pulmonary oxygen toxicity. *Am. J. Physiol.* **226**(b): 1401–1407.

Douglas, J. S., Curry, G., and Geffkin, S. A. (1977). Superoxide dismutase and pulmonary ozone toxicity. *Life Sci.* **20:** 1187–1192.

Fairshter, R. D., Rosen, S. M., Smith, W. R., Glausner, F. L., McRae, D. M., and Wilson, A. F. (1976). Paraquat poisoning: New aspects of therapy. *Q. J. Med.* **45:** 551.

Frischer, H., Chu, L. K., Ahmad, T., Justice, P., and Smith, G. F. (1981). Superoxide dismutase and glutathione peroxidase abnormalities in erythrocytes and lymphoid cells in Down Syndrome. In: *The Red Cell: Fifth Ann Arbor Conference.* Alan R. Liss, New York, pp. 269–283.

Giri, S. N., Hollinger, M. A., and Schedt, M. J. (1981). The effects of paraquat and superoxide dismutase on pulmonary vascular permeability and edema in mice. *Arch. J. Environ. Health* **36:** 149–154.

Harley, J. B., Carinspan, S., and Root, R. K. (1977). Paraquat suicide in a young woman: Results of therapy directed against the superoxide radical. *Yale J. Biol. Med.* **50:** 481–488.

Hunter, M. I. S., Brzeski, M. S., and deVane, P. J. (1981). Superoxide dismutase, glutathione peroxidase and thiobarbituric acid-reactive compounds in erythrocytes in Duchenne muscular dystrophy. *Clin. Chim. Acta* **115:** 93–98.

Kobayashi, Y., Ishigame, K., Ishigame, K., and Vsui, T. (1977). Superoxide dismutase activity of human blood cells. In: O. Hayaishi and K. Asada, *Biochemical and Medical Aspects of Active Oxygen.* Japan Scientific Society Press, Tokyo, Japan, pp. 261–274.

Legge, M., Brian, M., Winterbourn, C., and Carrell, R. (1977). Red cell superoxide dismutase activity in the newborn. *Aust. Paediatr. J.* **13:** 25–28.

McCord, J. M., Keele, B. B., Jr., and Fridovich, I. (1971). An enzyme-based theory of obligate anaerobiosis: The physiological function of superoxide dismutase. *Proc. Nat. Acad. Sci. U.S.A.* **68:** 1024–1027.

Menander-Huber, K. B., Edsmyr, F., and Huber, W. (1978). Orgotein (superoxide dismutase). A drug for the amelioration of radiation-induced side effects. *Urol. Res.* **6:** 255–257.

Michelson, A. M., Puget, K., Durosay, P., Bonneau, J. C., and Ropartz, C. (1977). Superoxide dismutase levels in human erythrocytes. In: O. Hayaishi and K. Asada, *Biochemical and Medical Aspects of Active Oxygen.* Japan Scientific Society Press, Tokyo, Japan, pp. 247–260.

Mustafa, M. G., Delucia, A. J., Hussain, M. Z., Chow, C. K., and Cross, C. E. (1974). American Thoracic Meeting, Cincinnati, Ohio.

Oberley, L. W. and Buettner, G. R. (1979). Role of superoxide dismutase in cancer: A review. *Cancer Res.* **39:** 1141–1149.

Petkav, A. (1978). Radiation protection by superoxide dismutase. *Photochem. Photobiol.* **28:** 765–774.

Petkav, A., Chelack, W. S., Kelly, K., Barefoot, C., and Monasterski, L. (1976). Tissue distribution of bovine ^{125}I-superoxide dismutase in mice. *Res. Commun. Chem. Pathol. Pharmacol.* **15:** 641–654.

Introduction

Petkav, A., Chelack, W. S., Pleskach, S. D., Meeker, B. E., and Brady, C. M. (1975). Radio protection of mice by superoxide dismutase. *Biochem. Biophys. Res. Comm.* **65**(3): 886–893.

Petkav, A., Chelack, W. S., and Pleskach, S. D. (1978). Protection by superoxide dismutase of white blood cells in X-irradiated mice. *Life Sci.* **22**: 867–882.

Reddy, K., Kimball, R., Peirce, T., Macres, S., and Cross, C. (1975). *Clin. Res.* **23**: 117A.

Rhodes, M. L. and Patterson, C. E. (1978). Effect of exogenous superoxide dismutase on paraquat toxicity. *Am. Rev. Respir. Dis.* **117**: 255 (abstract).

Rotilio, G., Rigo, A., Bracci, R. Bagnoli, F., Sargentini, I., and Brunori, M. (1977). Determination of red blood cell superoxide dismutase and glutathione peroxidase in newborns in relation to neonatal hemolysis. *Clin. Chim. Acta* **81**: 131–134.

Scudder, P., Stocks, J., and Dormandy, T. L. (1976). The relationship between erythrocyte superoxide dismutase activity and erythrocyte copper levels in normal subjects and in patients with rheumatoid arthritis. *Clin. Chim. Acta* **69**: 397–403.

Stone, D., Lin, P.-S., and Kwock, L. (1978). Radiosensitization of human erythrocytes by diethyldithiocarbamate. *Int. J. Radiat. Biol.* **33**: 393–396.

Teng, Y. S. and Lie-Injo, L. E. (1977). Erythrocyte superoxide dismutase in different racial groups in Malaysia. *Hum. Hered.* **36**: 231–234.

Wasserman, B. and Block, E. R. (1978). Prevention of acute paraquat toxicity in rats by superoxide dismutase. *Aviat. Space Environ. Med.* **49**: 805–809.

Welch, S. G. and Mears, G. W. (1972). Genetic variants of human indophenol oxidase in the Westray Island of the Orkneys. *Hum. Hered.* **22**: 38–41.

Ueda, K. and Ogata, M. (1978). Levels of erythrocyte superoxide dismutase activity in Japanese people. *Acta Med. Okayama* **32**: 393–397.

Yoshioka, T., Sugive, A., Shimada, T., and Utsumi, K. (1979). Superoxide dismutase activity in the maternal and cord blood. *Biol. Neonate* **36**: 173–180.

N. Porphyrias

1. INTRODUCTION

Metabolic disorders associated with the synthesis of porphyrins (usually in erythroid and liver cells) are collectively known as porphyrias. This group of closely related diseases is characterized by a marked increase in the synthesis and excretion of porphyrins or their precursors. A porphyrin is any one of a group of iron-free or magnesium-free cyclic tetrapyrrole derivatives. Porphyrins are vitally important molecules, some of which serve as intermediates in the synthesis of hemoglobin, the cytochromes, chlorophyll, and vitamin B_{12}. The porphyrin ring is shown in Figure 2-7. The scheme of heme biosynthesis is shown in Figure 2-8.

The excessive production of porphyrins leads to abnormally high levels of such substances in the urine and various tissues of the body. Porphyrins often have a photosensitizing activity that produces skin lesions, prickling, itching, and burning following exposure to light (Marver and Schmid, 1972). It is thought that these skin manifestations are influenced by chemicals produced from photochemical reactions that are initiated by

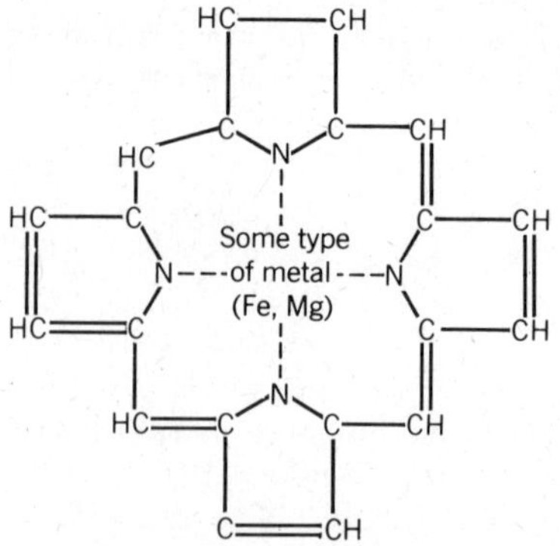

Figure 2-7. The chemical structure of the porphyrin ring.

the light absorbed by porphyrins. Rimington et al. (1967) have suggested that the release of lysosomal enzymes plays a role in the inflammation process.

Besides their photosensitizing properties, porphyrins produce a broad variety of toxic effects, such as vascular and intestinal spasms and neuropathologic changes (Marver and Schmid, 1972). It has been suggested that porphyrins may act to reduce the release of acetylcholine that usually occurs following ionic depolarization. Also, it appears that aminolevulinic acid (ALA), a porphyrin precursor, may inhibit brain tissue dependent ATPase. Thus, it seems that intermediates in heme biosynthesis may be neurotoxins (Marver and Schmid, 1972).

2. MAJOR TYPES OF PORPHYRIAS

a. Erythropoietic Porphyria

This is a metabolic disorder in erythroid cells of bone marrow that leads to excessive production of porphyrins. The disorder results in chronic skin photosensitivity and severe porphyrinuria associated with hemolytic anemia.

b. Hepatic Porphyria

The three hepatic porphyrias that are genetically transmitted (acute intermittent porphyria, variegate porphyria and coproporphyria) have several common characteristics:

1. The acute phase has increased levels of ALA and porphobilinogen (PBG); this is caused, in part, by increased activity of ALA synthetase (Marver and Schmid, 1972).

Pollutant Interactions

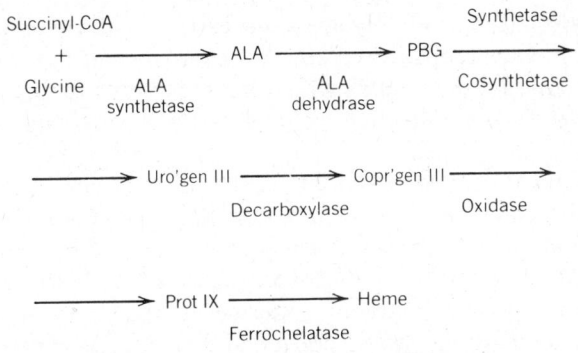

```
Succinyl-CoA                                    Synthetase
     +      ————→  ALA  ————————→  PBG  ————————→
  Glycine    ALA            ALA              Cosynthetase
          synthetase     dehydrase


         ————→  Uro'gen III  ————————→  Copr'gen III  ————————→
                  Decarboxylase                Oxidase


         ————→  Prot IX  ————————→  Heme
                        Ferrochelatase
```

Figure 2-8. Scheme of heme biosynthesis.

2. In many individuals, the metabolic disorder is in a latent form so that afflicted individuals are entirely free of symptoms or descriptions of nervousness. However, an acute attack may be precipitated by therapeutic doses of several drugs such as barbiturates, sulfonamides, general anesthetics, excessive amounts of ethanol, and chloraquine (Waldenström, 1957; Eales and Linder, 1962; Goldberg, 1968; Waldenström and Haeger-Aronsen, 1963). These drugs are thought to induce hepatic ALA synthetase (Marver and Schmid, 1972).

3. Clinical symptoms are associated with similar neurologic features (Marver and Schmid, 1972).

4. Often the first clinical and biochemical signs of the disease occur during late puberty. This may be associated with naturally occurring steroids that induce ALA synthetase (Granick, 1966). Clinical exacerbations have also been noted in association with pregnancy (Vine et al., 1957; Neilson and Neilson, 1958; Gould et al., 1961; Petrie and Mooney, 1962), and the menstrual cycle (Perlroth et al., 1965; Marver and Schmid, 1972), which significantly alters steroid hormone levels.

c. Protoporphyria

This is a mild disorder characterized by chronic solar eczema caused by the overproduction of protoporphyrin IXa in liver and erythroid cells.

3. POLLUTANT INTERACTIONS

A large outbreak of porphyria affecting several thousand people, occurred during 1956 in Turkey. The toxic effects were caused by chronic ingestion of hexachlorobenzene, which was used as a fungicidal agent

(Schmid, 1960; Cam and Nigogoysan, 1963). When the use of hexachlorobenzene was stopped in 1959, new cases of porphyria ceased. This disorder occurred in individuals who were not known to have any genetic predisposition toward developing porphyria (Cam and Nigogoysan, 1963).

Lead inhibits the activities of at least two enzymes (ALA dehydrase and ferrochelatase) concerned with the synthesis of heme in the red blood cells (Haeger-Aronsen, 1960; Schwartz et al., 1952; Lichtman and Feldman, 1963; Goldberg, 1968). Typical features of lead toxicity are usually indicated by hypochromic anemia, increased free erythrocyte protoporphyrin, and excessive urinary excretion of ALA and coproporphyrin with near normal concentrations of PBG. Tschudy and Collins (1957) have also demonstrated that the herbicide AT partially inhibits ALA dehydrase.

It is important to note that under normal conditions heme inhibits the activity of ALA synthetase, ALA dehydrase, and ferrochelatase. Thus, the scheme for heme biosynthesis operates under self-limiting controls (Marver and Schmid, 1972).

The presence of the genetic disorders that stimulate excessive activity of ALA synthetase and/or exposure to drugs, industrial chemicals, or hormones that induce this enzyme along with lead exposure, would be expected to produce higher than normal levels of ALA. In this instance, ALA is inhibited from being converted to heme by the lead, thereby preventing the ALA from seeking its natural product (heme). This would result in excessive quantities of ALA being both excreted and accumulating in various tissues of the body. Since the production of heme would be reduced in such a situation, the activity of ALA synthetase would not be inhibited by negative feedback from the heme.

4. FREQUENCY IN THE POPULATION

The frequency of the various porphyria metabolic disorders is uncertain, however their occurrence is considered rare (Marver and Schmid, 1972). With regard to acute intermittent porphyria, the incidence is about 1.5 per 100,000 in Sweden (Marver and Schmid, 1972), Denmark (With, 1963), Ireland (Fennelly et al., 1960), and western Australia (Saint and Curnow, 1962), while considerably rarer in Blacks (Lyon, 1968).

5. SCREENING TESTS FOR PORPHYRIAS

Of the various types of porphyrin metabolic disorders the most frequent type is the hepatic porphyrias; that is, acute intermittent porphyria and variegated porphyria.

a. Acute Intermittent Porphyria

The levels of PBG in the normal individual's urine are too slight to be determined with the conventional method (Hammond and Welcker, 1948) described by Watson and Schwartz (1941). However, during an acute attack of porphyria, the levels of urinary PBG are quite high (Ackner et al., 1961; With, 1963). Additionally, the levels of ALA also increase significantly during an acute attack (With, 1963; Mauzerall and Granick, 1956; Ackner et al., 1961). Thus, individuals experiencing acute intermittent porphyria can be detected. In latent porphyria, chromatographic analysis of the urine frequently (Stein et al. 1966), but not invariably (With, 1963), reveals increased excretion of ALA or PBG or both, thereby making confident diagnosis of asymptomatic individuals difficult (Watson, 1960; Waldenstrom and Haeger-Aronsen, 1963).

b. Variegate Porphyria

In patients with this disease, the concentrations of proto- and coproporphyrin and peptide conjugates of dicarboxylic porphyrins are greatly increased in the feces even when clinical symptoms are minimal (Dean and Barnes, 1959; Eales, 1963; Sweeney, 1963). Although the frequency of variegate porphyria is not generally known, research does indicate an estimated three cases per 1000 white South Africans (Dean, 1963).

6. SUMMARY

Individuals with the genetically determined condition collectively referred to as porphyrias are theoretically at increased risk to lead-induced toxicity principally as an aggravation of their condition. However, experimental evidence to support this theoretical proposal is lacking.

REFERENCES

Ackner, B., Cooper, J. E., Gray, C. H., Kelly, M., and Nicholson, D. C. (1961). Excretion of porphobilinogen and δ-aminolaevulinic acid in acute porphyria. *Lancet* **1:** 1256.

Cam, C. and Nigogoysan, G. (1963). Acquired toxic porphyria cutania tarda due to hexachlorobenzene. *J. Am. Med. Assoc.* **183:** 88.

Dean, G. (1963). *The Porphyrias: A Story of Inheritance and Environment.* Pitman Medical, London.

Dean, G. and Barnes, H. D. (1959). Porphyria in Sweden and South Africa. *S. Afr. Med. J.* **33:** 274.

Eales, L. (1963). Porphyria as seen in Cape Town: A survey of 250 patients and some recent studies. *S. Afr. J. Lab. Clin. Med.* **9:** 151.

Eales, L. and Linder, G. C. (1962). Porphyria—The acute attack. *S. Afr. Med. J.* **36**: 284.

Fennelly, J. J., Fitzgerald, O., and Hingerty, D. J. (1960). Observations on porphyria with special reference to Ireland. *Ir. J. Med. Sci.* **411**: 130.

Goldberg, A. (1968). Lead poisoning as a disorder of heme synthesis. *Semin. Hematol.* **5**: 424.

Gould, S., Allison, H. M., and Bellew, L. N. (1961). Acute porphyria complicated by pregnancy: Report of a case. *Obstet. Gynecol.* **17**: 109.

Granick, S. (1966). The induction *in vitro* of the synthesis of δ-aminolevulinic acid synthetase in chemical porphyria: A response to certain drugs, sex hormones and foreign chemicals. *J. Biol. Chem.* **241**: 1359.

Haeger-Aronsen, B. (1960). Studies on urinary excretion of δ-aminolaevulinic acid and other haem precursors in lead workers and lead-intoxicated rabbits. *Scand. J. Clin. Invest.* **12**(suppl.): 47.

Hammond, R. S. and Welcker, M. L. (1948). Porphobilinogen tests on a thousand miscellaneous patients in a search for false positive reactions. *J. Lab. Clin. Med.* **33**: 1254.

Lichtman, H. C. and Feldman, F. (1963). *In vitro* pyrrole and porphyrin synthesis in lead poisoning and iron deficiency. *J. Clin. Invest.* **42**: 830.

Lyon, C. J. (1968). Acute porphyria in American Negro. *N. Y. Med. J.* **68**: 2441.

Marver, H. S. and Schmid, R. (1972). The porphyrias: In: *The Metabolic Basis of Inherited Disease.* J. B. Stanbury, J. B. Wyngaarden, and D. S. Fredrickson (eds.). McGraw-Hill, New York, p. 1087.

Mauzerall, D. and Granick, S. (1956). The occurrence and determination of δ-aminolevulinic acid and porphobilinogen in urine. *J. Biol. Chem.* **219**: 435.

Neilson, D. R. and Neilson, R. P. (1958). Porphyrin complicated by pregnancy. *West. J. Surg.* **66**: 133.

Perlroth, M. G., Marver, H. S., and Tschudy, D. P. (1965). Oral contraceptive agents and the management of acute intermittent porphyria. *J. Am. Med. Assoc.* **194**: 1037.

Petrie, S. J. and Mooney, J. B. (1962). Porphyria with the complication of pregnancy. *Am. J. Obstet. Gynecol.* **83**: 264.

Rimington, C., Magnus, I. A., Ryan, E. A., and Cripps, D. J. (1967). Porphyria and photosensitivity. *Q. J. Med.* **36**: 29.

Saint, E. G. and Curnow, D. H. (1962). Porphyria in Western Australia. *Lancet* **1**: 133.

Schmid, R. (1960). Cutaneous porphyrin in Turkey. *N. Engl. J. Med.* **263**: 397.

Schwartz, S., Keprios, M., and Schmid, R. (1952). Experimental porphyria. II. Type produced by lead, phenylhydrazine and light. *Proc. Soc. Exp. Biol. Med.* **79**: 463.

Stein, J. A., Tschudy, D. P., Marver, H. S., Berard, C. W., Ziegel, R. F., Recheigl, M., Jr., and Collins, A. (1966). Acute intermittent porphyria: New morphologic and biochemical findings. *Am. J. Med.* **41**: 149.

Sweeney, G. D. (1963). Patterns of porphyrin excretion in South African patients. *S. Afr. J. Lab. Clin. Med.* **9**: 182.

Tschudy, D. P. and Collins, A. (1957). Effect of 3-amino-1,2,4-triazole on δ-aminolevulinic acid dehydratase activity. *Science* **126**: 168.

Vine, S., Shaffer, H. M., Pauler, G., and Margolis, E. J. (1957). A review of the relationship between pregnancy and porphyria and presentation of a case. *Ann. Intern. Med.* **47**: 834.

Waldenström, J. (1957). The porphyrias as inborn errors of metabolism. *Am. J. Med.* **22**: 758.

Waldenström, J. and Haeger-Aronsen, B. (1963). Different pattern of human porphyria. *Br. Med. J.* **2**: 272.

Watson, C. J. (1960). The problem of porphyria—some facts and conditions. *N. Engl. J. Med.* **263**: 1205.

Watson, C. J. and Schwartz, S. (1941). A simple test for urinary porphobilinogen. *Proc. Soc. Exp. Biol. Med.* **47**: 393.

With, T. K. (1963). Acute intermittent porphyria: Family studies on the excretion of PBG and delta-ALA with ion exchange chromatography. *A. Klin. Chem.* **1**: 134.

O. ALA Dehydratase Deficiency

This enzyme is known to be very sensitive to lead inhibition. Individuals with a deficient condition have been recently identified (Bird et al., 1979). The authors of such research speculated that those individuals may be at increased risk to lead poisoning within the red cell. This is not very likely since ALA dehydratase is not the rate limiting step in the synthesis of heme, and experimental animal studies have shown that decreasing ALA dehydratase activity by 98% did not alter hemoglobin production (Stopps, 1974).

REFERENCES

Bird, T. D., Hamerynik, P., Nutter, J. Y., and Labbe, R. F. (1979). Inherited deficiency of delta-aminolevulinic acid dehydratase. *Am. J. Hum. Genet.* **31**: 662–668.

Stopps, G. J. (1974). Is there a safe level of lead exposure? *J. Wash. Acad. Sci.* **61**(2): 103–120.

P. Rhodanese Variants

The enzyme rhodanese is known to be the principal enzyme involved in the detoxification of cyanide. Recent studies have revealed the presence of rhodanese variants in the population (Scott and Wright, 1980). To what extent these variants exist and whether those with this variant may have their capacity to detoxify cyanide modified is unknown but worthy of investigation.

REFERENCE

Scott, E. M. and Wright, R. C. (1980). Genetic polymorphism of rhodanese from human erythrocytes. *Am. J. Hum. Genet.* **32**: 112–114.

Q. Hb Variants

It has long been recognized that exposure to carbon monoxide (CO) will result in the formation of carboxyhemoglobin (COHb). That the formation of COHb occurs after exposure to CO is not unexpected since CO has several hundred times more affinity to Hb than O_2. The toxicity caused by exposure to elevated levels of CO is directly related to its capacity to prevent Hb from transporting oxygen to tissues for cellular respiration. The formation of COHb as a result of CO exposure occurs in all mammalian species studied. Since all mammals are dependent on Hb to convey O_2 to tissues for energy production, the fact that they form COHb as do humans, suggests the possibility that a variety of animal models may serve as reliable predictors of COHb formation in humans. The likelihood that such mammalian models may be effective predictors of humans with regard to COHb formation is strongly supported by a variety of observations which indicate that a large number of respiratory structural and functional variables are mathematical functions of body weight, which are consistent across the spectrum of mammalian species. Presumably, therefore, the different mammalian species should be exposed to the same amount of O_2 or CO for that matter as other mammals including humans.

Does this imply that mammalian animal models should be good predictors of the occurrence of CO-induced COHb formation in humans? Yes, it does, but the problem is somewhat more complex. At least one additional question that needs to be answered with regard to interspecies predictions for CO-induced formation is the relative CO affinity for Hb in the different species.

It is well-known that a large number of structurally different Hb exist. Many of these genetic variant forms of Hb are not known to have any adverse health consequences while others may—such as in the case of Hb S in sickle-cell trait and anemia. In light of the occurrence of structurally and functionally modified Hb molecules, it would not only be unexpected if differences in CO affinity for Hb differed between groups of humans but also on an interspecies basis.

While limited data exist on this topic, it has long been known that humans and mice displayed a different capacity for COHb formation. More specifically, human blood was found to be more easily saturated with CO than mouse blood across a wide range of CO exposures. For example, when human blood had approximately a 25% CO saturation, mice displayed approximately 15% COHb (Douglas et al., 1912).

These findings suggest that while animal models may be good qualitative predictors of human COHb formation, quantitative predictions may be difficult if there are differences in the affinity of the CO for Hb. In fact, Musselman et al. (1959) reported considerable variability in the COHb levels of rats, rabbits, and dogs that had been exposed to 50 ppm CO for 24 hr/day, 7 days per week for 3 months. The average COHb levels reached were 7.3% for dogs, 3.2% for rabbits, and only 1.8% for rats. Why there was so much interspecies variability in COHb levels among the species is uncertain although it would appear that any variation in the respiratory tract variables would not be able to account for such differences. Whether and to what extent differences in the relative affinities of Hb for CO may have played a role is also unknown. However, it would appear that this would be an interesting question to evaluate.

It is interesting to note that there are several types of human Hb [i.e., Hb Zurich (Hb Z) and Hb M-Chicago (Hb Mc)] that display a greater affinity for CO than normal human Hb. For example, the reaction rate of CO with Hb M and Hb Z has been found to be 20 and 10 times, respectively, greater than normal Hb (Giacometti et al., 1980; Winterhalter et al., 1969). Thus individuals with these forms of Hb may be considered at increased risk to the formation of COHb as a result of CO exposure. As is typical of other comparable situations, the question becomes, which group of humans does a possible model represent? In many respects the variability of affinities in the different animal models can be exploited by researchers who are trying to obtain models for normal and high-risk groups.

REFERENCES

Anson, M. L., Barcroft, J., Mirsky, A. E., and Oinuma, S. (1924). On the correlation between the spectra of various hemoglobins and their relative affinities for oxygen and carbon monoxide. *Proc. R. Soc. London Series B.* **97:** 61–83.

Douglas, C. G., Haldane, J. S., and Haldane, J. B. S. (1912). The laws of combination of haemoglobin with carbon monoxide and oxygen. *J. Physiol.* **44:** 276–304.

Giacometti, G., Brunori, M., Antonini, E., DiLorio, E. E., and Winterhalter, K. H. (1980). The action of hemoglobin Zurich with oxygen and carbon monoxide. *J. Biol. Chem.* **225**(13): 6160–6165.

Musselman, N. P., Groff, W. A., Yerich, P. P., Wilinski, F. T., Weeks, M. H., and Oberst, F. W. (1959). Continuous exposure of laboratory animals to a low concentration of carbon monoxide. *Aerosp. Med.* **30:** 524–529.

Winterhalter, K. H., Anderson, N. M., Amiconi, G., Antonini, E., and Brunori, M. (1969). Functional properties of hemoglobin Zurich. *Eur. J. Biochem.* **11:** 435–440.

3 Liver Metabolism

A. Defects in Conjugation Mechanisms

1. GLUCURONIDATION

a. Introduction

Since the late 1960s when PCBs were first recognized as a potentially widespread environmental contaminant, hundreds of research papers have been published concerning their structure, industrial uses, presence in the environment, toxicity to numerous animals from insects to man, tissue storage, metabolism, and excretion (Golberg, 1974; Selikoff, 1972; Calabrese and Sorenson, 1977). As a result of such data accumulation, regulatory agencies are now trying to approach the problem of standard setting with respect to PCBs. Standards for air, water, and food are necessary if the total human exposure is to be regulated and controlled. One important component in standard setting is a consideration of the individuals within the population who may be at high risk to the pollutant because of various genetic, physiological, psychological, and behavioral traits. Calabrese (1977) has indicated that individuals lacking the ability to detoxify and excrete PCBs represent such a high-risk group since they will be unable to prevent undue accumulation of possible toxic levels of PCBs within the body. The theoretical foundations upon which this high-risk group is based (Calabrese, 1977) will be subsequently discussed.

Mammals, including rabbits, dogs, and humans have been reported to excrete PCBs in part by conjugate glucuronidation and/or sulfonation (Golberg, 1974; Block and Cornish, 1959). The urinary excretion of biphenyl and 4-chlorobiphenyl has been studied in rabbits. Biphenyl glucosiduronic acid and 4-hydroxybiphenyl were isolated from the urine. The rabbits fed the 4-chlorobiphenyl excreted 4-(p-chlorobiphenyl)-phenol and 4-chlorobiphenyl glucosiduronide. Twice as much 4-chlorobiphenyl as biphenyl was excreted as the glucosiduronic acid derivative. It was suggested that other low chlorinated biphenyls are excreted in a similar manner.

In dogs injected with 2,4,4'-trichloro-2'hydroxydiphenyl ether, nearly 100% of the material excreted in their urine and feces over a five-day period appeared as the glucuronide or sulfate conjugate. Human adults similarly exposed excrete 65% in urine and 20% in feces as a free compound or as a glucuronide (Golberg, 1974). Other evidence suggests that some chlorinated biphenyls are hydroxylated by the rat and pigeon. No evidence of reductive dechlorination was observed in either the trout, rat, or pigeon (Hutzinger et al., 1972).

It is apparent that individuals lacking the ability to form either conjugate glucuronides or sulfates or both will be at high risk with respect to toxic compounds of a phenolic nature. The recognition of low levels of glucuronide formation in the cat has been suggested as an explanation of the widely observed phenomenon in veterinary literature that phenolic compounds should not be used on or around cats because of the increased incidence of toxicity in this species (Clarke and Clarke, 1967; Jones, 1965; Stecher, 1968; Wilkenson, 1968) (Table 3-1). Furthermore, intravenous injections of phenol in cats, pigs, dogs, and goats revealed that cats were at least 2 times more sensitive with respect to fatality caused by the phenol (Oehme, 1969). Toxicity in this study was related to the partial deficiency in the cat to conjugate phenol with glucuronic acid. The author concluded that the cat is more likely to be poisoned by acute doses of phenol and is more likely to become chronically affected with phenol toxicity due to the inability to rapidly form and excrete glucuronides. Biochemical studies with the cat have also indicated that conjugate glucuronidation is the major pathway for the detoxification and excretion of phenol (e.g., more than twice as effective as sulfonation) (Oehme, 1969).

b. Developmental Immaturity and Susceptibility

Because cats are hypersusceptible to phenolic-like compounds as a result of their diminished capacity to form glucuronides, it is strongly suspected

Glucuronidation

Table 3-1. The Toxicity of Several Phenols to the Cat Compared with Some Noncarnivorous Species[a]

Compound	Species	Route of Administration	Toxicity	Dose (mg/kg)
Phenol	Cat	sc	MLD	80
	Guinea pig			450–550
2-Methylphenol (o-cresol)	Cat	sc	LD	55
	Rabbit			450–500
	Rat			650
	Guinea pig			350–400
3-Methylphenol (m-cresol)	Cat	sc	LD	180
	Rabbit			500–600
	Rat			900
	Guinea pig			300–400
4-Methylphenol (p-cresol)	Cat	sc	LD	80
	Rabbit			300–400
	Rat			500
	Guinea pig			200–300
2-Nitrophenol	Cat	sc	LD	600
	Rabbit			1700
4-Nitrophenol	Cat	sc	MLD	197
	Rabbit			600
1-Naphthol	Cat	or	LD	100–150
	Rabbit			9000
2-Naphthol	Cat	or	LD	100–150
	Rabbit			3800

Source: Hirom et al. (1977).

[a] Data taken from Spector (1956). LD means lethal dose (the amount that kills an animal) and MLD means minimum lethal dose (the smallest of several doses which kills one of a group of test animals). sc means subcutaneous and or means oral.

that human embryos, fetuses, and neonates (2–3 months old), which also are deficient in a functional conjugate glucuronidation system, will also be predisposed to the toxic effects of phenolic compounds including PCBs (Gillette, 1967; Smith and Williams, 1966; Nyhan, 1961). Usually by the age of 2–3 months, adult levels of most enzyme systems are achieved in humans as previously suggested (Gillette, 1967; Smith and Williams, 1966; Nyhan, 1961). Unfortunately, this "developmental immaturity" in the unborn and the very young may predispose them to the toxic effects of certain substances since they may be unable to detoxify and excrete them as quickly as necessary (Nyhan, 1961). It has already been pointed out that human adults do partially excrete PCBs via conjugate glucuronidation. Clinical experience has shown that infants may respond differently to doses of drugs that are easily tolerated by older children and adults. Presumably, this is because older children and adults have fully functioning enzyme detoxification systems, while the neonate lacks such development.

The widespread occurrence of glucuronic acid conjugation may be due to the facility with which glucuronic acid can be produced in the body from carbohydrate sources and the variety of chemical groups to which glucuronic acid can be transformed enzymatically. The other conjugation mechanisms are more restricted in their occurrence probably because of the limited availability of the conjugating agents such as glycine and cysteine (via glutathione) and the small number of chemical groupings to which conjugating agents such as sulfate, glycine, cysteine, methyl, and acetyl can be transferred. The acitivity and amount of the transferring enzymes concerned in these conjugations are probably also limiting factors. Glycine conjugation is confined mainly to aromatic carboxyl groups; cysteine or glutathione conjugation to some aromatic hydrocarbons and halogenated hydrocarbons; methylation to certain hydroxyl and amino groups; and acetylation to some amino and hydrazine groups. However, glucuronic acid conjugation can occur with several types of hydroxyl, amino, carboxyl groups, and with sulfhydryl groups (Adamson and Davies, 1973; Dutton, 1961, 1962; Gillette, 1967).

The effect of conjugation of a substance with glucuronic acid is to produce a strongly acidic compound that is more water soluble at physiological pH values than the precursor. The majority of foreign substances are ultimately cleared from the body via their excretion in the urine and bile, most often in the form of polar conjugates such as glucuronides with the avenue of excretion of the glucuronide probably varying with the species of animal considered (Adamson and Davies, 1973; Dutton, 1962).

A tragic example of toxicity arising from the inability to form glucuronides has been reported. The drug chloramphenicol, which is known to be metabolized in humans by the action of the glucuronic pathway, caused the death of more than 30 premature babies who had been treated with the antibiotic for infectious diseases. Theoretically, the premature babies were not able to conjugate the drug with glucuronic acid, and thus the drug could only be slowly excreted and so tended to accumulate in the body to toxic proportions (Smith and Williams, 1966). Inefficient glucuronidative mechanisms in the very young have broad significance since the absence of such mechanisms prevents the prompt elimination of toxic substances from the body.

A further problem for the very young with regard to PCBs is that in addition to having a difficult time excreting PCBs, they may also consume more PCBs per unit of body weight than at any other time during their life span due to high levels in human breast milk (Calabrese, 1982).

An additional problem encountered by many neonates is that approxi-

Glucuronidation

mately 5% of the mothers of normal infants secrete milk that inhibits the activity of glucuronyl transferase (thus the glucuronidation process) by more than 20% via the action of a steroid (pregname 3α,20β-diol) present in the breast milk. Inhibition of this glucuronyl transferase has been reported for up to 49 days after birth. Clinically, these children have been reported to develop unusually severe neonatal jaundice. This condition develops because glucuronide formation that assists in the elimination of bilirubin (breakdown product of Hb) is partially inhibited. Cow's milk does not contain sufficient amounts of this steroid to affect a noticeable inhibition of the glucuronidation process. Consequently, about 5% of the breast fed neonates would be expected to have their ability to excrete PCBs further impaired (Garther and Arias, 1966). For a contrary view see Ramos et al. (1966). Administration of the antibiotic novobiocin has also been associated with unconjugated hyperbilirubinemia in infants (Sutherland and Keller, 1961). Novobiocin is a noncompetitive inhibitor of glucuronyl transferase activity *in vitro* (Lokietz et al., 1963). Thus, children receiving concomitant exposure of novobiocin and PCBs would be expected to have their ability to detoxify and excrete PCBs impaired (Calabrese, 1977).

c. Gilbert's Syndrome/Crigler-Najjar Syndrome

After the neonatal period, a broad range of conditions is associated with the improper or incomplete development of the glucuronide conjugation system. The usual physiological problem associated with these conditions is the inadequate detoxification and excretion of bilirubin. This spectrum extends from the frequently occurring "mild" condition known as Gilbert's syndrome to the very rare, but severe and often fatal, Crigler-Najjar syndrome (Lester and Schmid, 1964).

Since the clinical effects associated with these syndromes are considered to be caused by metabolic disturbances of the glucuronide conjugation scheme, it is expected that PCB elimination in these individuals would be impeded. The population incidence of Gilbert's syndrome has been variously reported as 1 in 200 males (Billing, 1970), 7% based on examinations of 100 medical students (Kornberg, 1942) and 6% of 252 healthy National Blood Transfer Service donors, and 47 healthy medical students (197 males and 102 females) with no difference in frequency between males and females (Owens and Evans, 1975).

Concomitant exposure to PCBs and other druglike chemicals may potentiate the toxic effects of PCBs. For example, rats and monkeys given SKF525A (β-diethylaminoethyl-2, 2-diphenylpentanoate, a nonspecific inhibitor for many of the microsomal enzymes, especially hepatic microsomal enzymes) during the initial 24 hr of exposure to PCB succumbed rapidly as compared to the control group (Allen, 1975). Of possible significance is the fact that SKF525A inhibits the proper functioning of the glucuronic pathway (Smith and Williams, 1966).

Individuals with liver infections may also be at high risk with respect to PCBs. For example, depression of glucuronide synthesis has been observed in humans with infectious hepatitis (Smith and Williams, 1966).

Even though PCB-like pollutants require a functional conjugate glucuronidation system for their excretion, this does not imply that such substances are merely passive molecules in this process. For example, synthesis of microsomal enzymes such as glucuronyl transferase can be stimulated by drugs such as phenobarbital. Phenobarbital treatments have been successfully employed in persons with a partial defect in bilirubin conjugation in order to reduce the levels of bilirubin in the blood (Yaffe et al., 1966; Smith et al., 1967; Arias et al., 1969; Kreek and Sleisenger, 1968). Thompson et al. (1969) demonstrated similar results with oral administration of DDT. Although it has been suggested that the beneficial effects of phenobarbital or DDT are caused by their induction of glucuronyl transferase in the liver (Remmer, 1965), conclusive evidence still remains to be demonstrated (Kreek and Sleisenger, 1968; Whelton et al., 1968; Marver and Schmid, 1968).

With regard to public health implications, it appears that people with a partial glucuronyl transferase deficiency may not accumulate significant quantities of PCB-like substances. These substances seem capable of inducing microsomal enzymes that lead to their (PCBs) metabolism and excretion. This constitutes a built-in safety feature that should assist these individuals from accumulating PCB-like chemicals in their bodies. In contrast, individuals with the Crigler-Najjar syndrome seem to have an absolute deficiency of glucuronyl transferase, and thus do not have such a built-in compensatory safety system. However, it is not known how effective such a system is in those with the partial enzyme deficiency. It should also be noted that substances which require the activity of glucuronyl transferase for metabolism and excretion and which do not affect microsomal enzyme induction would be expected to accumulate within the body of individuals who have any type of deficiency of glucuronyl transferase (Calabrese, 1977).

d. Identification of Hyperbilirubinemia: From the Crigler-Najjar Syndrome to Gilbert's Syndrome

Since cases of hyperbilirubinemia are often identified in families, it has been suggested that it is of genetic origin. The precise modes of inheritance for the various disorders, including the Crigler-Najjar syndrome, are uncertain. However, Arias et al. (1969) have suggested that in cases of intermediate hyperbilirubinemia, the genetic transmission is according to an autosomal dominant gene with incomplete penetrance. Difficulty in determining the basis of inheritance is compounded by the problem of identifying carriers of the trait who do not exhibit hyperbilirubinemia. According to Arias (1962) and Arias et al. (1969), the oral menthol tolerance test, even though widely used, is not successful in differentiating between icteric patients and anicteric family members thought to be carrying the abnormal gene. Thus, the lack of differentiability by the menthol tolerance test severely restricts the degree of its diagnostic usefulness. Dutton (1966) indicated that the discrepancies with the extent of glucuronide synthesis with menthol and with bilirubin may not be unusual at all in light of the suspected variety of enzymes that may catalyze glucuronide synthesis.

With regard to mild cases of hyperbilirubinemia, there may be several unrelated causes such as infectious hepatitis (Arias, 1962) in addition to any genetic basis (Gilbert's syndrome). Individuals with Gilbert's syndrome are usually identified by satisfying the following criteria (Powell et al., 1967): chronic, mild unconjugated hyperbilirubinemia with normal values for direct-reacting bilirubin; normal erythropoiesis and red blood cell survival; and absence of histologic or functional abnormalities of the liver and biliary tree. Based on limited familial studies, the data suggest an autosomal dominant type of inheritance (Powell et al., 1967).

REFERENCES

Adamson, R. H. and Davis, D. S. (1973). Comparative aspects of absorption, distribution, metabolism and excretion of drugs. *Comp. Pharmacol.* **2:** 851.

Allen, J. R. (1975). Response of the non-human primate to polychlorinated biphenyl exposure. *Fed. Proc.* **34**(8): 1675.

Arias, I. M. (1962). Chronic unconjugated hyperbilirubinemia without overt signs of hemolysis in adolescents and adults. *J. Clin. Invest.* **41:** 2233.

Arias, I. M., Gartner, L. M., Cohen, M., Ezzer, J. B., and Levi, A. J. (1969). Chronic nonhemolytic unconjugated hyperbilirubinemia with glucuronyl transferase deficiency. *Am. J. Med.* **47:** 395.

Billing, B. H. (1970). Bilirubin metabolism and jaundice with special reference to unconjugated hyperbilirubinemia. *Ann. Clin. Biochem.* **7:** 69–74.

Block, W. D. and Cornish, H. H. (1959). Metabolism of biphenyl and 4-chlorobiphenyl in the rabbit. *J. Biol. Chem.* **234**(12): 3301.

Calabrese, E. J. (1977). Insufficient conjugate glucuronidation: A possible cause of PCB toxicity. *Med. Hypoth.* **2**(4): 257–283.

Calabrese, E.J. (1982). Human breast milk contamination in the U.S. and Canada by chlorinated hydrocarbon insecticides and industrial pollutants: current status. *Journ. Amer. College Toxicol.* **1**(3): 91–98.

Calabrese, E. J. and Sorenson, A. (1977). The health implications of PCBs with particular emphasis on high risk groups. *Rev. Environ. Health* **32**(2): 131–147.

Clarke, E. G. C. and Clarke, M. L. (1967). *Garner's Veterinary Toxicology.* 3rd ed. Williams and Wilkins, Baltimore, pp. 156–158.

Dutton, G. J. (1961). The mechanism of glucuronide formation. *Biochem. Pharmacol.* **6:** 65.

Dutton, G. J. (1962). Glucuronide conjugation. *Proc. First Int. Pharmacol. Meeting* **6:** 39.

Dutton, G. J. (1966). Free and combined. In: *Glucuronic Acid.* G. J. Dutton (ed.). Academic, New York, pp. 222, 230.

Garther, L. M. and Arias, I. M. (1966). Studies of prolonged neonatal jaundice in the breast-fed infant. *J. Pediatr.* **68**(1): 54.

Gillette, J. R. (1967). Individually different responses to drugs according to age, sex and functional or pathological states. *Drug Responses in Man.* Churchill, London, p. 24.

Golberg, L. (ed.). (1974). The toxicity of polychlorinated polycyclic compounds and related chemicals. *Crit. Rev. Toxicol.* **2**(4): 445.

Hirom, P. C., Idle, J. K., and Millburn, P. (1977). Comparative aspects of the biosynthesis and excretion of xenobiotic conjugates by non-primate mammals. In: *Drug Metabolism: From Microbe to Man.* D. V. Parke and R. L. Smith (eds.). Taylor and Francis, London, pp. 299–329.

Hutton, J. J. (1971). Genetic regulation of glucose-6-phosphate dehydrogenase activity in the inbred mouse. *Biochem. Genet.* **5:** 315.

Hutzinger, O., Nash, D. M., Safe, S., Defreitas, A. S. W., Nortstrom, R. J., Wildish, D. V., and Zitko, V. (1972). Polychlorinated biphenyls: Metabolic behavior of pure isomers in pigeons, rats and brook trout. *Science* **178:** 312.

Jones, L. M. (1965). *Veterinary Pharmacology and Therapeutics.* 3rd ed. Iowa State University, Ames, pp. 442–447.

Kornberg, A. (1942). Latent liver disease in persons recovered from catarrhal jaundice and in otherwise normal medical students as revealed by the bilirubin excretion test. *J. Clin. Invest.* **21:** 299–308.

Kreek, M. J. and Sleisenger, M. H. (1968). Reduction of serum unconjugated bilirubin with phenobarbitone in adult congenital non-haemolytic unconjugated hyperbilirubinaemia. *Lancet* **1:** 73.

Lester, R. and Schmid, R. (1964). Bilirubin metabolism. *N. Engl. J. Med.* **270**(15): 779.

Lokietz, H., Dowben, R. M., and Hsia, D. Y. (1963). Studies on the effect of novobiocin and glucuronyl transferase. *Pediatrics* **32:** 47.

Marver, H. S. and Schmid, R. (1968). Biotransformation in the liver: Implications for human disease. *Gastroenterology* **55:** 282.

Nyhan, W. L. (1961). Toxicity of drugs in the neonatal period. *J. Pediatr.* **59**(1): 1.

Oehme, F. M. (1969). A comparative study of the bio-transformation and excretion of phenol. Ph.D. dissertation, University of Missouri.

Owens, D. and Evans, J. (1975). Population studies on Gilbert's Syndrome. *J. Med. Genet.* **12:** 152–156.

Powell, L. W., Hemingway, E., Billing, B. H., and Sherlock, S. (1967). Idiopathic unconjugated hyperbilirubinemia (Gilbert's syndrome): A study of 42 families. *N. Engl. J. Med.* **277:** 1108.

Ramos, A., Silverberg, M., and Stern, L. (1966). Pregnanediols and neonatal hyperbilirubinemia. *Am. J. Dis. Child.* **111:** 353.

Remmer, H. (1965). The fate of drugs in the organism. *Rev. Pharmacol.* **5:** 405.

Selikoff, I. J. (ed.). (1972). Polychlorinated biphenyl-environmental impact: A review of the panel on hazardous trace substances. *Environ. Res.* **5:** 249.

Smith, P. M., Middleton, J. E., and Williams, R. (1967). Studies on the familial incidence and clinical history of patients with chronic unconjugated hyperbilirubinaemia. *Gut* **8:** 449.

Smith, R. L. and Williams, R. T. (1966). Implication of the conjugation of drugs and other exogenous compounds. In: *Glucuronic Acid.* G. J. Dutton (ed.). Academic, New York, p. 457.

Spector, S. C. (1956). *Handbook of Biological Data.* Saunders, Philadelphia, p. 371.

Stecher, P. G. (ed.). (1968). *The Merck Index.* 8th ed. Merck, Rahneay, New Jersey, p. 810.

Sutherland, J. M. and Keller, W. H. (1961). Novobiocin and neonatal hyperbilirubinemia. *Am. J. Dis. Child.* **101:** 447.

Thompson, R. P. G., Stathers, G. M., Pilcher, C. W. T., McLean, A. E. M., Robinson, J., and Williams, R. (1969). Treatment of unconjugated jaundice with dicophane. *Lancet* **1:** 4.

Whelton, N. H., Krustev, L. P., and Billing, B. H. (1968). Reduction in serum bilirubin by phenobarbital in adult unconjugated hyperbilirubinemia. Is enzyme induction responsible? *Am. J. Med.* **45:** 160.

Wilkenson, T. T. (1968). A review of drug toxicity in the cat. *J. Small Anim. Pract.* **9:** 21.

Yaffe, S. J., Levy, G., Matsuzawa, T., and Baliah, T. (1966). Enhancement of glucuronide-conjugating capacity in a hyperbilirubinemic infant due to apparent enzyme induction by phenobarbital. *N. Engl. J. Med.* **275:** 1461.

2. SULFATION

Tyramine-Induced Migraine Headache: A Possible Defect in Sulfate Conjugation

In 1967, Hanington reported that nearly ⅓ of patients with classical migraine headaches can have their symptoms precipitated by consumption of tyramine and/or tyramine-containing foods (e.g., especially cheeses). Tyramine is thought to be principally degraded via oxidative deamination that is catalyzed by monoamine oxidase (MAO) (Asatoor et al., 1963; Lemberger et al., 1966). While one may suspect that those susceptible to tyramine-induced headache may have a defect in oxidative deamination of this agent, Youdim et al. (1971) could not find any generalized *in vivo* defect in oxidative deamination as measured by the level of urinary *p*-hydroxy-phenylacetate, the principal metabolite of tyramine in migraine subjects given tyramine loading.

Failing to explain their enhanced susceptibility to migraine based on defects in oxidative deamination of tyramine, Youdim et al. (1971) sought their answer in the role of conjugation of tyramine, which is thought to be a relatively minor route of tyramine metabolism. In a tyramine loading study with eight female migraine patients and 12 normal volunteers (seven male and five female), these authors found that there was a significantly greater increase in free tyramine and a decrease in conjugated (i.e., either sulfate or glucuronic acid conjugates) tyramine in the migraine versus control subjects. They speculated that the possible enzyme defect may be one of sulfate conjugation, since these data with migraine patients showed considerably reduced urinary metabolites of total metadrenalines and 4-hydroxy-3-methoxyphenylglycol, which are normally excreted conjugated in humans as sulfate (LaBrosse and Mann, 1960; Axelrod et al., 1959). In 1974, Sandler et al. reported that chocolate, the most common trigger of migraine, contains substantial quantities of phenylethylamine and that persons susceptible to chocolate (and phenylethylamine) induced migraine have a low activity of blood platelet MAO activity that is needed to metabolize the phenylethylamine.

These data collectively suggest that there are metabolic differences that are presumably genetically based between normal individuals and those with cheese- and chocolate-induced migraines. It is of interest that the tyramine-sensitive migraine patients reported in the 1971 study of Youdim were not thought to have a defect in MAO metabolism of tyramine based on urinary metabolite measurement but to have a conjugation defect. In contrast, the chocolate sensitive patients were found to have a phenylethylamine oxidizing defect as indicated by human platelet MAO activity. Could it be that there are different metabolic subgroups that are susceptible to food agent induced migraine? These findings suggest this may be possible. While the evidence is far from complete with respect to both genetic and biochemical mechanisms, the data are sufficiently encouraging to recommend further research in this area of ecogenetics.

REFERENCES

Asatoor, A. M., Levi, A. J., and Milne, M. D. (1963). *Lancet* **2**: 733.
Axelrod, J., Kopin, I. J., and Mann, J. D. (1959). *Biochim. Biophys. Acta* **36**: 576.
Hanington, E. (1967). *Br. Med. J.* **2**: 550–551.
LaBrosse, E. H. and Mann, J. D. (1960). *Nature* **185**: 40.
Lemberger, L., Klutch, A., and Kuntzman, R. (1966). *J. Pharmacol. Exp. Ther.* **153**: 183.
Sandler, M. (1972). Migraine: A pulmonary disease? *Lancet* **1**: 618–619.
Sandler, M., Youdim, M. B. H., and Hanington, E. (1974). A phenylethylamine oxidizing defect in migraine. *Nature* **25**: 335–337.
Youdim, M. B. H., Carter, S. B., Sandler, M., Hanington, E., and Wilkinson, M. (1971). Conjugation defect in tyramine-sensitive migraine. *Nature* **230**: 127–128.

3. ACETYLATION

Fast Versus Slow Acetylation and Susceptibility to Arylamine Induced Bladder Cancer

Introduction

Acetylation, a type of phase II conjugation, is a common pathway for the metabolism of compounds possessing a primary amino group. Smith and Williams (1974) have identified five types of amino groups that can be conjugated by humans and other primate species: the aromatic and aliphatic amino groups, the sulphonamide group, the amino group of amino acids, and the hydrazine group. Interspecies studies have revealed

Acetylation

the dog to be a poor excreter of acetylated aromatic amines while the rabbit acetylates such compounds efficiently.

Susceptibility to Drugs

Humans display a genetic polymorphism with respect to acetylation with the population consisting of fast and slow acetylators. Population genetics studies have revealed that the hepatic N-acetyltransferase is under Mendelian genetic regulation as an autosomal recessive trait with an approximate 50:50 polymorphic distribution in North American Caucasians and Blacks, while among Japanese there are nine fast acetylators to one slow one (Harris et al., 1958). Karim et al. (1981) have summarized the data in numerous surveys of acetylator phenotype frequency in Europe, North and South America, Africa, Asia, and the Pacific countries and clearly support the previous reports of polymorphic distribution. Numerous reports in the literature indicate that the ability to acetylate is associated with susceptibility to toxicity for a number of acetylatable nitrogen compounds. Individuals with the slow acetylator phenotype have been found to be at considerably increased risk (Table 3-2) to the development of peripheral neuropathies associated with isoniazid (Hughes et al., 1954), phenelzine (Evans et al., 1965), hydralazine (Perry et al., 1967), and salicylazosulfapyridine (Das et al., 1973) presumably because of the differential ability to detoxify these substances by N-acetylation. There have been situations when acetylated metabolites have proven to be more toxic than the parent compound and in such cases the fast acetylator is the individual at increased risk (e.g., hepatotoxicity caused by isoniazid (Table 3-2) (Black et al., 1975; Mitchell et al., 1975).

Susceptibility to Carcinogenic Aromatic Amines

It is broadly recognized that the carcinogenic aromatic amines require metabolic activation to chemically reactive metabolites (i.e., electrophilic agents) by oxidative enzyme systems and that substrate specificity and functional capacity of enzyme systems involved in activation and detoxification frequently are the principal factors affecting species and tissue susceptibility (Clayson and Garner, 1976; Miller and Miller, 1969; Weisburger and Weisburger, 1973; Lower et al., 1979). Susceptibility to bladder cancer in the dog has been directly associated with the capacity of the liver microsomal enzymes to N-hydroxylate the aromatic amine to proximate N-hydroxyalkylamines. Aromatic amines that become N-hydroxylated (e.g., β-naphthalamine) exhibit strong carcinogenic potential in the dog while structurally very similar aromatic amines that are not efficiently N-hydroxylated (α-naphthalamine) are thought to be noncarcinogenic in the dog (Radomski and Brill, 1970; Radomski et al., 1971). In addition, species (e.g., guinea pig, steppe lemming) (Parke,

Table 3-2. N-Acetyltransferase Phenotype as a Determinant of Human Susceptibility to the Dose-Related Toxicities of Polymorphically Acetylated Nitrogen Compounds

Polymorphically Acetylated Nitrogen Compounds	Dose-Related Toxicities (Susceptible Phenotype)[a]
Isoniazid	Peripheral neuropathies (S)
Phenelzine	Drowsiness, dizziness, nausea (S)
Hydralazine	Peripheral neuropathies (S) (Lupus erythematosus-like syndrome)
Salicylazosulfapyridine (sulfapyridine)	Cyanosis, hemolysis, and reticulocytosis (S)
Isoniazid	Hepatitis (R)
Procainamide	Systemic lupus erythematosus (S)
?	"Spontaneous" systemic lupus erythematosus (S)
?	Diabetic neuropathy (S)

Source: Lower et al. (1979).
[a] S—slow, R—rapid.

1968) that are not able to N-hydroxylate aryl nitrogen compounds are not susceptible to aromatic amine induced cancer (Miller et al., 1964).

Lower et al. (1979) have indicated that such direct relationships as illustrated above in explaining the biochemical foundations for interspecies differences and tissue sensitivity to aromatic amine induced cancer are frequently overshadowed by the complexity of competing metabolic interrelationships. This may be caused by the

relationships of metabolic pathways involved in arylamine activation and detoxification, and the conversion of arylamines to differing ultimate electrophiles with differing tissue specificities and toxicities. Thus, while the proximate carcinogenic metabolites involved in urinary bladder carcinogenesis are evidenced to be nonacetylated N-hydroxyarylamines (arylhydroxylamines), the proximate carcinogenic metabolites involved in hepatocarcinogenesis are evidenced to be acetylated N-hydroxylarylacetamides (arylhydroxamic acids), with both of these proximate carcinogenic metabolites being derived from arylamines and arylacetamides by virtue of parallel divergent metabolic pathways (Poirier et al., 1963; Lower and Bryan, 1973).

For example, administration of a variety of carcinogenic arylamines to dogs has resulted in the formation of only urinary bladder tumors, the liver apparently being refractory in keeping with an apparent total deficiency in the capacity to N-acetylate arylamines, observations providing an indication that N-acetylation is not required for bladder carcinogenesis (Lower and Bryan, 1973). In contrast, administration of structurally analogous carcinogenic arylacetamides to dogs has resulted in the unequivocal formation of both urinary bladder tumors and hepatomas, with susceptibility to bladder carcinogenesis being correlated with the substrate specificity of arylacetamide deacetylase enzyme systems, observations providing an indication that removal of the acetyl group is required for bladder carcinogenesis (Lower and Bryan, 1976).

In other words, N-hydroxylating enzyme systems can be viewed as components of activation pathways with respect to both arylamine bladder carcinogenesis and arylacetamide hepatocarcinogenesis, while N-acetyltransferase enzyme systems can be viewed as components of activation pathways with respect to arylamine hepatocarcinogenesis and as components of the detoxication pathways with respect to arylamine bladder carcinogenesis. Similarly, arylacetamide deacetylase enzyme systems can be viewed as components of detoxication pathways with respect to arylacetamide hepatocarcinogenesis and as components of activation pathways with respect to arylacetamide bladder carcinogenesis. These components appear to be generally applicable to mammalian systems with the potentially rate-limiting role of a given enzyme system being a relative one, depending on species-specific metabolic capabilities. Thus, it is probable that activating and detoxifying enzyme systems might also act as partial determinants of human susceptibility to bladder carcinogenesis by arylamines, and it becomes important to clarify the role of these metabolic factors from a human frame of reference (Lower and Bryan, 1976).

Humans are able to acetylate aromatic amines but, as noted previously, this capacity exists as a genetic polymorphism with there being fast and slow acetylators (Lower et al., 1979). Persons who are fast acetylators have about 9–10 times more acetylase activity than the "slow" individuals for benzidine, β-naphthylamine and methylene bis-2-choroaniline (MOCA), while having slightly greater and less activity for 2-aminofluorene and α-naphthylamine, respectively (Table 3-3) (Glowinski et al., 1978). Consequently, it would appear that the dog tends to simulate the slow acetylator more closely than the fast acetylator phenotype. However, no evidence exists as to whether fast and slow phenotypes differ in their capacity to deacetylate. The susceptibility of the dog to aromatic amine induced bladder cancer can be explained in part by its lack of ability to acetylate these compounds and its ability to deacetylate acetylated parent compounds. Lower et al. (1979) hypothesized that humans with the slow acetylator phenotype would be at increased risk to develop aromatic amine induced bladder cancer. Preliminary epidemiological findings tend to support this hypothesis.

Lower et al. (1979) reported that a population of urban urinary bladder cancer patients in Denmark exhibited a 13.4% excess ($p = 0.065$) of individuals with the slow acetylation phenotype as compared to a control group. This amounted to an increased relative risk of 1.74. As expected, no significant difference in bladder cancer cases occurred between fast and slow acetylator phenotypes in rural Sweden because of a presumed lower exposure to carcinogenic aromatic amines. According to the authors, these data are consistent with the hypothesis that slow acetylators

Table 3-3. N-Acetyltransferase Activity for Arylamine Substrates in Liver From Rapid and Slow INH Acetylator Rabbits and Humans

	Rabbit[a]			Human[b]		
	Activity			Activity		
Substrate	Rapid	Slow	Ratio	Rapid	Slow	Ratio
	nmoles/min/mg			nmoles/min/mg		
Sulfamethazine	8.62 (72)	0.030 (66)	290	0.837	0.226	3.7
	8.26 (72)	0.030 (49)	275			
	3.02 (18)	0.029 (49)	104			
	7.43 (58)	0.016 (49)	464			
	4.71 (58)	0.078 (63)	60			
p-Aminobenzoic acid	0.71 (18)	0.34 (49)	2	0.150	0.157	0.96
	1.61 (58)	0.32 (49)	5			
	1.10 (58)	0.83 (63)	1.5			
Carcinogenic Arylamines						
α-Naphthylamine	12.40 (72)	0.023 (66)	540	1.17	0.225	5.2
Benzidine	8.88 (58)	0.016 (49)	560	0.173	0.019	9.0
	4.14 (18)	0.022 (49)	190			
β-Naphthylamine	8.73 (72)	0.28 (66)	310	0.231	0.026	8.9
2-Aminofluorene	7.53 (72)	0.013 (49)	580	0.285	0.021	13.0
MOCA				0.144	0.015	9.6

Source: Glowinski et al. (1978).
[a] Numbers in parentheses refer to individual rabbits.
[b] One rapid acetylator human (#4) and one slow acetylator human (#9) were used for initial rate determinations.

have enhanced risk to developing an arylamine induced bladder cancer. Considering the fact that the selection of patients did not involve persons occupationally exposed to arylamines, the findings of Lower et al. (1979) were impressive since they did not presumably investigate the most ideal population to test truly this theory. However, it must be emphasized that the study had important methodologic limitations in that potential confounding variables such as smoking and occupation were not controlled nor was it stated whether the N-acetyltransferase phenotype was determined blindly (Epidemiology Resources, 1982).

A recent British epidemiological study reported by Cartwright et al. (1982) has offered further support for the hypothesis of Lower et al. (1979) that a slow acetylator phenotype is a predisposing factor for the recurrence of arylamine induced bladder cancer. Particularly striking in their findings was the occurrence of 22 persons with the slow acetylator phenotype out of 23 bladder cancer patients with a past history of chemical exposure in the dye-manufacturing industry. Considering that slow acetylators occur about one in every two Caucasians, this observation was quite striking ($P < 0.000048$). Further research revealed that susceptibility to bladder cancer is not equal among those generally grouped as "slow acetylators" but that susceptibility is greater as the acetylation capacity decreases. Consequently, future studies should consider acetylation-ratio subgroups as well as the typical fast and slow acetylator phenotype comparisons.

When one considers that 50% of the North American Caucasian and Black populations are slow acetylators, the sheer magnitude of those at potential increased risk is striking. The present situation is that the theory is well founded in cancer research with a variety of animal models. However, what is needed are epidemiologic studies of industrial bladder cancer populations so that it may be possible to determine if there is a different relative risk for bladder cancer depending on one's ability to acetylate arylamines.

Screening Tests

Urine tests for phenotyping slow and fast acetylators have been developed since 1970 in order to deal with the potential medical problem of slow acetylators being at enhanced risk of developing adverse reactions to isoniazid, an antitubercular treatment. The procedure is straightforward and simple, displaying an excellent capability to distinguish fast from slow acetylators and can be performed either with or without the use of a spectrophotometer depending on the capability of the laboratory (Jessamine et al., 1975). More recently, Cartwright et al. (1982) have utilized a technique involving the acetylation of dapsone that is appropriate for population studies since large numbers can be quickly screened. The dapsone method was advantageous since it does not require urine collection, the timing of collection is not critical, the acetylated dapsone is thought to be stable, and the ratios are highly reproducible.

REFERENCES

Black, M., Mitchell, J. R., Zimmerman, H. J., Ishak, K. G., and Epler, G. R. (1975). Isoniazid-associated hepatitis in 114 patients. *Gastroenterology* **69**: 289.

Cartwright, R. A., Glashan, R. W., Rogers, H. J., Ahmad, R. A., Barham-Hall, D., Higgins, E., and Kahn, M. A. (1982). Role of *N*-acetyltransferase phenotypes in bladder carcinogenesis. A pharmacogenetic epidemiological approach to bladder cancer. *Lancet* **2**: 842–845.

Clayson, D. B. and Garner, R. C. (1976). Carcinogenic aromatic amines and related compounds. In: *Chemical Carcinogens*. C. E. Searle (ed.). *Am. Chem. Soc. Monog.* 173. American Chemical Society. p. 366.

Das, K. M., Eastwood, N. A., McManus, J. P. A., and Sircus, W. (1973). Adverse reactions during salicylazousulfapyridine therapy and the relation with drug metabolism and acetylator phenotype. *N. Engl. J. Med.* **289**: 491.

Epidemiology Resources, Inc. (1982). Report for the U.S. Office of Technology Assessment. Chestnut Hill, Massachusetts.

Evans, D. A. P., Davidson, K., and Pratt, R. T. C. (1965). The influence of acetylator phenotype on the effects of treating depression with phenelzine. *Clin. Pharmacol. Therap.* **6**: 430.

Glowinski, I. R., Radtke, H. E. and Weber, W. W. (1978). Genetic variation in *N*-acetylation of carcinogenic arylamines by human and rabbit liver. *Mol. Pharmacol.* **14**: 940–949.

Harris, H. W., Knight, A., and Selin, M. J. (1958). Comparison of isoniazid concentrations in the blood of people of Japanese and European descent—therapeutic and genetic implications. *Am. Rev. Tuber. Dis.* **78**: 944.

Hughes, H. B., Biehl, J. P., Jones, A. P., and Schmidt, L. H. (1954). Metabolism of isoniazid in man as related to the occurrence of peripheral neuritis. *Am. Rev. Tuber.* **70**: 266.

Jessamine, A. G., Hodkin, M. M., and Eidus, L. (1975). Urine tests for phenotyping slow and fast acetylators. *Can. J. Public Health* **65**: 119–123.

Karim, A. K. M. B., Elfellah, M. S., and Evans, D. A. P. (1981). Human acetylator polymorphism: Estimate of allele frequency in Libya and details of global distribution. *J. Med. Genet.* **18**: 325–330.

Lower, G. M. (1982). Concepts in causality: Chemically induced human urinary bladder cancer. *Cancer* **49**: 1056–1066.

Lower, G. M., Jr. and Bryan, G. T. (1973). Enzymatic *N*-acetylation of carcinogenic aromatic amines by liver cytosol of species displaying different organ susceptibilities. *Biochem. Pharmacol.* **22**: 1581–1588.

Lower, G. M. and Bryan, G. T. (1976). Enzymatic deacetylation of carcinogenic arylacetamides by tissue microsomes of the dog and other species. *J. Toxicol. Environ. Health* **1**: 421.

Lower, G. M., Jr., Nilsson, T., Nelson, L. E., Wolf, H., Gamsky, T. E., and Bryan, G. T. (1979). *N*-acetyltransferase phenotype and risk in urinary bladder cancer: Approaches in molecular epidemiology. Preliminary results in Sweden and Denmark. *Environ. Health Perspect.* **29**: 71–79.

Miller, E. C., Miller, J. A., and Enomoto, M. (1964). The comparative carcinogenicities of 2-acetylaminofluorene and its *N*-hydroxy metabolite in mice, hamsters, and guinea pigs. *Cancer Res.* **23**: 2018.

Miller, J. A. and Miller, E. C. (1969). The metabolic activation of carcinogenic aromatic amines and amides. *Prog. Exptl. Tumor Res.* **11**: 273–301.

Mitchell, J. R., Thoregeirsson, U. P., Black, M., Timbrell, J. A., Snodgrass, W. R., Potter, W. Z., Jollow, D. J., and Keiser, H. R. (1975). Increased incidence of isoniazid hepatitis in rapid acetylators: Possible relation to hydrazine metabolites. *Clin. Pharmacol. Ther.* **18**: 70.

Parke, D. V. (1968). *The Biochemistry of Foreign Compounds*. Pergamon, New York.

Perry, H. M., Jr., Sakamoto, A., and Tan, E. M. (1967). Relationship of acetylating enzyme to hydralazine toxicity. *J. Lab. Clin. Med.* **70**: 1020.

Poirier, L. A., Miller, J. A., and Miller, E. C. (1963). The *N*- and ring-hydroxylation of 2-acetylaminofluorene and the failure to detect *N*-acetylation of 2-aminofluorene in the dog. *Cancer Res.* **23**: 790.

Radomski, J. L. and Brill, E. (1970). Bladder cancer induction by aromatic amines. Role of *N*-hydroxy metabolites. *Science* **167**: 992.

Radomski, J. L., Brill, E., Deichmann, W. B., and Glass, F. M. (1971). Carcinogenicity testing of *N*-hydroxy and other oxidation and decomposition products of 1- and 2-naphthylamine. *Cancer Res.* **31**: 1461.

Smith, R. L. and Williams, R. T. (1974). Comparative metabolism of drugs in men and monkeys. *J. Med. Primatol.* **3**: 138–152.

Weisburger, J. H. and Weisburger, E. K. (1973). Biochemical formation and pharmacological, toxicological and pathological properties of hydroxylamines and hydroxaminic acids. *Pharmacol. Rev.* **25**: 1.

B. Oxidation Center Defects

1. INTRODUCTION

Numerous drugs and environmental pollutants are metabolized in part via hydroxylation. The metabolic significance of this process is profound since hydroxylation may result in a metabolite either more or less toxic (or carcinogenic) than the parent compound and/or metabolite more capable of further metabolism via conjugation reactions such as glucuronidation.

Interspecies differences in the ability to hydroxylate various xenobiotics have resulted in profound differences in toxicity and carcinogenic responses. For example, the lack of ability to *N*-hydroxylate aromatic amines by guinea pigs is thought to explain its lack of susceptibility to developing cancer from these compounds. Another example of species specificity may be seen with a study of the metabolism of aniline. Parke (1968) reports that cats, dogs, and other carnivores tend to favor the conversion of aniline to *O*-aminophenol, while herbivores including rabbits produce greater amounts of *p*-aminophenol. Interestingly, the carnivores are much more susceptible than herbivores to aniline toxicity, presumably because of the enhanced toxicity of the *O*-aminophenol metabolite. In a similar fashion, Parke (1968) reported that β-naphthylamine is preferentially converted by carnivores (i.e., cats and dogs) into 2-amino-1-naphthol while rats and rabbits excrete only 5% as this isomer with the major metabolic end-product being 2-amino-6-naphthol.

Among humans, interindividual variations exist with respect to the oxidation of carbon centers in several drugs including debrisoquine, guanoxan, and phenacetin. Studies concerning debrisoquine interindividual metabolism have been the primary form of research in this area and will be emphasized here.

2. GENETIC DIFFERENCES IN THE METABOLISM OF DEBRISOQUINE

Idle and Smith et al. (1979) have reported that the quanidine-based antihypertensive drug debrisoquine is metabolized by humans principally via alicyclic oxidation to yield 4-hydroxydebrisoquine (Figure 3-1). The metabolic process is under the control of a single autosomal gene (Mahgoub et al., 1977; Evans et al., 1980). It has been suggested by Mbanefo et al. (1980) that at least two alleles occur in the population that affect the extent of occurrence of this reaction. The two alleles have been designated D^H (extensive hydroxylation of debrisoquine) and D^L (low hydroxylation). Persons who display homozygosity for the D^L allele show a significantly reduced capacity to 4-hydroxylate the drug and excrete the substance, for the most part, unchanged. Those persons homozygous for the D^L allele make up what is called by Mbanefo et al. (1980), the poor metabolizing (PM) phenotype. Individuals who are designated as genotypically homozygous D^H or heterozygous are known to readily hydroxylate debrisoquine and they constitute what is called the extensive metabolizing (EM) phenotype.

As indicated in the previous paragraph, humans may oxidize debrisoquine to 4-hydroxydebrisoquine to a variable extent with varying amounts of unchanged drug and metabolite being eliminated in the urine of different persons. The quotient, percent dose eliminated as unchanged debrisoquine/percent dose excreted as 4-hydroxydebrisoquine in the urine after a single 10-mg oral dose is bimodally distributed and referred to as the "metabolic ratio." The EM phenotype exhibits metabolic ratios in the range of 0.01–10.0, while the PM phenotype displays metabolic ratios in the range of 18–200 (Figure 3-2). If one considered the extreme range of values given for the metabolic ratios of 0.01–200, a difference of 20,000-fold metabolic variations is seen. In fact, such differences in metabolism help to explain the marked variation in the optimal dose requirement of debrisoquine (20–440 mg/day) to control blood pressure in hypertensive patients (Idle and Smith, 1979).

The occurrence of genetic polymorphism of debrisoquine hydroxylation has been shown in a number of ethnic groups including Egyptians, Ghanaians, Nigerians, Saudis, and United Kingdom Caucasians. A

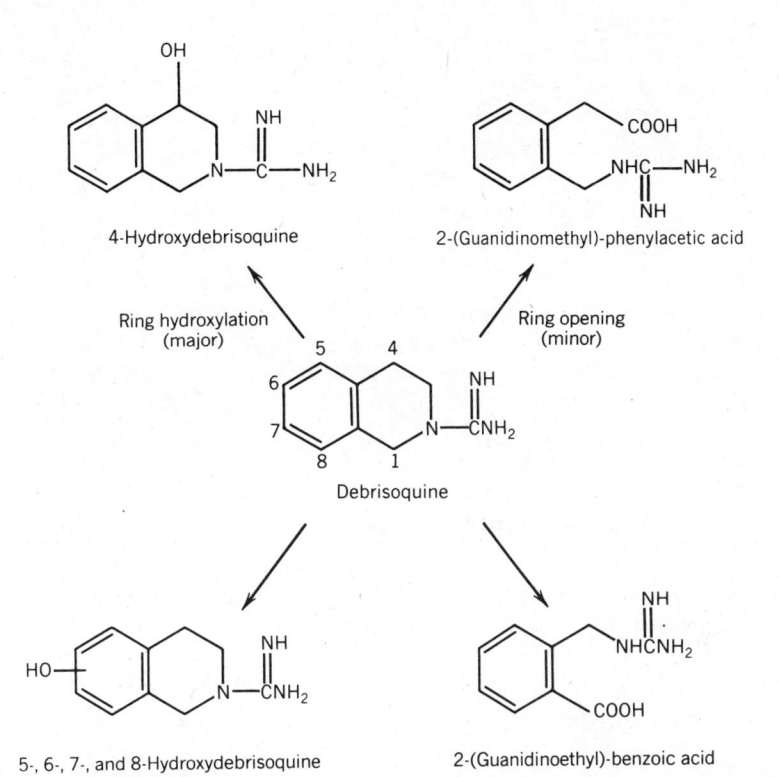

Figure 3-1. Metabolic disposition of debrisoquine in man. *Source:* Idle and Smith (1979).

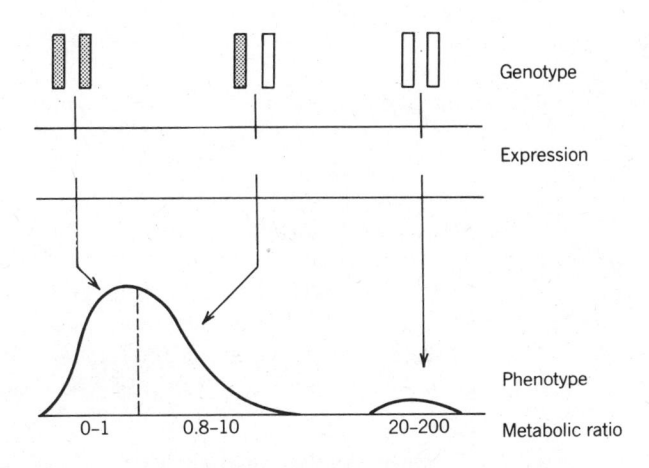

Figure 3-2. The genetic model in which three genotypically distinct allele pairs give rise to two distinct phenotypes. *Source:* Idle and Smith (1979).

nearly ninefold difference in the frequency of the PM phenotype among these ethnic groups has been reported. For example, the 4-hydroxylation of debrisoquine among the Saudis was found to be polymorphic with a PM frequency of approximately 1%. Other ethnic groups that have been studied display the following PM frequencies: British Caucasians, 8.9% (Evans et al., 1980); Nigerians, 8.1% (Mbanefo et al., 1980); Ghanaians, 6.3% (Woolhouse et al., 1979); and Egyptians, 1.4% (Mahgoub et al., 1979).

Because the PM frequency in the Saudi population is lower than in British Caucasians, the Saudis as a population will be faster hydroxylators of drugs such as debrisoquine than British Caucasians. However, a number of other pharmacological and toxicological implications emerge as well from these data.

3. TOXICOLOGICAL IMPLICATIONS

Islam et al. (1980) have summarized the broad thrust of such pharmacological and toxicological implications as follows:

Drugs,* typified by debrisoquine, which are derived from plasma by metabolic oxidation, often by a first passage through the liver, and which owe their activity to the parent drug, may be less effective in Saudis and Egyptians, at doses found to be effective in clinical trials in European white volunteers. . . . In contrast, drugs which owe their therapeutic value to an oxidative metabolite, such as the anticancer drug cyclophosphamide (Dixon, 1968) may be more effective in such Arab patients or show a higher incidence of adverse drug reactions. It may also be postulated that environmental or dietary carcinogens which are oxidized to chemically reactive metabolites in the tissues, such as aflatoxin B_1 (Swenson et al., 1974), benzo(a)pyrene (Sims et al., 1974), dimethylnitrosamine (Montesano and Magee, 1970) or estrogole (Drinkwater et al., 1976), may have measurably different toxicities in more rapidly oxidizing populations such as Saudi or Egyptian Arabs.

In 1981, Idle et al. evaluated the hypothesis of Islam et al. (1980) by exploring whether there was an association between genetically determined oxidation status and the occurrence of primary hepatoma and tumors of the gastrointestinal tract. These carcinomas were selective since they are often widely assumed to be caused by environmental agents such as aflatoxin. Consequently, the study involved a comparison

---

*It should be noted that individuals phenotypically classified as EM or PM for the metabolic ratios of debrisoquine 4-hydroxylation responded in an identical manner with respect to the aromatic hydroxylation of guanoxan and the oxidative de-ethylation of phenacetin (Idle and Smith, 1979), and suggests the generalizableness of these findings.

of the genetically determined debrisoquine oxidation status of 123 normal, healthy Nigerians with 59 Nigerian patients with carcinoma of the liver and gastrointestinal tract. Consistent with the hypothesis of Islam et al. (1980), Idle et al. (1981) found that the cancer group displayed a statistically significantly larger number of volunteers who were EM phenotypes (i.e., 16% greater than expected) as compared to the controls.

The authors concluded by suggesting that there is an association between genetically determined metabolic oxidation status of Nigerians and susceptibility to develop liver and gastrointestinal tract cancer. Since numerous environmental carcinogens require bioactivation, a genetic predisposition to carry out such bioactivation may well be a critical factor in the overall susceptibility to such cancers.

4. SUMMARY

The evidence presented here indicates that one's genetic condition affects the ability to oxidize molecules of dissimilar structures such as alicyclic, aliphatic, and aromatic carbon centers. Whether the defect extends to oxidation of nitrogen and sulfur centers in xenobiotics is of considerable importance and is under study by Idle and Smith, 1979.

The occupational and/or environmental health implications of these findings are enormous. For example, many known animal model/human mutagens and/or carcinogens require an initial bioactivation step via an oxidative process. To what extent humans differ in their ability to bioactivate toxic and/or carcinogenic compounds may contribute significantly in explaining variation in population responses to such agents. In addition, the number of the potentially affected persons is enormous.

Idle and Smith (1979) suggested that this knowledge could be put to very practical use in the testing of new drugs. They recommended that panels of known oxidation phenotypes be identified for drug metabolism studies. It would also seem reasonable to develop cell cultures from humans of such different phenotypes for metabolism studies of various environmental and industrial substances. Finally, Nadir et al. (1982) have reported a rat model displaying some of the metabolic and genetic characteristics of this human polymorphism. Further development of an *in vivo* animal model as described by Nadir et al. (1982) is also likely to markedly enhance progress in this area.

REFERENCES

Dixon, R. L. (1968). *J. Pharm. Sci.* **57**: 1351.
Drinkwater, N. R., Miller, E. C., Miller, J. A., and Pitot, H. C. (1976). *J. Nat. Cancer Inst.* **57**: 1323.

Evans, D. A. P., Mahgoub, A., Sloan, T. P., Idle, J. P., and Smith, R. L. (1980). A family and population study of the genetic polymorphism of debrisoquine oxidation in a British white population. *J. Med. Genet.* **17:** 102.

Idle, J. R., Mahgoub, A., Sloan, T. P., Smith, R. L., Mbanefo, C. O., and Bababunmi, E. A. (1981). Some observations on the oxidation phenotype status of Nigerian patients presenting with cancer. *Cancer Letters* **11:** 331–338.

Idle, J. R. and Smith, R. L. (1979). Polymorphisms of oxidation at carbon centers of drugs and their clinical significance. *Drug Metabo. Rev.* **9(2):** 301–317.

Islam, S. I., Idle, J. R., and Smith, R. L. (1980). The polymorphic 4-hydroxylation of debrisoquine in a Saudi Arab population. *Xenobiotica* **10:** 819–826.

Mahgoub, A., Idle, J. R., Dring, L. G., Lancaster, R., and Smith, R. L. (1977). Polymorphic hydroxylation of debrisoquine in man. *Lancet* **2:** 584–587.

Mahgoub, A., Idle, J. R., and Smith, R. L. (1979). A population and familial study of the defective alicyclic hydroxylation of debrisoquine among Egyptians. *Xenobiotica* **9:** 51–56.

Mbanebo, C., Barabunmi, E. A., Mahgoub, A., Sloan, T. P., Idle, J. R., and Smith, R. L. (1980). A study of the debrisoquine hydroxylation polymorphism in a Nigerian population. *Xenobiotica* **10:** 811–818.

Montesano, R. and Magee, P. N. (1970). *Nature* **228:** 173.

Nadir, H. H., Al-Dabbagh, S. G., and Idle, J. R. (1982). Elevated serum cholesterol in drug-oxidation-deficient rats. *Biochem. Pharm.* **31:** 1665–1668.

Parke, D. V. (1968). *The Biochemistry of Foreign Compounds.* Pergamon, New York.

Sims, P., Grover, P. L., Swaizland, A. J., Pal, K., and Hewer, A. (1974). *Nature* **252:** 326.

Sloan, T. P., Mahgoub, A., Lancaster, R., Idle, J. R., and Smith, R. L. (1978). Polymorphism of carbon oxidation of drugs and clinical implications. *Br. Med. J.* **2:** 655–657.

Swenson, D. H., Miller, E. C., and Miller, J. A. (1974). *Biochem. Biophys. Res. Commun.* **60:** 1036.

Woolhouse, N. M., Andoh, B., Mahgoub, A., Sloan, T. P., Idle, J. R., and Smith, R. L. (1979). Debrisoquine hydroxylation polymorphism among Ghanaians and Caucasians. *Clin. Pharmacol. Ther.* **26:** 584–591.

C. Alcohol Dehydrogenase Variants

The causes of alcoholism within a society are extremely complex with social, cultural, and biological influences tightly interwoven. While the predominant extent of research has focused on the social determinants of alcoholism, there has been a slowly growing recognition that a genetic predisposition for developing alcoholism may also exist.

The case for a genetic basis for alcoholism has been summarized by Schuckit and Rayses (1979). They indicated that considerable evidence has emerged to suggest that alcoholism is a multifactorial genetically influenced condition. To support this contention they stated that:

1. There is a 25–50% lifetime risk for alcoholism in the sons and brothers of severely alcoholic men.
2. Research with animal models indicates that alcohol preferences can be selectively bred in various mouse strains.

3. There is a 55% or higher concordance rate for alcoholism in identical twins with only a 28% rate for same-sex fraternal twins.
4. Studies using either the half-sibling method in the United States or adoption procedure in Scandinavia are the most convincing support for a genetic factor at present. These investigations have shown a fourfold or higher increase in alcoholism for the children of alcoholics as compared to controls, even when the children have been separated from their paternal and maternal parents soon after birth and raised without knowledge and interaction of the biological parents.

1. INTERRACIAL STUDIES

Other study methodologies have tried to assess the role of genetic factors in the metabolism of ethanol. Of particular importance in this respect have been interracial comparisons of ethanol metabolism. In the first such comparison, Fenna et al. in 1971 studied the metabolism of intravenously administered ethanol among individuals of three races (i.e., Canadian Inuit, Indians, and Caucasians). While the three races were comparable in the amount of ethanol needed to reach and sustain a blood level of 125 mg/100 mL, they differed markedly in the rate of metabolism based on the rate of disappearance of ethanol from the blood with the Caucasian being significantly fastest. In contrast, Bennion and Li (1976) noted that American Indians from Phoenix, Arizona metabolized ethanol in a similar fashion as Caucasians (e.g., 92–93 mg/kg hr). However, it should be pointed out that the Indian subjects in this study were heavier than their Caucasian comparison groups (82.2 vs. 70.0 kg). If the Indians were more obese than the Caucasians, Propping (1978) has stated that the Indians should have displayed a higher rate of metabolism per kilogram of lean body mass.* Ewing et al. (1974) reported that Caucasians and Orientals in the United States did not significantly differ in their blood ethanol and acetaldehyde concentrations following oral intake. Adding more apparent confusion was a report by Reed et al. (1976) that indicated a greater metabolic rate in Chinese and Canadian Indians than in Caucasians.

That the results of such studies may differ among each other is not surprising since a number of potentially important methodological differences are apparent as well as a general deficiency of each study to consider the known major factors that may affect ethanol metabolism. For

---

*Since ethanol distribution is essentially confined to body water additional fat increases the total body weight and thereby decreases the fraction of body weight that is related to ethanol metabolism. Consequently, the leaner individual will have a higher calculated metabolic rate in milligrams per kilogram hour. (Reed et al., 1976).

example, Fenna et al. (1971) gave the ethanol by intravenous while the others administered it orally. The doses varied considerably among the various studies ranging from 0.3 to 1.5 mg/kg. The weight of subjects also markedly differed between study groups in the Bennion and Li (1976) study. Finally, there may well be markedly different dietary factors which affect the metabolism of ethanol that are not known.

2. TWIN STUDIES

While the interracial studies are equivocal in their answer to the question of whether alcoholism is affected by genetic determinants, reasonably firm biological evidence exists from twin studies that genetic factors are of considerable importance in the metabolism of ethanol. Vesell et al. (1971) compared the metabolism of ethanol in healthy monozygous and dizygous twins. The twins were not receiving other drugs and were not restricted in any way. Figure 3-3 illustrates three typical comparisons between identical and fraternal twins. These findings clearly reveal a striking similarity between the identical twins with much greater variability between the fraternal twins. These findings strongly implicate genetic factors in the metabolism of ethanol. These results are not unique in the sense that previous findings also illustrated the predominant influence of genetics over environmental factors in the metabolism of phenylbutazone (Vesell and Page, 1968), antipyrine (Vesell and Page, 1968a), and bishydroxycoumarin (Vesell and Page, 1968b). More specifically, Vesell (1973) stated that the contribution of heredity to interindividual variations of identical twins in the half-life of these three substances plus alcohol were calculated to be 0.99, 0.98, 0.97, and 0.99, respectively. These studies were designed to maximize the likelihood for genetic differences to occur between mono- and dizygous twins. Considerable environmental similarity is needed when twin populations are studied for the purpose of differentiating the influence of heredity factors as they were in these examples. A recent experiment on ethanol metabolism was conducted in which environmental heterogeneity occurred with regard to several important variables (Kopun and Propping, 1977) and lower heritability indices for ethanol were obtained as compared to the striking findings of Vesell et al. (1971). Nevertheless, Vesell (1973) asserted that even though significant environmental differences existed among the twins in the Kopun and Propping (1977) study, they still found that about half the total variability in ethanol disposition could still be explained by genetic factors.

3. GENETIC VARIANTS FOR ALCOHOL METABOLISM

That individuals may differ in the genetic ability to metabolize ethanol was first established by Von Wartburg et al. in 1965 who reported an

Genetic Variants for Alcohol Metabolism

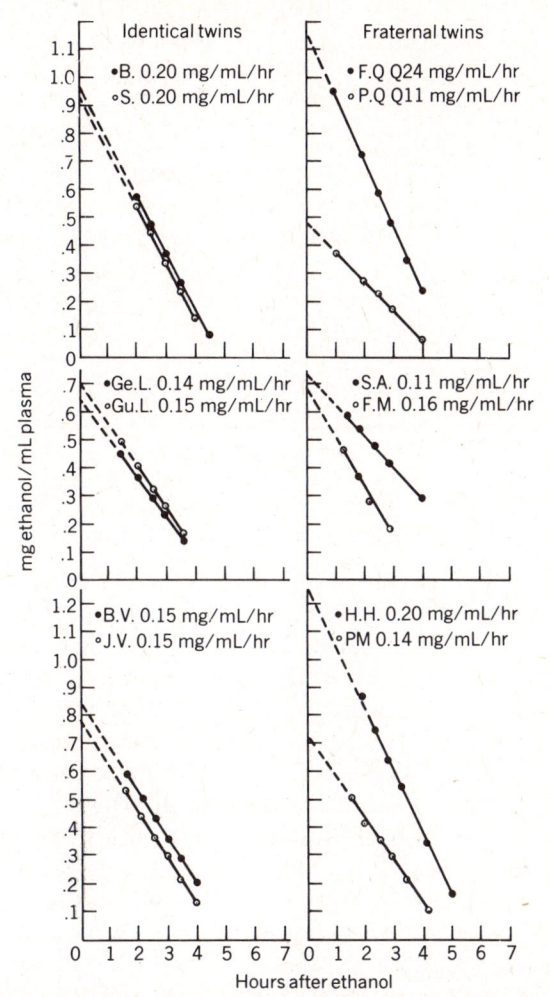

Figure 3-3. Close similarity of identical twins in rates of ethanol removal and large intrapair differences between fraternal twins. *Source:* Vesell et al. (1971).

atypical alcohol dehydrogenase in 2 out of 32 human livers. The purified enzyme differed from the normal one regarding (1) pH rate profile, (2) substrate specificity, and (3) sensitivity to metal-binding agents. Thiourea inhibited the variant enzyme but it activated the normal human liver alcohol dehydrogenase. Total alcohol dehydrogenase activity was considerably higher in livers containing the atypical enzyme. The pH optimum for the typical enzyme is 10.8 while for the atypical enzyme it is 8.5.

A series of *in vitro* experiments evaluated the metabolism of several drugs by liver homogenates containing normal or atypical human alcohol dehydrogenase (ADH) (Table 3-4). They found that β-pyridyl carbinol is oxidized at comparable rates by atypical and normal ADH although the

Table 3-4. Turnover of Some Drugs by Liver Homogenates Containing Normal or Atypical Human Alcohol Dehydrogenase[a]

Substrate	Concentration (mM)	Normal Liver I.U./g	Normal Liver Percent of EtOH	Atypical Liver I.U./g	Atypical Liver Percent of EtOH
Ethanol	17	2.84	100	25.60	100
β-Pyridyl carbinol	1.7	2.86	101	3.71	15
1,2-Propandiol	17	0.47	16	0.46	2
3-o-Toloxy-1,2-propandiol	0.33	0.31	11	0.40	2
Chloral hydrate	3.3	6.10	214	8.70	34
Acetaldol	1.7	15.30	537	77.80	304

Source: von Wartburg and Schürch (1968).
[a]Conditions: Na pyrophosphate, 3.3×10^{-2} M, pH 8.8; NAD, 1.6×10^{-2} M; substrate concentrations as indicated; change in A_{300} determined at 25°C.

metabolism relative to ethanol is very dissimilar. A comparable situation pertains to 1,2-propandiol and its derivative 3-O-toloxy-1,2-propandiol. As for chloral hydrate, the normal liver reduces this substrate about twice as fast as it does ethanol, while the rate is only ⅓ with the atypical liver ADH. However, the absolute rates of reduction are almost the same for both enzymes. In contrast, acetaldol (i.e., β-hydroxybutyraldehyde) and ethanol are reduced considerably faster by the atypical enzyme than by normal ADH. The enhanced metabolism of ethanol by the atypical ADH was subsequently supported in an *in vivo* study in which drinking experiments were performed on a normal and an atypical subject (Figure 3-4). The frequency of the atypical ADH* variant in the population has been determined by several research teams. For example, Von Wartburg et al. (1965) reported that 5% of the English and 20% of the Swiss population had the atypical form of ADH. A frequency of 10% of the atypical enzyme was determined among 166 English subjects by Smith et al. (1971a). Reports of German subjects also closely mirror the frequencies from those of other Europeans (Klein et al., 1962; Harada et al., 1978). Fukui and Wakasuyi (1972) and Stamatoyannopoulos et al. (1975) reported that the Japanese have a considerably different frequency of the enzyme with about 90% of those tested showing the atypical ADH vari-

---

*The two principal enzymes responsible for ethanol oxidation in humans are ADH and aldehyde dehydrogenase (ALDH), both located in the liver. Three autosomal gene loci are concerned (ADH_1, ADH_2, ADH_3) in determining the structure of ADH in humans. Locus ADH_1 and ADH_3 are primarily active during fetal life; locus ADH_2 is expressed during embryogenesis and becomes more active so that in adults this locus is responsible for most of the liver ADH activity (Smith et al., 1971). In adults the enzyme is expressed only in the liver and kidneys.

Genetic Variants for Alcohol Metabolism

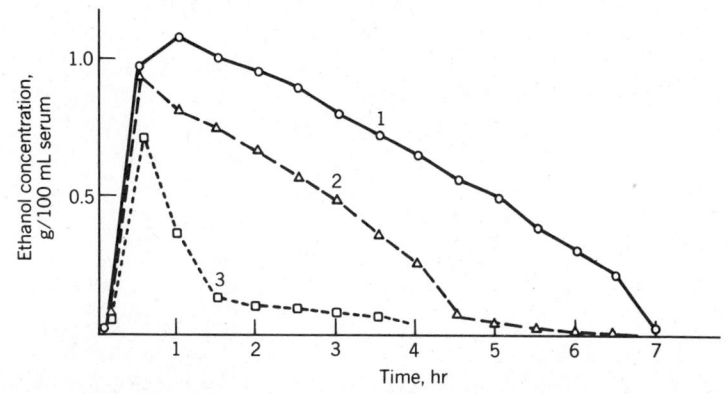

Figure 3-4. Blood alcohol concentrations after peroral administration of ethanol. 1. Normal individual; 2. individual with atypical alcohol dehydrogenase; 3. infant with glycogenesis (glucose–6–phosphatase deficiency): Ethanol dose at time 0:0.7 g/kg of body weight. Blood alcohol concentration determined enzymatically (ADH method). *Source:* von Wartburg and Schürch (1968).

ant. Table 3-5 summarizes the gene frequencies of the atypical (i.e., ADH_2) locus in Europeans and Japanese.

If the difference in the population frequency of the atypical variant was of biological significance, then there should be some degree of interracial differences in susceptibility to the effects of alcohol. Some reports supporting that hypothesis have in fact been published. For example, Wolff (1972, 1973) has reported that the vast majority of Mongoloids studied display a rapid, intense flushing of the face with symptoms of

Table 3-5. Gene Frequencies at ADH_2 Locus in Europeans and Japanese[a]

Origin	Number of Livers	Normal	Atypical	References
England	118	0.95	0.05	Smith et al. (1971)
England	50	0.98	0.02	Von Wartburg and Schurch (1968)
England	23	0.96	0.04	Edwards and Evans (1967)
Switzerland	59	0.89	0.11	Von Wartburg and Schurch (1963)
Germany	35	0.97	0.03	Klein et al. (1962
Germany	46	0.96	0.04	Harada et al. (1978)
Japan	62	0.31	0.69	Fukui and Wakasuyi (1972)
Japan	40	0.39	0.61	Stamatoyannopoulos et al. (1975)

Source: Propping (1978).

[a] The data of Klein et al. are based on activity determinations only, since the atypical variant was not known at that time.

alcohol intoxication following consumption of doses of alcohol that seem to have no effect in most Caucasians. Stamatoyannopoulos et al. (1975) hypothesized that the flushing response is the result of the initially high amount of the ethanol metabolite acetaldehyde produced via the greater activity of the atypical ADH. This idea has been supported by observations indicating that alcohol intoxication symptoms in Mongoloids mimic effects of disulfiram, which inhibits acetaldehyde dehydrogenase thereby resulting in higher acetaldehyde levels. In addition, Reed et al. (1976) and Ewing et al. (1974) reported higher acetaldehyde levels in Mongoloids as compared to Caucasians after oral alcohol loading. Based on these findings, Propper (1978) speculated that Orientals may be metabolically protected from developing alcoholism. He claimed that the initially greater level of discomfort presumably experienced by Orientals may, in fact, be protective by creating a type of avoidance phenomenon. It is well-known that Orientals have a lower incidence of alcoholism than Caucasians. Previously it was thought that cultural influences were the primary explanatory factors. The occurrence of genetic differences in the ability to metabolize ethanol may well play a role in the overall developmental unfolding of the disease of alcoholism. Clearly, these metabolism studies support the earlier genetic research of ethanol within identical and fraternal twins.

4. SCREENING TESTS

A simple screening test is available to determine the type of ADH present in crude human liver homogenates. However, there is only a trace of ADH activity in normal serum and hemolyzed erythrocytes. According to Von Wartburg et al. (1965) this amount of activity is insufficient to assess the activity levels needed for screening.

REFERENCES

Armussen, E., Hold, J., and Larson, V. (1948). The pharmacological action of acetaldehyde on the human organism. *Acta Pharmacol.* **4**: 311–320.

Bennion, C. J. and Li, T. K. (1976). Alcohol metabolism in American Indians and whites. *N. Engl. J. Med.* **294**: 9–13.

Edwards, J. A. and Evans, D. A. P. (1967). Ethanol metabolism in subjects possessing typical and atypical liver alcohol dehydrogenase. *Clin. Pharmacol. Ther.* **8**: 824–829.

Ewing, J. A., Rouse, B. A., and Pellizzari, E. D. (1974). Alcohol sensitivity and ethnic background. *Am. J. Psychiatry* **131**: 206–210.

Farris, J. J. and Jones, B. M. (1978). Ethanol metabolism in male American Indians and whites. *Alcoholism Clin. Exp. Res.* **2**: 77–81.

Fenna, D., Mix, L., Schaefer, O., and Gilbert, J. A. L. (1971). Ethanol metabolism in various racial groups. *Can. Med. Assoc. J.* **105**: 472–475.

References

Fukui, M. and Wakasuyi, C. (1972). Liver alcohol dehydrogenase in the Japanese population. *Jpn. J. Leg. Med.* **26**: 46–51.

Hanna, J. M. (1978). Metabolic responses of Chinese, Japanese and Europeans to alcohol. *Alcoholism Clin. Exp. Res.* **2**: 89–92.

Harada, S., Agarwood, D. P., and Goedde, H. W. (1978). Human liver alcohol dehydrogenase isoenzyme variations. *Humn. Genet.* **40**: 215–220.

Klein, H., Fahrig, H., and Wolf, H. P. (1962). Die Bestimmung der Alkoholdehydrogenase- und Glutaminsarre-Oyalussigsarre Transaminose Aktivitat der menschlichen. Leber-nach dem Tode. *Dtsch. Z. Gesamte Gerichtl. Med.* **52**: 615–629.

Kopun, M. and Propping, P. (1977). The kinetics of ethanol absorption and elimination in twins and supplementary repetitive experiments in singleton subjects. *Eur. J. Clin. Pharmacol.* **11**: 337–344.

Omenn, G. S. (1975). Alcoholism: A pharmacogenetic disorder. In: *Genetics and Psychopharmacology*. J. Mendlewicz (ed.). Karger, New York, pp. 12–22.

Omenn, G. S. and Motulsky, A. G. (1972). A biochemical and genetic approach to alcoholism. *Ann. N.Y. Acad. Sci.* **197**: 16–23.

Omenn, G. S. and Motulsky, A. G. (1978). "Eco-genetics": Genetic variation in susceptibility to environmental agents. In: *Genetic Issues in Public Health and Medicine*. Thomas, Springfield, Illinois, pp. 83–111.

Propping, P. (1978). Alcohol and alcoholism. In: Human Genetic Variation in Response to Medical and Environmental Agents. Pharmacogenetics and Ecogenetics. *Hum. Genet.* Suppl. 7, pp. 91–99.

Reed, T. E., Kalant, H., Gibbins, R. J., Kapus, B. M., and Rankin, J. G. (1976). Alcohol and acetaldehyde metabolism in Caucasians, Chinese, and American Indians. *Can. Med. Assoc. J.* **115**: 851–855.

Sauter, A. M., Boss, D., and Wartburg, J. P. (1977). Re-evaluation of the disulfuram-alcohol reaction in man. *J. Stud. Alcohol* **38**: 1680–1695.

Schuckit, M. A. and Rayses, V. (1979). Ethanol ingestion; difference in blood acetaldehyde concentration in relatives of alcoholics and controls. *Science* **203**: 54–55.

Smith, M., Hopkinson, D. A., and Harris, H. (1971). Developmental changes and polymorphism in human alcohol dehydrogenase. *Am. J. Hum. Genet.* **35**: 243–253.

Smith, M., Hopkinson, D. A., and Harris, H. (1971a). Developmental changes and polymorphism of human-liver alcohol dehydrogenase. *Eur. J. Biochem.* **24**: 251–271.

Stamatoyannopoulos, S., Chen, S. H., and Fukui, M. (1975). Liver alcohol dehydrogenase in Japanese: High population frequency of atypical form and its possible role in alcohol sensitivity. *Am. J. Hum. Genet.* **27**: 789–796.

Von Wartburg, J. P., Papenberg, J., and Aebi, H. (1965). An atypical human alcohol dehydrogenase. *Can. J. Biochem.* **43**: 889–898.

Von Wartburg, J. P. and Schurch, P. M. (1968). A typical human liver alcohol dehydrogenase. *Ann. N.Y. Acad. Sci.* **151**: 936–946.

Vesell, E. S. (1973). Advances in Pharmacogenetics. *Prog. Med. Genet.* **9**: 291–367.

Vesell, E. S. and Page, J. G. (1968). Genetic control of drug levels in man: Phenylbutazone. *Science* **159**: 1479–1480.

Vesell, E. S. and Page, J. G. (1968a). Genetic control of drug levels in man: Antipyrine. *Science* **161**: 72–73.

Vesell, E. S. and Page, J. G. (1968b). Genetic control of dicoumarol levels in man. *J. Clin. Invest.* **47**: 2657–2663.

Vesell, E. S., Page, J. G., and Passananti, G. T. (1971). Genetic and environmental factors affecting ethanol metabolism in man. *Clin. Pharmacol. Ther.* **12**: 193–201.

Wolff, P. H. (1972). Ethnic differences in alcohol sensitivity. *Science* **175**: 449–450.

Wolff, P. H. (1973). Vasomotor-sensitivity to alcohol in diverse mongoloid populations. *Am. J. Hum. Genet.* **25**: 193–199.

D. Gout

Gout is a characteristically human clinical disorder that principally affects the joints and produces a specific type of acute and chronic arthritis associated with the presence of sodium urate crystals. The predominant biochemical feature is hyperuricemia. Wyngaarden and Kelley (1972, 1978) indicate that in primary gout, the hyperuricemia results from one of a variety of possible inborn errors of metabolism. However, in secondary gout, the hyperuricemia is an outgrowth of an acquired disorder or by exposure to certain diets, drugs, or environmental agents. These authors presented a classification of the various types of gout, its accompanied metabolic disturbance, and proposed modes of inheritance (Table 3-6).

The hyperuricemia seen in persons with primary gout is asymptomatic at first. In fact, a certain percentage of affected individuals may remain symptomless their entire lives. In others, however, the genetic predisposition may become clinically evident by repeated attacks of acute gouty arthritis and/or *renal lithiasis*. Over the course of time, the arthritic condition may become chronic as a result of destructive effects of tissue urate deposits that tend to form in the proximity of the joints of the extremities, especially the feet and in cartilaginous areas. During the course of the disease, most patients with primary gout develop renal disease involving the tubular and interstitial tissue that often contains deposits of cystalline urate. Additional features frequently seen are glomerular and renal vascular changes and hypertension. The cause of events in secondary gout, like that of primary gout, may also pass through the stage of asymptomatic hyperuricemia, recurrent acute gouty arthritis and/or renal lithiasis, and chronic gouty arthritis.

1. PREVALENCE AND INCIDENCE

The prevalence of primary gout varies widely in different parts of the world. In 1960, Lawrence reported a prevalence of 0.3% in Europe. At the same time, Wyngaarden (1960) estimated a similar prevalence in the United States of 275/100,000 or 0.27%. Studies on the long-standing Framingham (Massachusetts) Heart Disease Epidemiology Study reported a prevalence of gouty arthritis of 0.2% in a population of 5127 (2283 men and 2844 women) aged 30–59 years (mean age 44). Fourteen years later the prevalence had jumped to 1.5% of the population now with a mean age of 58. The prevalence in the men was 2.9%, while only 0.4% in

Table 3-6. Classification of Hyperuricemia and Gout

Type	Metabolic Disturbance	Inheritance
Primary		
Idiopathic		
Normal excretion (75–80% of primary gout)	Overproduction of uric acid and/or underexcretion of uric acid (specific defects undefined)	Polygenic Autosomal dominant forms?
Overexcretion (20–25% of primary gout)	Overproduction of uric acid (specific defects undefined)	
Associated with specific enzyme defects		
Glucose-6-phosphatase: deficiency or absence	Overproduction plus underexcretion of uric acid; glycogen storage disease, type I (von Giercke)	Autosomal recessive
Hypoxanthine-guanine phosphoribosyltransferase: deficiency, partial or "virtually complete" (Lesch-Nyhan syndrome)	Overproduction of uric acid	X linked
Glutamine-PP-ribose-P-amidotransferase: feedback resistance	Overproduction of uric acid	Unknown
GSH reductase variant: increased activity	Overproduction of uric acid suspected, but undetermined	Autosomal dominant
Secondary		
Associated with increased nucleic acid turnover	Overproduction of uric acid	
Associated with decreased renal excretion of uric acid	Reduced renal functional mass	
	Inhibited tubular secretion of uric acid	
	Enhanced tubular reabsorption of uric acid	

Source: Wyngaarden and Kelley (1972).

the women (Hall et al., 1967). Other surveys have revealed even higher prevalence amongst Filipino males in northwestern North America (Decker et al., 1962), and considerably higher values among the Chamorros and the Carolinians in the Mariana Islands (Burch et al., 1966), and in the Maori of New Zealand who had a prevalence of 10% in males (Prior and Rose, 1966).

2. RELATIONSHIP OF LEAD TO GOUT

As far back as 1859 Garrod postulated an association between lead and gout. He stated that 75% of the gout cases in his hospital had a hereditary basis, while the remaining 25% had lead involved in the etiology of this disease. These early observations of Garrod were subsequently verified by other investigators. For example, a high incidence of gout has been found in patients with chronic lead nephropathy (Emmerson, 1963), with at least some of this observed increased incidence of gout thought to be caused by a defective renal excretion of the uric acid (Emmerson, 1965). Additional cases of gout have been reported to occur in patients with a history of appreciable lead exposure with the occurrence of nephropathy (Richet et al., 1964; Morgan et al., 1966).

Rapado (1969) indicated that 70 of 450 patients with gouty arthritis in Spain displayed a history of direct and prolonged contact with leaded gasoline, thereby suggesting that subclinical lead intoxication may be important. Consistent with these findings were more recent reports of Campbell et al. (1977) that indicated that joint pains occur in 25% of patients with chronic lead poisoning in the United Kingdom, and frequently associated with hyperuricemia. In addition, a study by Campbell et al. (1978) assessed the blood lead levels in 32 gouty patients with no history of overt lead exposure with age/sex matched nongout controls who were also neither hypertensive nor had raised serum urate levels nor known overt lead exposure. Interestingly, the gouty patients displayed highly significantly greater blood lead values than the controls. While these specific findings do not prove that lead caused the gout, taken in context with other findings that associated lead with symptomless hyperuricemia (Campbell et al., 1977), a causal connection may be reasonably hypothesized.

In 1981, a paper by Tak et al. offered a new proposed mechanism to explain how lead may affect the occurrence of gout without causing nephropathy. They reported that lead urate can spontaneously crystallize from solutions with low urate concentration and that lead urate crystals can serve as nucleation sites for monosodium urate. More specifically, these authors reported that lead urate will spontaneously nucleate at urate levels as low as 5 mg % in the presence of 10 ppm of lead. Spontaneous nucleation of lead urate also occurred at 10 mg % urate and 1 ppm of lead. The levels of lead employed in the studies of Tak et al. (1981), while exceeding ambient blood levels, fall within the range of blood lead levels found with acute lead intoxication. These findings suggest an alternative way of explaining how lead may precipitate gout. This interpretation would be strengthened by the detection of lead in naturally occurring gout urate crystals. The findings of Tak et al. (1981) not only suggest a possible mechanism of secondary gout caused by lead exposure, but they also imply that genetic conditions that result in the formation of hyperuricemia may be at increased risk to the occurrence of lead induced gout urate crystals.

Many cases of lead ingestion associated with the occurrence of gout have been found in the "moonshine belt" of the southeast United States. The home brewing of whiskey had been done in stills often haphazardly assembled with lead solder materials that commonly include automobile radiators. The lead exposure in such drinking incidents has been shown to be extraordinarily high. Levels exceeding 1 ppm were not uncommon in samples of moonshine from Alabama stills (Ball and Morgan, 1968). Despite this high exposure to lead, it must also be recognized that hyperuricemia may also be induced by excessive alcohol consumption (Lieber, 1965). The mechanism by which alcohol affects the occurrence of hyperuricemia has been speculated by Lieber (1965) to be by increasing the blood lactate concentration, a phenomenon he experimentally demonstrated in terms of temporal association. Consequently, in those consuming lead tainted moonshine, both alcohol and lead were present. This leads to the possibility of a multiple factor interaction in the affected cases.

While it is known that lead and alcohol as well as drugs such as certain diuretics (e.g., thiazide, ethacrynic acid, salicylates) may lead to the development of hyperuricemia, no information is available on whether those with genetic predispositions to developing gout are at even greater risk and how much if exposed to these agents. This represents an important area of further study.

REFERENCES

Ball, G. and Morgan, J. M. (1968). Chronic lead ingestion and gout. *South. Med. J.* **61**: 21–24.

Ball, G. V. and Sorensen, L. B. (1969). Pathogenesis of hyperuricemia in saturnine gout. *N. Engl. J. Med.* **280**: 1199–1202.

Burch, T. A., O'Brien, W. M., Reed, R., and Kurland, L. T. (1966). Hyperuricemia and gout in Mariana Islands. *Ann. Rheum. Dis.* **25**: 114.

Campbell, B. C. and Baird, A. W. (1977). *Br. J. Indust. Med.* **34**: 298.

Campbell, B. C., Beattie, A. D., Moore, M. R., Goldberg, A., and Reid, A. G. (1977). Renal insufficiency associated with lead exposure. *Br. Med. J.* **1**: 482–485.

Campbell, B. C., Moore, M. R., Goldberg, A., Hernandez, L. A., and Dick, W. C. (1978). Subclinical lead exposure: A possible cause of gout. *Br. Med. J.* **2**: 1403.

Decker, J. L. and Lane, J. J., Jr. (1959). Gouty arthritis in Filipinos. *N. Engl. J. Med.* **261**: 805.

Decker, J. L., Lane, J. J., Jr., and Reynolds, W. E. (1962). Hyperuricemia in a Filipino population. *Arthritis Rheum.* **5**: 144.

Emmerson, B. T. (1963). Chronic lead nephropathy: the diagnostic use of calcium EDTA and the association with gout. *Aust. Ann. Med.* **12**: 310.

Emmerson, B. T. (1965). The renal excretion of urate in chonic lead nephropathy. *Aust. Ann. Med.* **14:** 295–303.

Emmerson, B. T. (1968). The clinical differentiation of lead gout from primary gout. *Arthritis Rheum.* **11**(5): 623–634.

Garrod, A. B. (1859). *The Nature and Treatment of Gout and Rheumatic Gout.* 2nd ed. Walton and Moberly, London.

Hall, A. P., Barry, P. E., Dawber, T. R., and McNamara, M. (1967). Epidemiology of gout and hyperuricemia: A long term population study. *Am. J. Med.* **42:** 27.

Lawrence, J. S. (1960). Heritable disorders of connective tissue. *Proc. R. Soc. Med.* **53:** 522.

Lieber, C. (1965). Hyperuricemia induced by alcohol. *Arthritis Rheum.* **8**(5): 786–798.

Morgan, J. M., Hartley, M. W., and Miller, R. E. (1966). Nephropathy in chronic lead poisoning. *Arch. Intern. Med.* **118:** 17.

Prior, I. A. M. and Rose, B. S. (1966). Uric acid, gout and public health in the South Pacific. *N. Z. Med. J.* **65:** 295.

Rapado, A. (1969). Gout and saturnism. *N. Engl. J. Med.* **281:** 851.

Richet, G., Albahary, C., Ardaillov, R., Sultan, C., and Morel-Maroger, A. (1964). Le Rein du Saturnisme Chronique. *Rev. Fr. Etud. Clin. Biol.* **9:** 188.

Tak, H-K., Wilcox, W. R., and Cooper, S. M. (1981). The effect of lead upon urate nucleation. *Arthritis Rheum.* **24**(10): 1291–1295.

Wyngaarden, J. B. and Kelley, W. N. (1972, 1978). Gout. In: *The Metabolic Basis of Inherited Disease*, pp. 889–968 (1972); pp. 916–1010 (1978). J. B. Stanbury, J. B. Wyngaarden, and D. S. Fredrickson (eds.). McGraw-Hill, New York.

Wyngaarden, J. B. (1960). Gout. In: *Metabolic Basis of Inherited Disease.* 1st. ed. J. B. Stanbury, J. B. Wyngaarden, and D. S. Fredrickson (eds.). McGraw-Hill, New York, p. 679.

E. Animal Models for the Study of Hyperuricemia

Hyperuricemia is a metabolic condition unique to humans and is characterized by serum urate levels greater than 6.5 mg/100 mL. This elevated level of urate may result from (1) an overproduction of uric acid, and (2) diminished urinary excretion of uric acid, or (3) a combination of both (Stavric and Nera, 1979).

It is important to realize that uric acid is the final product in the degradative metabolism of purines in humans and is thus excreted. In contrast, all other mammals with the notable exception of some primate species have allantoin as the end product of purine metabolism. Allantoin is more completely oxidized than uric acid and is much more soluble than uric acid in water. The enzyme that converts uric acid to allantoin is uricase and it is located primarily in the liver of most mammals. However, as implied above, humans lack this enzyme possibly as a result of an evolutionary change (Stavric and Nera, 1979).

The presence of uricase in almost all mammals including those typically employed in toxicology studies offers extraordinary difficulties in trying to find an experimental model. This has lead to a consideration of those nonhuman primates that, like humans, lack uricase. However, the limited number of such animals, their cost, the inherent need for more selective use of experimental apes, and their generally relatively low serum urate levels has stimulated the need for alternative animal models.

The search for acceptable animal models to study hyperuricemia has produced an interesting array of potential models. For example, the Dalmatian dog, in contrast to the mongrel, has high serum urate levels and elevated urinary levels of uric acid. This difference in metabolism has been related to a defect in liver metabolism (Duncan et al., 1961; Kuster et al., 1972; Mudge et al., 1968; Berger and Yü, 1970). Austic and Cole (1972) have described the use of inbred chickens with an inherited decreased capacity for renal excretion of uric acid for a possible model. Fitzgerald (1974) mentioned another possible model for acute gout. This involved using urate-induced paw swelling in rats, mice, and dogs. A variety of investigators have attempted to induce hyperuricemia in rats, rabbits, and dogs by intravenous administration of uric acid (Fitzgerald, 1974; Duncan et al., 1961; Smith and Lee, 1957; Dunn and Polson, 1926). This procedure proved to be unacceptable since no increased serum urate levels or significant renal structural changes could be observed with prolonged low-level exposures. Any changes that did occur were caused by rapidly flooding the kidneys with large quantities of uric acid, a situation not comparable to the human state in which hyperuricemia is a chronic condition with periodic elevations of serum urate.

While the previous attempts to develop animal models investigated (1) naturally occurring states (i.e., primates, Dalmatian dogs), (2) highly inbred strains (i.e., chickens), or (3) the loading of uric acid, a different approach was subsequently employed. This involved the development of competitive inhibitors of uricase. The inhibitors were representative of the S-triazines that appear similar to the pyrimidine of the purine ring system of uric acid. The most potent competitive inhibitor of uricase was found to be the potassium salt of oxonic acid (Figure 3-5) (Fridovich, 1965).

Based on the work of Fridovich (1965) with uricase inhibitors, Johnson et al. (1969) and Stavric et al. (1969) developed a biochemical procedure for creating an animal model for hyperuricemia and hyperuricosuria. They hypothesized that chemical inhibition of uricase by potassium oxonate would create a chronic hyperuricemia state applicable to the human condition. In fact, this research indicated that a hyperuricemic condition could be maintained following just a few days of treatment and that it could be sustained if treatment was continued (Johnson et al., 1969). While the original work was accomplished with rats (Johnson et al., 1969), a comparable K-oxonate induced hyperuricemia state was created in mice (Gralla et al., 1975), pigs (Hatfield et al., 1974), rabbits (Wildman and Philp, 1970), and mongrel dogs (Yu et al., 1971).

Stavric and Nera (1979) reported that maintenance of rats on a diet

Smith, J. F. and Lee, Y. C. (1957). Experimental uric acid nephritis in the rabbit. *J. Exp. Med.* **105**: 615.

Stavric, B., Johnson, W. J., and Grice, H. C. (1969). Uric acid nephropathy: An experimental model. *Proc. Soc. Exp. Biol. Med.* **130**: 512.

Stavric, B. and Nera, E. A. (1979). Use of the uricase-inhibited rat as an animal model in toxicology. In: *Toxicology Annual. Vol. 3*. C. L. Winek and S. P. Shanor (eds.). Marcel Dekker, New York, pp. 47–74.

Tak, H-K., Wilcox, W. R., and Cooper, S. M. (1981). The effect of lead upon water nucleation. *Arthritis Rheum.* **24**(10): 1291–1295.

Wildman, R. A. and Philp, R. B. (1970). Effect of allopurinol on platelets and uric acid in oxonate-induced hyperuricemic rabbits. *Proc. Can. Fed. Biol. Sci.* **13**: 9.

Yü, T. F., Gutman, A., Berger, L., and Kaung, C. (1971). Low uricase activity in the Dalmatian dog simulated in mongrels given oxonic acid. *Am. J. Physiol.* **220**: 973.

F. Sulfite Oxidase Deficiency*

1. INTRODUCTION

In recent years the identification of individuals who may be at increased risk to the toxic effects of environmental pollutants has played an important role in the process of standard setting for air and drinking water contaminants within the United States by the EPA (Calabrese, 1978). For example, the increased susceptibility of infants to the development of nitrate- and/or nitrite-induced methemoglobinemia led to the adoption of a drinking water standard that was sufficiently restrictive to protect infants from nitrates. Similarly, the increased susceptibility of those with cardiorespiratory disorders has been considered in the formation of ambient air standards for ozone, carbon monoxide, nitrogen dioxide, and sulfur dioxide (SO_2). The inclusion of the knowledge of high-risk groups in the standard setting process is made with the understanding that these groups are the first to experience morbidity and/or mortality and if they can be protected from adverse health effects then the entire population could in turn be protected.

Evaluation will be made of the controversial hypothesis, originally proposed by Hickey et al. (1976), that individuals with a sulfite oxidase deficiency are at increased risk to SO_2, sulfite, and bisulfite toxicity. The following will be shown:

1. SO_2 is converted to sulfite within the blood stream.
2. Acute sulfite toxicity is inversely related to hepatic sulfite oxidase activity in various animal models.
3. Experimentally induced hepatic sulfite oxidase deficiency enhances acute SO_2 toxicity.

*This section was previously published in *Medical Hypotheses* **7**: 133–145, 1981 and is reprinted by permission of the publisher.

4. Sulfite is mutagenic in several test systems.
5. Hepatic sulfite oxidase deficiencies (both intermediate and complete) exist in the human population.

In light of these interrelated observations it is proposed that persons with an extreme or intermediate hepatic sulfite oxidase deficiency may be at increased risk to SO_2-induced systemic mutagenicity and other insidious systemic toxicological effects of sulfite. Finally, an animal model for evaluating this hypothesis will be presented.

2. SO₂ AND HEALTH EFFECTS

SO_2 still remains a major pollutant in urban areas and may prove to be an even more serious future threat in the United States, which has abundant reserves of high-sulfur coal. It is, therefore, important that the effects of SO_2 pollution on health be examined thoroughly, especially the effects of chronic exposure.

Respiratory abnormalities have been the most thoroughly investigated of the health effects induced by chronic levels of SO_2. The potency of SO_2 as a lung and upper respiratory irritant is due to its conversion to sulfuric acid and particulate sulfates in the atmosphere (Amdur, 1971). For example, 1 ppm (1.3 mg/m^3) SO_2 produces an increase of about 15% in lung flow resistance, whereas the same amount of SO_2 converted to sulfuric acid causes an increase in resistance of approximately 60% (Amdur, 1971). Generally, various animal studies have revealed that elevated levels of SO_2 exposure cause the development of narrowed air passages, impairment of normal pulmonary mechanics, thickening of alveoli, lymphatic engorgement, edema, alveolar hyperplasia, and the acceleration of the onset of precancerous conditions and lung tumors in susceptible mice (Balchum, 1959; Alarie et al., 1972; Wakisaka, 1976).

Information concerning pulmonary toxicity due to SO_2 in experimentally exposed human subjects is consistent with the animal model studies. For example, volunteers exposed to sulfuric acid aerosols containing from 0.35 to 0.5 mg/m^3 developed an increased respiration rate and a concomitant decrease in the maximum inspiratory and expiratory flow rates (Amdur et al., 1972). Exposure to higher concentrations increased airway resistance and initiated intense coughing, wheezing, and long-lasting bronchitic symptoms (Sim and Pattle, 1957).

3. FATE OF SO₂ IN MAMMALIAN TISSUES

Sulfur oxides may enter the body by inhaling SO_2 or by consuming foods treated with sulfite or bisulfite salts. In both cases, the SO_2 is absorbed into the bloodstream to form sulfite (SO_3^{2-}) or bisulfite (HSO_3^-), the hydrated form. Once in the blood the bisulfite is distributed to all body tissues including the brain (Hickey et al., 1974).

Two laboratory groups have independently studied the fate of sulfur oxides by exposing animals to ^{35}S-labeled SO_2. Yokoyama et al. (1971) exposed dogs to 20–24 ppm $^{35}SO_2$ for 1 hr or less and found that the bulk of the whole blood radioactivity was contained in the plasma with $\frac{2}{3}$ being associated with the α-globulin and albumin proteins. Two-thirds of the red blood cell associated ^{35}S was intracellular; it was not determined if this affected red blood cell function. Although most of the inhaled gas was eventually excreted as sulfate most of the SO_2 dose was still present in the animal 3 hr after exposure. During the postexposure period, plasma SO_2 levels tended to fall whereas red blood cell levels increased. When rabbits were exposed to the same level of SO_2 for up to 200 continuous hours, plasma radioactivity levels steadily rose throughout the exposure and at 62 hr all of the ^{35}S in the plasma was in the form of nondialyzable S-sulfonate (R-S-SO_3^-) or small dialyzable molecules such as S-sulfocysteine. Although sulfite was not detected in the blood even during initial exposure *in vitro*, investigators have demonstrated that SO_2 is rapidly hydrated to bisulfite in an aqueous medium (Gunnison and Beaton, 1971; Petering and Shis, 1975). Under physiological conditions, the equilibrium position of the reaction of bisulfite with compounds containing disulfide linkages lies far to the right, i.e., in the combined state (Petering and Shis, 1975):

$$\text{R-S-S-R} + HSO_3 \rightleftharpoons \text{R-S-}SO_3^- + \text{RSH}$$

The clearance time of the ^{35}S-sulfonate from the plasma depended upon the route of entry. Following the ingestion of 26 μmol sulfite/mL or inhalation of 10 ppm SO_2 for 10 days, the mean half-lives for exogenous S-sulfonate in plasma were 1.3 ± 0.3 days and 3.2 ± 2.3 days, respectively (Gunnison and Palmes, 1973). Long-term ingestion and inhalation exposures drastically lengthened the clearance time compared with acute sulfite exposure by injection. The reason for the extreme variations in clearance time of exogenous S-sulfonate could not be explained.

Experimental findings with human subjects exposed to SO_2 were in accord with animal exposure investigations (Gunnison and Palmes, 1974). A highly significant direct linear relationship was found between the level of SO_2 inhaled and the concentration of S-sulfonates found in the blood. Subjects were exposed to SO_2 concentrations of 0.3, 1.0, 3.0, 4.2, and 6.0 ppm. [The EPA standard for SO_2 in community air is 0.03 ppm for an annual arithmetic mean and 0.14 ppm for a 24-hr maximum. The U.S. Occupational Health Standard for an 8-hr daily exposure is 5 ppm (Calabrese, 1978).] An increase of 1.1 $n$moles of S-sulfonate per milliliter of plasma for each part per million increase in SO_2 concentration was observed and was estimated to be $\frac{1}{3}$ of the response observed in rabbits to the same level of SO_2.

Ultimately the absorbed SO_2 is excreted mainly in the form of inorganic sulfate (Yokoyama et al., 1971). When ^{35}S-labeled sulfite was ingested by rats, mice, and monkeys, 70–95% of the label was absorbed from the intestine and eliminated as inorganic sulfate within 24 hrs (Gibson and Strong, 1973). It is believed that the enzyme sulfite oxidase, which has been found in mammalian tissues, is responsible for the conversion of sulfite to sulfate (Johnson and Rajogopalan, 1976; Cohen and Fridovich, 1971). Sulfite oxidase controls the last step of the metabolic pathway which catabolizes sulfur amino acids to inorganic sulfate that is excreted in the urine.

4. SULFITE OXIDASE DEFICIENCY

A genetically transmitted deficiency of sulfite oxidase in humans was first documented in 1967 by Mudd et al. A severely retarded 32-month-old infant with a previously unrecognized metabolic disorder was found to have no detectable kidney, liver, or brain sulfite oxidase activity. Elevated levels of S-sulfo-L-cysteine, sulfite, and thiosulfates were found in the deficient patient's urine while no inorganic sulfate could be detected. In the absence of sulfite oxidase activity, excess sulfite is converted to thiosulfate (Segal and Johnson, 1963) via the action of β-metacaptopyruvate sulfurtransferase, and to S-sulfo-cysteine (Sorbo, 1957) by a yet unknown pathway. Oral feeding of L-cysteine to a diet-controlled sulfite oxidase deficient patient increased the sulfite concentration slightly in the urine and increased the plasma concentration to more than 10-fold the diet control level (Shih et al., 1977). Also found associated with sulfite oxidase deficiency were bilateral dislocation of the lenses and many neurological abnormalities, including severe mental retardation (Shurman et al., 1970). Several similar cases have since been reported (Shih et al., 1977; Shurman et al., 1970).

5. RELATIONSHIP BETWEEN SULFITE OXIDASE AND THE TOXICITY OF SO_2 AND BISULFITE

An inverse correlation has been found between the levels of sulfite oxidase activity in the liver and kidney and bisulfite toxicity in five species of laboratory animals (Tejnarova, 1977, 1978). Sulfite oxidase activity was found to be species specific; activity values decreased in the following order: rat, mice, guinea pig, hamster, and rabbit. Enzyme activities varied considerably with rats having more than five times the activity of rabbits (Tejnarova, 1978). Consistent with this, the LD_{50} for sodium bisulfite (N_2HSO_3) was three times higher for rats than for rabbits. This study further supports the hypothesis that sulfite oxidase is a main participant in the mechanism by which exogenous sulfite is detoxified.

6. MOLYBDENUM AND SULFITE OXIDASE DEFICIENCY

Sulfite oxidase deficiency has been produced in rats by supplying a low molybdenum diet and concomitantly administering tungsten, a competitive antagonist of molybdenum (Cohen et al., 1973). In another study by the same laboratory group administration of tungsten from 1 to 100 ppm produced progressively greater decreases in the enzyme activity (Johnson and Rojagopalan, 1974). No physical abnormalities were observed in the tungsten-treated animals except that sulfite oxidase and xanthine oxidase activities were decreased. Similarly, no health effects were noted in the progeny of female rats fed 100 ppm tungsten in drinking water during gestation (Cohen et al., 1974). The animals retained approximately 2–5% residual sulfite oxidase activity in the liver.

Sulfite oxidase and xanthine oxidase constitute the major molybdenum-containing enzyme in animals and humans. Earliest investigations on the effects of molybdenum antagonists were limited to xanthine oxidase, which is involved in the terminal metabolism of free purine bases to form uric acid (Westerfield and Richert, 1953). It had been established that molybdenum was required for the maintenance of normal levels of xanthine oxidase in rat tissues (Higgens et al., 1956). Xanthine oxidase activity was depleted by using a molybdenum-free diet along with a competitive inhibitor of molybdenum, sodium tungstenate. Except for the lack of enzyme activity no other indications of molybdenum deficiency were observed. Control and experimental groups showed comparable weight gains and there were no differences in the urinary output of uric acid. It was concluded that the rat has a remarkably low requirement for molybdenum and tissue xanthine oxidase activity (Higgens et al., 1956). Similarly, humans with deficiencies of xanthine oxidase show no impairment of health (Watts et al., 1964).

In contrast to xanthine oxidase, it was found that sulfite oxidase activity was required for normal body function for rats as well as for humans. When female rats were fed 400 ppm tungsten, no impairment of health (i.e., weight gain, urinary output, and general appearance and behavior) was observed except that hepatic sulfite oxidase activity was less than 1%. However, the litters from these rats showed poor weight gain and poor survival rates, similar to that observed in human sulfite oxidase deficiency (Cohen et al., 1974).

It has been demonstrated via the rat model that sulfite oxidase is involved in the oxidative metabolism and consequent detoxification of SO_2 and bisulfite (Cohen et al., 1973). The rats were maintained on a low molybdenum diet supplemented with deionized water containing 100 ppm tungsten for 3–5 weeks before toxicity studies were performed. The treatment resulted in a dose-dependent decrease in both sulfite oxidase and xanthine oxidase activities along with a loss of hepatic molybdenum. Sulfite oxidase activity of the kidney, intestinal ileum, heart, and lung also decreased (Cohen et al., 1973). The LD_{50} for bisulfite in tungsten-treated rats was between 148 and 221 mg $NaHSO_3$/kg of body weight compared to 394–569 mg/kg for rats fed a normal diet. Similarly, sulfite oxidase deficient rats exposed to 925 and 2350 ppm SO_2 exhibited significant decreases in survival times as compared to the appropriate controls at the respective concentrations (Johnson and Rojagopalan, 1974).

Of particular interest was the finding that sulfite oxidase deficient animals appeared to be dying of bisulfite-induced neural toxicity while at the same bisulfite concentration the normal rats died because of respiratory damage. Presumably, the rats with the low levels of sulfite oxidase activity were not as capable of converting bisulfite to sulfate (Cohen et al., 1973). However, when the levels of SO_2 were reduced to 590 ppm, slight differences between the groups were noted. Insufficient amounts of bisulfite are presumably formed *in vivo* at the 590 ppm exposure to cause obvious central nervous system toxicity in the deficient animals, since both deficient and control animals ultimately died from respiratory damage (Cohen et al., 1973). It is important to realize that the 925 ppm level of SO_2 that was required to cause enhanced toxicity in the deficient rat is about 200 times greater than levels even rarely approached (5 ppm) in urban centers. While such findings cause one to question whether persons with an intermediate deficiency of sulfite oxidase may be at increased risk to ambient levels of SO_2, it must be recognized that the rat studies were concerned with acute respiratory toxicity resulting from 590 to 925 ppm SO_2 exposure and not with the more insidious chronic effects that sulfite may exert by its potentially mutagenic effect and its disruption of critical sulfhydryl protein bonds in cellular membranes.

Livers of tungsten treated rats have been shown to contain inactive protein that is immunologically identical to antibody prepared against native rat liver sulfite oxidase (Johnson et al., 1974). In addition it was demonstrated that when the enzyme activity is lowered by the use of tungsten, the rats continue to synthesize inactive sulfite oxidase, and tungsten is incorporated into 35% of the molybdenum-free molecules.

Purified human liver sulfite oxidase has been compared to the rat liver enzyme (Johnson and Rojagopalan, 1976). Both consist of two subunits with one metal and one cytochrome b_5-type heme per subunit. The geometry of the molybdenum center, which is essential for enzyme activity, was extensively investigated by electron paramagnetic resonance. The center was found to have the same unique sensitivity to alterations in pH and ionic strength as those of rat, chicken, and bovine enzymes (Johnson and Rojagopalan, 1976). The enzymes differ in that the human subunits had a larger molecular weight (61,100 as compared to 57,200 daltons) and a higher negative charge. The specific activity of the rat liver sulfite oxidase was more than 10 times that found in human liver.

7. SULFITE/BISULFITE TOXICITY

Bisulfite is known to react with numerous biomolecules, essentially acting as a nucleophile attracted to electrophilic centers (Morrison and Boyd, 1973). The relative concentrations of bisulfite reaction products that form in the body may be determined by the relative concentrations of the reactants and their respective equilibrium constants. Administration of ^{35}S to laboratory animals has revealed that much of the bisulfite reacts with the disulfide bonds of plasma proteins, α-globulin, and albumin to form S-sulfonates (Yokoyama et al., 1971).

Bisulfite reacts reversibly with aldehydes and ketones, most notably in 5 and 6 carbon sugars (Petering and Shis, 1975). In addition, it also reacts reversibly with the coenzyme, nicotinamide adenine dinucleotide (NAD^+). However, bisulfite reacts irreversibly with thiamine (Hermus, 1969; Joslyn and Leichter, 1968), and epinephrine (Yang, 1973), while tryptophan is destroyed by sulfite during aerobic oxidation in the presence of Mn^{2+} (Yang, 1973). None of the previously mentioned reactions possesses a large enough equilibrium constant to be considered a major component of bisulfite toxicity. The products that form would necessarily reflect the prevalence of the reactants. Consequently, the interaction of bisulfite with proteins would seem to be the most prominent because of the presence of innumerable disulfide bonds in blood proteins.

8. TARGETS OF BISULFITE TOXICITY

Several studies have indicated that SO_2 exposure may interfere with normal immunological mechanisms, especially those concerning the removal of microorganisms and the production of antibodies and agglutinins. In fact, as early as 1908, it was noted that exposure of rabbits and guinea pigs to 500 ppm of SO_2 for a month reduced their resistance to disease (Ardelean et al., 1966). More recently rabbits exposed to as little as 7 ppm SO_2 for 113 days exhibited a reduction in agglutinin formation (Schneider and Calkins, 1970).

Since lymphocytes recognize a foreign body or antigen through their specific membrane structure, an alteration in the membrane morphology may result in the loss of immunological specificity for antigens and leave the organism more susceptible to disease. By the use of electron paramagnetic resonance (EPR) and a lymphocyte membrane spin label, changes in structure of the lymphocyte membrane were examined using increasing concentrations of SO_2 and N_2HSO_3. It was observed that the turnover of lymphocyte membrane proteins was abnormally high, which would indicate that denatured or damaged proteins were constantly being replaced on the cell surface due to SO_2 interaction (Gause and Rowlands, 1975).

It is known that bisulfite ions can selectively deplete red bloods cells of 2,3-diphosphoglycerate by reversibly increasing the metabolism of glucose to lactate (Parker, 1969). Several authors have proposed that 2,3-diphosphoglycerate is intimately involved in the transport of oxygen from the blood to the tissues (Benesch and Benesch, 1967; Chanutin and Curnish, 1967). Therefore, bisulfite may potentially deplete the body tissues of oxygen by altering the concentration of 2,3-diphosphoglycerate.

Another possible effect of bisulfite in the blood is the inhibition of platelet aggregation. Platelet aggregation was observed to be inhibited by a sulfite–bisulfite mixture in platelet-rich plasma (Kikugaqa and Kazuhiro, 1972). The addition of other blood cells such as erythrocytes to the mixture had little influence on the effects of bisulfite.

9. MUTAGENICITY

Potentially the most serious property of SO_2 is its mutagenic capability, that is, its ability to deaminate cytosine to uracil during DNA replication. The mutagenic activity of sodium bisulfite was first demonstrated using phage (Hayatsu and Muira, 1970), and *Escherichia coli* (Mukai et al., 1970). Later, mutagenesis was found to occur in the T_4 system causing C·G T·A transitions (Mukai et al., 1970; Summers and Drake, 1971). Chromatid aberrations in *Tradescantia* pollen tubes are significantly enhanced by contact with humid air containing 0.05–0.1 ppm SO_2 for a period of 18–20 hr (Ma et al., 1973).

The biochemical mechanisms for the alteration in cytosine by bisulfite is the same as the mutagen hydroxylamine, which is also cytosine specific. When the dinucleoside monophosphate C·A was treated with bisulfite, 5,6-dehydrocytidine-6-sulfonate $C(SO_3^-)$·A was produced; if left for longer periods, 5,6-dehydrouridine-6-sulfonate $[U(SO_3^-)]$ was formed (Hayatsu et al., 1970). The action of the bisulfite-mediated transition is single-strand specific. While C → U transition can be demonstrated with poly (C), poly (C) · poly (I) does not react with bisulfite under comparable conditions (Braverman et al., 1975). These bisulfite-induced modifications have been demonstrated in bacterial systems and cause disruption of protein synthesis (Braverman et al., 1975), which suggests that bisulfite alters DNA residues. However, at lower concentrations, which is the case with physiologic exposure, the deamination of cytosine has been found to occur at a rate proportional to the first order of bisulfite concentration, which implies that mutagenesis is possible under physiological conditions (Hayatsu et al., 1970).

Several investigations have indicated the potential mutagenic effects of SO_2 on mammalian systems. Exposure of human lymphocytes cultured *in vitro* with 5.7 ppm SO_2 revealed reduced cell enlargement, depression of DNA synthesis, and an increase in chromosome abnormalities compared with control cultures (Schneider and Calkins, 1970). The SO_2 seemed to preferentially affect cells that were preparing for DNA synthesis by causing clumping of the chromosomes. SO_2 and its sodium salts cause depression of DNA synthesis and chromosome abnormalities in established cell lines including HeLa (Kuroda, 1975).

10. HYPOTHESIS

It is hypothesized that humans with an intermediate deficiency of sulfite oxidase are at increased risk to the mutagenic and other potential systemic effects (e.g., disruption of platelet aggregation, depression of immunological response, etc.) of SO_2 exposure. This is supported by (1) an inverse relationship of bisulfite toxicity and sulfite oxidase activity in five species (Tejnarova, 1977, 1978); (2) an increase in bisulfite toxicity as sulfite oxidase activity decreased in rats with experimentally induced enzyme deficiency (Cohen et al., 1973); (3) the formation of S-sulfonates in animal models and humans following SO_2 inhalation; (4) the mutagenic and toxicological capabilities of SO_2 and sulfite in several test systems.

The hypothesis that a sulfite oxidase deficiency enhances susceptibility to SO_2, sulfite, and bisulfite has been strongly criticized by Alarie (1976) because a total sulfite oxidase deficiency is thought to be exceedingly rare and to have serious clinical implications of its own, making it irrelevant from a public health standpoint. However, the issue, as Hickey et al. (1976) have pointed out, is not with individuals with a total sulfite oxidase deficiency but with the heterozygotes (i.e., intermediate deficiency) who may be at an increased risk to the chronic toxicity resulting from prolonged exposure to sulfite. Intermediate levels of sulfite oxidase activity have been reported in humans as shown by data on liver and kidney tissues and the mode of inheritance is believed to be autosomal recessive (Mudd et al., 1967; Shih et al., 1977). Rojagopalan and Johnson (1977) have postulated that SO_2 does not pose a genetic threat to normal individuals based on calculations of sulfite oxidase activity in the lung and liver tissue. However, the reports of Gunnison and Palmes (1973, 1974) illustrated that SO_2 exposures as low as 0.3 ppm for over 120 hr results in the formation of S-sulfonates. S-sulfonates may liberate HSO_3^- (sulfite) in the germ cells or other tissues by reacting with thiol groups (Petering and Shih, 1975). Thus, S-sulfonate formation amply demonstrated that defense mechanisms do not completely protect normal individuals even at low SO_2 concentration and these S-sulfonate levels would be expected to be even greater in individuals with an intermediate deficiency.

At the present time the frequency of the heterozygote within the population is not known. However, it should be realized that even if only 1 in 25,000 persons is homozygous for the deficiency, then 1 in 80 would be heterozygous carriers using the Hardy–Weinberg equation (Gardner, 1968). Consequently, the extreme rareness of the homozygous recessive does not preclude a sizeable population of carriers.

11. TESTING THE HYPOTHESIS

The hypothesis that SO_2 may present a genetic hazard to individuals with an intermediate sulfite oxidase deficiency may be evaluated via an animal model study using hamsters. The levels of sulfite oxidase activity in normal humans are approximately equal to the levels found in the hamster (Cohen and Fridovich, 1971; Johnson and Rojagopalan, 1974). Intermediate levels of sulfite oxidase could be probably induced, as in rats, to duplicate the range of intermediate human deficiencies, although this needs to be evaluated as well. Hamsters could be exposed by the administration of bisulfite/sulfite by various routes, for example, inhalation, ingestion, or injection and by the inhalation of SO_2 at concentrations that simulate the ambient environment.

A Chinese hamster model may be of particular value to this study because hamsters have a low chromosome number and because the chromosomes are easily distinguishable from one another (Renner, 1979). Using this model several key questions concerning the possible mutagenicity of SO_2 could be explored. These include (1) does SO_2 reach the vicinity of the genetic material; if so to what extent, (2) what is the reactivity of SO_2 or its physiological product with the genetic material, (3) to what extent does its reactivity interfere with the normal molecular genetics of the cell, and (4) to what extent would the genetic alterations harm germ cell chromosomes.

As previously indicated, the actual proportion of individuals with intermediate levels of sulfite oxidase deficiencies is unknown. An indication of the extent of intermediate levels could be determined by genetic screening of the general population for sulfite oxidase activity in skin fibroblasts (Olney et al., 1975) or possibly by urine screening for sulfite (Levy and Kennedy, 1975).

12. CONCLUSION

It is proposed that exposure to SO_2 derived sulfite may cause a potential genetic risk to the general public and that this risk may be markedly enhanced in persons with a sulfite-oxidase deficiency.

In light of the projected increased and widespread use of high sulfur fuels, the evaluation of this hypothesis is particularly relevant.

REFERENCES

Alarie, Y. C. (1976). Rebuttal: health effects of atmospheric sulfur dioxide and dietary sulfides. *Arch. Environ. Health* **31:** 110–112.

Alarie, Y. C., Krumm, A. A., Busey, W. M., Ulrich, C. E., and Kantz, R. J. (1975). Long-term exposure to sulfur dioxide, sulfuric acid mist, fly ash, and their mixtures. *Arch. Environ. Health* **30:** 254–262.

Alarie, Y., Urich, C. E., Busey, W. M., Krumm, A. A., and MacFarland, H. N. (1972). Long-term continuous exposure to sulfur dioxide in cynomolgus monkeys. *Arch. Environ. Health* **24:** 115–128.

Amdur, M. O. (1971). Aerosols formed by the oxidation of sulfur dioxide. *Arch. Environ. Health* **23:** 459–468.

Amdur, M. O., Silverman, L., and Drinker, P. (1972). Inhalation of sulfuric acid mist by human subjects. *Arch. Ind. Hyg. Occup. Med.* **6:** 305–313.

Ardelean, I., Cucu, M., Andronacle, E., and Bodarian, S. (1966). Immunological changes in animals exposed to low sulfur dioxide concentrations. *Fiziol. Norm. Patol.* **12:** 12–15.

Balchum, O. J. (1959). Absorption and distribution of $^{35}SO_2$ inhaled through the nose and mouth of dogs. *Am. J. Physiol.* **197**(6): 1317–1321.

Benesch, R. and Benesch, R. E. (1967). The effect of organic phosphates from the human erythrocyte on the allosteric properties of hemoglobin. *Biochem. Biophys. Res. Commun.* **26:** 167.

Braverman, B., Shapiro, R., and Szer, W. (1975). Modification of *E. coli* ribosomes and coliophage MS2 RNA by bisulfite: effects of ribosomal binding and protein synthesis. *Nucleic Acids Res.* **2**(4): 501–507.

Calabrese, E. J. (1978). *Methodological Approaches to Deriving Standards for Environmental and Occupational Health Pollutants.* Wiley, New York.

Chanutin, A. and Curnish, R. R. (1967). Effect of organic and inorganic phosphates on the oxygen equilibrium of human erythrocytes. *Arch. Biochem. Biophys.* **121:** 96.

Cohen, H. J., Drew, R. T., Johnson, J. L., and Rojagopalan, K. V. (1973). Molecular basis of the biological function of molybdenum: the relationship between sulfite oxidase and the acute toxicity of bisulfite and SO_2. *Proc. Nat. Acad. Sci.* **70**(12): 3655–3659.

Cohen, H. J. and Fridovich, I. (1971). Hepatic sulfite oxidase: purification and properties. *J. Bio. Chem.* **246**(2): 359–366.

Cohen, H. J., Johnson, J. L., and Rojagopalan, K. V. (1974). Molecular basis of the biological function of molybdenum. Developmental patterns of sulfite oxidase and xanthine oxidase in the rat. *Arch. Biochem. Biophys.* **164:** 440–446.

Gardner, E. J. (1968). *Principles of Genetics.* Wiley, New York.

Gause, E. M. and Rowlands, J. R. (1975). Effects of sulfur dioxide and bisulfite ion upon human lymphocyte membranes. *Environ. Lett.* **9**(3): 292–305.

Gibson, W. B. and Strong, F. M. (1973). Metabolism and elimination of sulphite by rats, mice, and monkeys. *Fed. Cosmet. Toxicol.* **11:** 185–198.

Gunnison, A. F. and Benton, A. W. (1971). Sulfur dioxide: sulfite. *Arch. Environ. Health* **22:** 381–388.

Gunnison, A. F. and Palmes, E. D. (1973). Persistence of plasma S-sulfonates following exposure of rabbits to sulfite and sulfur dioxide. *Toxicol. Appl. Pharmacol.* **24:** 266–278.

Gunnison, A. F. and Palmes, E. D. (1974). S-sulfonates in human plasma following inhalation of sulfur dioxide. *J. Am. Ind. Hyg. Assoc.* **35:** 288–291.

Hayatsu, H. and Muira, A. (1970). The mutagenic action of sodium bisulfite. *Biochem. Biophys. Res. Commun.* **39:** 156–160.

References

Hayatsu, H., Wataya, Y., Kai, K., and Iida, S. (1970). Reaction of sodium bisulfide with uracil, cytasine and their derivatives. *Biochemistry* **9:** 2858.

Hermus, R. J. J. (1969). Sulfite-induced thiamine cleavages. Effect of storage and preparation of minced meat. *Int. J. Vit. Res.* **39:** 175.

Hickey, R. J. (1971). Air pollution. In: *Environment. Resources, Pollution and Society*, pp. 189–212. W. W. Murdoch (ed.). Sinauer Associates, Stanford.

Hickey, R. J., Clelland, R. C., Bowers, E. J., and Boyce, D. E. (1976). Health effects of atmospheric sulfur dioxide: dietary sulfites. *Arch. Environ. Health* **31:** 108–110.

Hickey, R. J., Clelland, R. C., and Boyce, D. E. (1974). Atmospheric sulfuric dioxide, nitrogen dioxide, and lead as mutagenic hazards to human health. *Mutat. Res.* **26:** 445–446.

Higgens, E. S., Richert, D. A., and Westerfield, W. W. (1956). Molybdenum deficiency and tungstate inhibition studies. *J. Nutr.* **59:** 539–559.

Higuchi, T. and Schroeter, L. C. (1960). Kinetics and mechanism of formation of sulfonate from epinephrine and bisulfite. *J. Am. Chem. Soc.* **82:** 1904.

Johnson, J. L., Cohen, H. J., and Rojagopalan, K. V. (1974). Molecular basis of biological function of molybdenum. *J. Biol. Chem.* **249**(16): 5046–5055.

Johnson, J. L. and Rojagopalan, K. V. (1974). Effect of tungsten on xanthine oxidase and sulfite oxidase in the rat. *J. Biol. Chem.* **249**(16): 859–866.

Johnson, J. L. and Rojagopalan, K. V. (1976). Purification and properties of sulfite oxidase from human liver. *J. Clin. Invest.* **58:** 543–550.

Joslyn, M. A. and Leichter, J. (1968). Thiamine instability in experimental wet diets containing commercial casein with sulfur dioxide. *J. Nutr.* **96:** 89.

Kikugaqa, K. and Kazuhiro, I. (1972). Inhibition of platelet aggregation by bisulfite-sulfite. *J. Pharm. Sci.* **61**(12): 1904–1907.

Kuroda, Y. (1975). Protective effect of vitamin E on reduction in colony formation of cultured human cells by bisulfite. *Exp. Cell Res.* **94:** 442.

Levy, H. L. and Kennedy, J. P. (1975). *Genetic Screening for Inborn Errors of Metabolism.* U.S. Department of Health, Education and Welfare. Rockville, Maryland.

Ma, T., Isbande, D., Khan, S. H., and Tseng, Y. (1973). Low level SO_2 enhanced chromatid aberrations in tradescantia pollen tubes and seasonal variation in the aberration rates. *Mutat. Res.* **21:** 93–100.

Morrison, R. T. and Boyd, T. N. (1973). *Organic Chemistry*, 3rd ed. Allyn and Bacon, Boston.

Mudd, S. H., Irrevere, F., and Laster, L. (1967). Sulfite oxidase deficiency in man: demonstration of the enzymatic defect. *Science* **156:** 1599–1602.

Mukai, F., Hawryluk, I., and Shapiro, R. (1970). The mutagenic specificity of sodium bisulfite. *Biochem. Biophys. Res. Commun.* **39:** 983–988.

Olney, J. W., Misra, C. H., and de Gubareff, T. (1975). Cysteine-S-sulfate brain damaging metabolite in sulfite oxidase deficiency. *J. Neuropathol. Exp. Neurol.* **34:** 167–177.

Parker, J. C. (1969). Influence of 2,3-diphosphoglycerate metabolism on sodium potassium permeability in human red blood cells: studies with bisulfite and other redox agents. *J. Clin. Invest.* **48:** 117–125.

Petering, D. H. and Shis, N. T. (1975). Biochemistry of bisulfite-sulfur dioxide. *Environ. Res.* **9:** 55–65.

Rojagopalan, K. V. and Johnson, J. L. (1977). Biological origin and metabolism of SO_2. In: *Biochemical Effects of Environmental Pollutants*, p. 307. S. D. Lee (ed.). Ann Arbor Science, Ann Arbor.

Renner, H. W. (1979). Monitoring of genetic environmental risk with new mutagenicity tests. *Ecotoxicol. Environ. Safety* **3:** 122–125.

Ronzani, E. (1908). Uber den Einfluss der Einat Mungenreizender Gase der Industrian auf

Verleidigiengokrafte der Organismur gegenuber den Infektiven Krankheiten. *Arch. Hyg.* **67**: 287–366.

Schneider, L. K. and Calkins, C. A. (1970). Sulfur dioxide-induced lymphocyte defects in human peripheral blood cultures. *Environ. Res.* **3**: 473–483.

Segal, I. H. and Johnson, M. J. (1963). Synthesis and characterization of sodium cysteine-S-sulfate monohydrate. *Anal. Biochem.* **5**: 330.

Shapiro, R., Braverman, B., Louis, J. B., and Servis, R. E. (1973). Nucleic acid reactivity and conformation. *J. Biol. Chem.* **248**(11): 4060–4064.

Shih, V. E., Abroms, I. F., Johnson, J. L., Carney, M., Mandel, R., Robb, R. M., Cloherty, J. P., and Rojagopalan, K. V. (1977). Sulfite oxidase deficiency. *N. Engl. J. Med.* **297**(19): 1022–1028.

Shurman, J. A., Gaull, G., and Rocha, N. C. R. (1970). Absence of cystathionase in human fetal liver: is cysteine essential? *Science* **169**: 74–76.

Sim, V. M. and Pattle, R. E. (1957). Effect of possible smog irritants in human subjects. *J. Am. Med. Assoc.* **165**: 1908–1913.

Sorbo, B. (1957). Enzymatic transfer of sulfur from mercaptopyruvate to sulfate or sulfinoates. *Biochem. Biophys. Acta* **24**: 324.

Summers, G. A. and Drake, J. W. (1971). Bisulfite mutagenesis in bacteriophage T_4. *Genetics* **68**: 603–607.

Tejnarova, I. (1977). The toxicity of sulphite anion in five species of laboratory animals. *C. Hyg.* **22**: 64–68.

Tejnarova, I. (1978). Sulfite oxidase activity in liver and kidney tissue in five laboratory animal species. *Toxicol. Appl. Pharmacol.* **44**: 251–256.

Wakisaka, I. (1976). Sensory irritation by sulfite aerosols on the upper respiratory tract in mice. *Jpn. J. Hyg.* **31**(2): 347–352.

Watts, R. W., Engelman, K., Klinenberg, J., Seegmiller, J. E., and Sjoerdsma, A. (1964). Enzyme defect in a case of xanthinuria. *Nature (London)* **201**: 395–396.

Westerfield, W. W. and Richert, D. A. (1953). Distribution of the xanthine oxidase factor (molybdenum) in foods. *J. Nutr.* **51**: 85.

Yang, S. E. (1973). Distribution of tryptophan during the aerobic oxidation of sulfite ions. *Environ. Res.* **6**: 395–402.

Yokoyama, E., Yoder, R. E., and Frank, N. R. (1971). Distribution of ^{35}S in the blood and its excretion in the urine of dogs exposed to $^{35}SO_2$. *Arch. Environ. Health* **22**: 389–395.

G. Paraoxonase Variants

The human serum has been found to contain an enzyme (i.e., paraoxonase) that hydrolyzes the cholinesterase inhibitor paraoxon, which is the oxidized metabolite of the organophosphate insecticide parathion (Figure 3-6). Paraoxonase displays considerable interindividual variability while its activity remains constant within a given subject (Geldmacher von Mallinckrodt et al., 1973; Playfer et al., 1976). Geldmacher von Mallinckrodt et al. (1973). have reported a diallelic model with the occurrence of a trimodal distribution of enzyme activity with a simple mode of inheritance. Persons classified in the low activity group have about 25% of the activity of the high group. The gene frequency of the low activity allele is determined to be 0.7. According to Geldmacher von Mallinckrodt (1978),

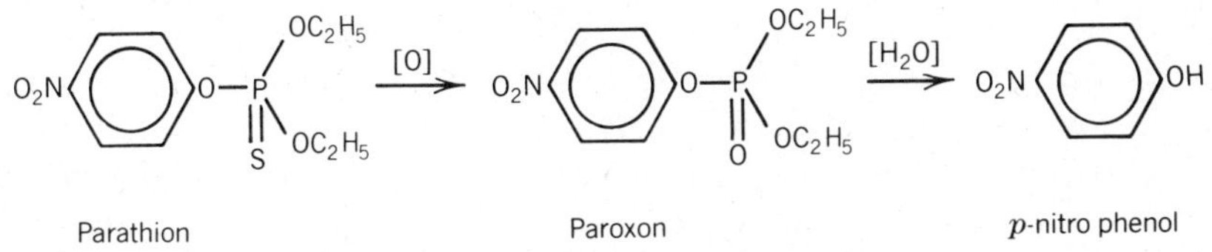

Ornithine Carbamoyl Transferase Deficiency

Figure 3-6. Biotransformation of parathion and paroxon. Parathion is converted to paroxon by means of microsomal oxidation and paroxon is hydrolyzed to yield *p*-nitrophenol. The reaction is speeded by the plasma enzyme, paroxonase. *Source:* Playfer et al. (1976).

an individual with low paraoxonase activity would be expected to be at increased risk to parathion toxicity although there is no substantiation of this hypothesis.

REFERENCES

Geldmacher-v-Mallinckrodt, M. (1978). Polymorphism of human serum paraoxonase. *Hum. Genet.* Suppl. 1, 65–68.

Geldmacher-v-Mallinckrodt, M., Lindorf, H. H., Petenyi, M., Flugel, M., Fischer, Th., and Hiller, Th. (1973). Genetisch determinierter Polymorphismus der menschlichen serum-paraoxonase. *Humangenetik* **17**: 331–335.

Playfer, J. R., Eze, L. C., Bullen, M. F., and Evans, D. A. P. (1976). Genetic polymorphism and interethnic variability of plasma paraxonase activity. *J. Med. Genet.* **13**: 337.

H. Ornithine Carbamoyl Transferase Deficiency

There have been several reports in the clinical literature associating the occurrence of toxic encephalopathy with the use of the common insect repellent, N,N-diethyltoluamide (DET) (Gryboski et al., 1961; Zadikoff, 1979; Heick et al., 1980). The authors of the most recent report, Heick et al. (1980), have indicated that their findings supported the premise that their patient was a heterozygote for a hepatic ornithine carbamoyl transferase (OCT) deficiency and that this condition may have predisposed her to DET-induced illness.

This hypothesis was based on the (1) observation that suggested an OCT deficiency such as a lifelong aversion to protein, elevated urinary orotic acid levels, low maternal levels of hepatic OCT, and previous behavioral symptomatic episodes consistent with OCT deficiency; and (2) the time course of events that precipitated toxic symptoms occurred only after extensive exposure to DET.*

The authors concluded that "chemicals and drugs that are considered

---

*Heick et al. (1980) reported that the effect of DET on the uric acid cycle has not been reported in the literature.

safe in the normal population may produce catastrophic effects in patients with an unrecognized genetic susceptibility. Our experience with this patient indicates the need for care in prescribing drugs in patients who may have metabolic disease." While these concluding words are well worth considering, it must be realized that the evidence indicating that an OCT heterozygote deficiency predisposes those affected to DET is far from conclusive.

REFERENCES

Gryboski, J., Weinstein, D., and Ordway, N. K. (1961). Toxic encephalopathy apparently related to the use of an insect repellent. *N. Engl. J. Med.* **264:** 289.

Heick, H. M. C., Shipman, R. T., Norman, M. G., and James, W. (1980). Reye-like syndrome associated with use of insect repellent in a presumed heterozygote for ornithine carbamoyl transferase deficiency. *J. Pediatr.* **97**(3): 471–473.

Zadikoff, C. M. (1979). Toxic encephalopathy associated with use of insect repellent. *J. Pediatr.* **95:** 140.

I. Carbon Disulfide Metabolism

Stokinger and Scheel (1973) have reported that there is an inherited susceptibility to carbon disulfide (CS_2) induced toxicity and that it satisfies the (i.e., their) prerequisites for hypersusceptibility testing in that it (1) has a relatively high prevalence in the worker population; (2) concerns substances commonly occurring in industry; (3) is compatible with an apparently normal life until industrial exposure occurs (i.e., the condition should be one such that it would not result in premature death early in life and does not prevent daily employment); and (4) the predictive test should be inexpensive and simple to perform, thereby permitting large-scale use.

According to Stokinger and Scheel (1973), the test for hypersusceptibility determines the rate of metabolism of a CS_2- containing substance [tetraethylthiuram disulfide (TETD), Disulfiram, Antabuse], one of the end products of which is CS_2 (Skalicka, 1967). Stokinger and Scheel (1973), commenting on Djuric et al. (1972), stated that those who had a diminished utility to metabolize TETD were thought to be more susceptible to CS_2 (Figure 3-7). This conclusion was based on the subsequent study.

Djuric et al. (1972) gave a 0.5-g tablet of TETD to three groups of workers and measured urinary dithiocarb excretion. The groups of workers were:

1. Eighteen persons previously exposed at relatively low levels of CS_2 (i.e., below industrial air limit of 60 mg/m^3).

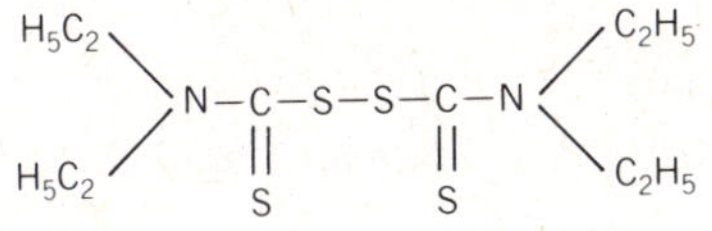

$$\mathrm{H_5C_2} \diagdown \mathrm{N-C-S-S-C-N} \diagup \mathrm{C_2H_5}$$
$$\mathrm{H_5C_2} \diagup \quad \underset{S}{\|} \quad \underset{S}{\|} \quad \diagdown \mathrm{C_2H_5}$$

TETD

$$2 \;\; \mathrm{H_5C_2} \diagdown \mathrm{N-C-S-H} \longrightarrow 2 \;\; \mathrm{H_5C_2} \diagdown \mathrm{NH} + 2\,CS_2$$
$$\mathrm{H_5C_2} \diagup \quad \underset{S}{\|} \quad\quad\quad \mathrm{H_5C_2} \diagup$$

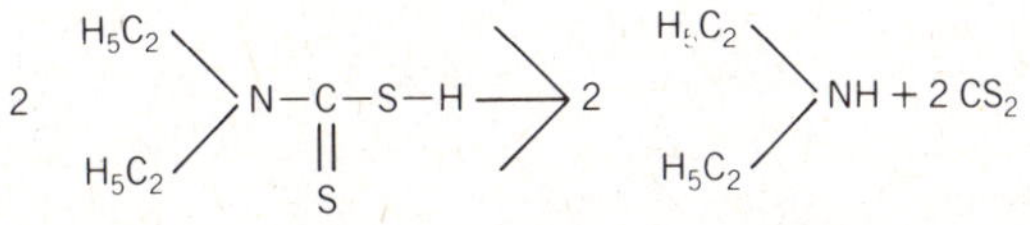

Figure 3-7. Tetraethylthiuram disulfide (TETD) and metabolites. *Source:* Stokinger, H. E. and Scheel, L. D. (1973). Hypersusceptibility and genetic problems in occupational medicine. A concensus report. *J. Occup. Med.* **15**: 564–573.

2. Twenty-one persons previously exposed at higher levels of CS_2 but nevertheless had not displayed any indication of CS_2 intoxication and thus were thought to be "resistant."
3. Thirty-three persons who displayed polyneuritis or other indications of overexposure and had been removed from exposure and consequently were labeled as "susceptible."

The results of this study indicate a statistically significant difference ($P < 0.05$) between the labeled susceptible and resistant groups to metabolize TETD. However, there was a wide range of metabolizing capacity in the control (i.e., resistant) group that partially overlapped values of the susceptible subgroup thereby diminishing the likelihood of employing the test. To get around this problem of overlap, Djuric et al. (1972) suggested that a prospective study of persons before and after exposure to quantifiable levels of CS_2 should be made. According to Stokinger and Scheel (1973) the proposed modification of Djuric et al. fulfilled the prerequisites mentioned above. It should be emphasized that the hereditary basis of such variation in the metabolism of TETD is not known. It is possible that the cause of the variation may not be genetic per se. Other factors including nutritional status such as pyridoxine levels and certain drugs (Calabrese, 1980) may also affect CS_2 intoxication. The available information, while highly suggestive of there being a metabolic basis for the occurrence of differential susceptibility to CS_2 toxicity, should be regarded as only preliminary and worthy of further study of both an industrial and familial nature.

Finally, Stokinger and Scheel (1973) reported that a test for hypersusceptibility to CS_2 using the iodide–azide test was reported in Argentina (see Stokinger et al., 1968). However, they were not aware of any similar tests in Japan, the Soviet Union, or in the industrialized parts of Australia, India, or South Africa.

REFERENCES

Calabrese, E. J. (1980). *Nutrition and Environmental Health. Volume I: The Vitamins.* Wiley, New York.

Djuric, D. et al. (1972). Antabuse as an indicator of human susceptibility to CS_2. Presented at XVII International Congress on Occupational Health. Buenos Aires, Argentina Sept. 17–24. (Cited in Stokinger and Scheel, 1973.)

Skalicka, B. (1967). Sirne Latky. V. moci po podavani: TETD. *Prac. Lek.* **19:** 408.

Stokinger, H. E., Mountain, J. T., and Scheel, L. D. (1968). Pharmacogenetics in the detection of the hypersusceptible worker. *Ann. N.Y. Acad. Sci.* **151:** 968. (Art. 2.)

Stokinger, H. E. and Scheel, L. D. (1973). Hypersusceptibility and genetic problems in occupational medicine—a consensus report. *J. Occup. Med.* **15:** 564–573.

J. Wilson's Disease

Mountain et al. (1953) reported that long-term feeding of vanadium (V) compounds to rats caused a reduction in the cystine content of rat hair. Subsequent research by Mountain et al. (1955) indicated that the fingernails of workers exposed to dusts containing V exhibited significantly lower amounts of cystine as compared to unexposed control populations. The decrease in cystine levels was correlated with the amount of V exposure. Mountain (1963), commenting on Keenan (unpublished observation), indicated that during studies of the V content of liver, there was found an inverse relationship between copper (Cu) and V levels.

Based on these observations of apparent antagonism between V and Cu, it was suggested by Mountain (1963) that certain toxic effects caused by V might be influenced by an induced Cu deficiency. The relationship between Cu and V was further clarified by the following observations: (1) copper is an essential catalyst for keratinization (Marston, 1954) and (2) the increased neutral sulfur in the urine after rats ingested V (Mountain et al., 1959) is similar to the cystinuria related to Wilson's disease where there is a genetically determined lack of normal protein for Cu (Stein et al., 1954). Table 3-7 indicates the similarity of the effects of V and Wilson's disease.

Based on this information, Mountain (1963) concluded that concurrent V exposure in a person with Wilson's disease may have at least additive and possibly synergistic effects with severe metabolic disturbances. This prediction is extremely difficult to confirm because Wilson's disease (homozygous condition) is so rare as to affect about one person in a million. However, Mountain (1963), commenting on Leff and Klendshij (unpublished observation), reports that such a case was found in which autopsy and other studies indicated that the genetic abnormalities were exacerbated by the environmental exposure of V. With regard to carriers (heterozygous condition) of the trait, it is estimated that there are 400,000 carriers in the United States. Particularly high incidences of the

Table 3-7. Relationship of Vanadium to Wilson's Disease

Vanadium
1. Inverse relationship between liver vanadium and copper content
2. Vanadium lowers cystine levels in rat hair following feeding
3. Increased urinary neutral sulfur in rats fed vanadium

Wilson's Disease
1. Display a lack of normal copper transport protein
2. Low fingernail cystine
3. Excessive excretion of urinary cystine (10 times normal)
4. Vanadium aggravates Wilson's disease

Source: Mountain (1963).

gene are found in Jews from Eastern Europe and non-Jews from the Mediterranean regions, especially Sicily (Greenblatt, 1974). Heterozygote carriers do not show clinical symptoms of the disease.

SCREENING

Attempts at identifying asymptomatic carriers of Wilson's disease have produced ambivalent results. For example, a decreased ceruloplasmin level, a common biochemical abnormality in people with Wilson's disease, has been present in some heterozygotes and not in others (Bearn, 1953, 1972; Sternlieb et al., 1961). A procedure for identifying heterozygotes without a liver biopsy has been suggested by Sternlieb et al. (1961). The proposed method relies on the ratio of Cu^{64} incorporation into ceruloplasmin at 48 hr to the level attained at 1–2 hr after oral ingestion of Cu^{64}. Ratios of Cu^{64} at 48 hr to Cu^{64} at 1–2 hr of < 0.559 and >1.253 indicate the heterozygote and homozygote, respectively, with 99% confidence. However, the ratio in the control group seems to be correlated with age and makes the determination of the heterozygote less certain (Bearn, 1972).

REFERENCES

Bearn, A. G. (1953). Genetic and biochemical aspects of Wilson's Disease. *Am. J. Med.* **15:** 442.

Bearn, A. G. (1972). Wilson's Disease. In: *Metabolic Basis of Inherited Disease*, 3rd ed. J. B. Stanbury, J. B. Wyngaarden, and D. S. Fredrickson (eds.). McGraw-Hill, New York, p. 1033.

Greenblatt, A. (1974). *Heredity and You*. Coward, McCann and Geoghegan, New York, pp. 155–166.

Marston, R. H. (1954). *Physiology and Biochemistry of the Skin*. S. Rothman (ed.). University of Chicago Press, Chicago.

Mountain, J. T. (1963). Detecting hypersusceptibility to toxic substances. *Arch. Environ. Health* 6: 357.

Mountain, J. T., Delker, L. L., and Stokinger, H. E. (1953). Studies in vanadium toxicology: I. Reduction in the cystine content of rat hair. *Arch. Indust. Hyg. Occ. Med.* 8: 406.

Mountain, J. T., Stockell, F. R., Jr., and Stokinger, H. E. (1955). Studies in vanadium toxicology: III. Fingernail cystine as an early indicator of metabolic changes in vanadium workers. *A.M.A. Arch. Indust. Hyg.* 12: 494.

Mountain, J. T., Stockell, F. R., and Stokinger, H. E. (1956). Effects of ingested vanadium on cholesterol and phospholipid metabolism in the rabbit. *Proc. Soc. Exp. Biol.* 92: 582–587.

Mountain, J. T., Wagner, W. D., and Stokinger, H. E. (1959). Effects of vanadium on growth, cholesterol metabolism, tissue components in laboratory animals on various diets. *Fed. Proc.* 18(pt. 1): 1678.

Stein, W. H., Bearn, A. G., and Moore, S. (1954). The amino acid content of the blood and urine in Wilson's Disease. *J. Clin. Invest.* 33: 410.

Sternlieb, I., Morell, A. G., Baver, C. C., Combes, B., Sternberg, S., and Scheinberg, I. H. (1961). Detection of the heterozygous carrier of the Wilson's disease gene. *J. Clin. Invest.* 40: 707.

K. Metabolic Predisposition to Cigarette Smoking–Induced Bladder Cancer

While the link between cigarette smoking and lung cancer has received great attention, smoking has also been causally linked to the occurrence of bladder cancer. Just how cigarette smoking may affect the occurrence of bladder cancer has been the subject of several research efforts focusing on the metabolism of nicotine. Nicotine is metabolized in cigarette smokers to cotinine and nicotine-1'-N-oxide (Bowman et al., 1959; Booth and Boyland, 1970), by the alternative pathways of oxidative metabolism involving nitrogen and 2-carbon oxidation (Gorrod et al., 1971). It is interesting to note that these compounds that are the principal basic metabolites of nicotine in humans, have been implicated in the production of tumors (Booth and Boyland, 1970; Gorrod et al., 1971; Truhaut et al., 1964). Follow-up studies by Gorrod et al. (1974) revealed that the urinary ratios of cotinine to nicotine-1'-N-oxide were significantly higher in cancer patients than in the controls (Figure 3-8). While this evidence suggests that bladder cancer patients and controls may metabolize nicotine differentially, considerably more research is needed to establish the causality of such metabolic differences with bladder cancer. According to Gorrod et al. (1971), "if these findings can be substantiated in larger populations, the excretion of these nicotine metabolites may be a useful indicator of the development of this neoplastic disease."

References

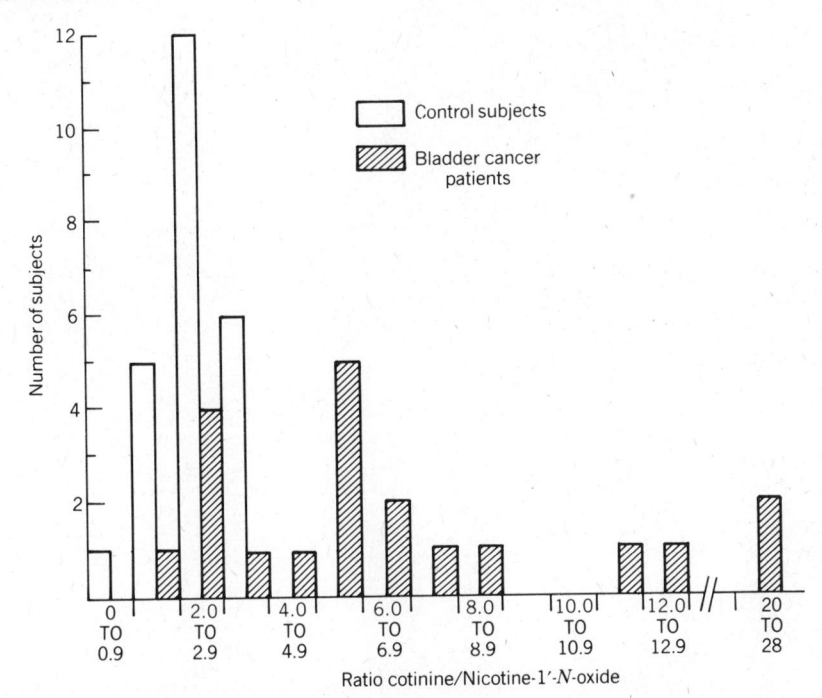

Figure 3-8. Distribution of cotinine/nicotine-1′-N-oxide ratios between the control group and patients with cancer of the urinary bladder. *Source:* Gorrod et al. (1971).

REFERENCES

Booth, J., and Boyland, E. (1970). The metabolism of nicotine as two optically active stereoisomers of nicotine-1-N oxide by animal tissues *in vitro* and by cigarette smokers. *Biochem. Pharmacol.* **19**: 733–742.

Bowman, E. R., Turnbull, L. B., and McKennis, H., Jr. (1959). Metabolism of nicotine in the human and excretion of pyridine compounds by smokers. *J. Pharmacol. Exp. Ther.* **127**: 92–95.

Boyland, E. (1968). The possible carcinogenic action of alkaloids of tobacco and betel nut. *Planta Med.* Suppl. 13–23.

Gorrod, J. W., Jenner, P., Keysell, G. R., et al. (1971). Selective inhibition of alternative pathways of nicotine metabolism *in vitro*. *Chem. Biol. Interact.* **3**: 269–270.

Truhaut, R., de Clercq, M. and Loisillier, F. (1964). Sur le toxicites aigue et chronique de la cotinine, et sur effet cancerigene chez le rat. *Pathol. Biol. (Paris)* **12**: 36–42.

4 Serum Variants

A. Albumin Variants

Of considerable potential toxicological significance is the extent to which drugs bind with serum proteins. The serum protein binding of drugs is an integral component of the pharmacokinetic appraisal of the drug with the test organism. It is well established that highly and tightly bound substances display more prolonged biological half-lives than those substances exhibiting a lower degree of plasma protein binding. The biomedical implications of such differences in serum protein binding are of such importance in the development of new therapeutic agents that it has assumed a prominent place in drug pharmacology.

In my recent book, *Principles of Animal Extrapolation* (1983), the extent to which interspecies differences exist with respect to serum protein binding of drugs and a limited number of pollutants was evaluated. This assessment was made with the expressed intention of evaluating the predictive utility of animal models vis-à-vis human responses. It was shown that a considerable amount of interspecies variety with respect to serum protein binding of drugs exists depending on the specific substance. Table 4-1 illustrates the extent of such interspecies variations.

Of particular concern is the question of whether such variability in serum protein binding of drugs and/or pollutants may have a defined toxicological significance. Let us consider several examples of the

Table 4-1. How the Human Compares With Common Animal Models in Protein Binding Characteristics

Rat vs. Human	
Of 22 direct comparisons:	13 drugs bound less in the rat
	7 drugs bound more in the rat
	2 drugs bound equally in both species
Result: Human markedly more efficient	
Monkey vs. Human	
Of 14 direct comparisons:	7 drugs bound less in the monkey
	7 drugs bound more in the monkey
Result: Human is similar to monkey	
Dog vs. Human	
Of 22 direct comparisons:	16 drugs bound less in the dog
	6 drugs bound more in the dog
Result: Human markedly more efficient	
Rabbit vs. Human	
Of 16 direct comparisons:	10 drugs bound less in the rabbit
	6 drugs bound more in the rabbit
Result: Human more efficient	
Mouse vs. Human	
Of 8 direct comparisons:	7 drugs bound less in the mouse
	1 drug bound more in the mouse
Result: Human markedly more efficient	

Source: Calabrese (1983).

toxicological significance of such interspecies differences. The markedly lower plasma albumin binding of the anticancer drug campothecitin in mice than humans was thought to have led to an underestimation of the required human dosages of this substance in initial clinical trials. Secondly, the enhanced susceptibility of dogs to copper has been postulated as being caused in part by their diminished ability to bind that substance to plasma albumin (Calabrese, 1983).

Since plasma proteins such as albumin may differ both in quantity and in structure between species, it is not unreasonable to assume that human genetic variants of plasma albumin exist as do variants of many other human proteins such as Hb. Rare electrophoretic variants of albumin have been reported in the sera of Europeans and Americans of European descent by Scheurlen (1955) who coined the condition "bisalbuminemia." This term described the two albumin peaks visible on paper electrophoresis in the cases of the affected individuals. In normal individuals, only a single peak is seen. Initial familial investigations supported the idea that this variant (called albumin B) was inherited as a simple autosomal codominant trait, and that bisalbuminemic persons were actually heterozygotes for a rare gene Al^B and the gene which determines the occurrence of common albumin (Al^A) (Wuhrmann, 1959). Subsequent studies with other families have also detected the presence of albumin B with the mode of inheritance following a simple autosomal pattern (Weitkamp et al., 1967; Melartin, 1967).

Additional albumin variants have also been uncovered. For example, separate families displaying fast-moving albumin variants (albumin Reading, albumin Gent, albumin Maku), and a slow-moving variant distinct from albumin B have been reported (Weitkamp et al., 1967; Weitkamp and Chagnon, 1968). These albumin variants were found to be relatively rare in the European populations where they were detected [i.e., 1 in 4750 (Laurell and Wilehn, 1966); 1 in 1015 (Efremov and Braend, 1964)]. No cases of homozygotes for these rare genes were found.

In contrast to the relatively rare occurrence of the above-mentioned variants, Melartin and Blumberg (1966) reported a fast-moving albumin variant (i.e., albumin Naskapi) that was found in a high frequency in the Naskapi Indians of the Ungava region of Quebec. More specifically, approximately 25% of the 203 individuals in the village were heterozygous for this trait and one was homozygous. Another example of an albumin variant of a relatively high frequency was found in Mexican Indians and identified as Al^{me} (i.e., albumin Mexico) (Melartin et al., 1967). According to Johnson et al. (1969), the albumin Naskapi and the albumin Mexico were found in sufficiently high frequency in several populations and are considered as "polymorphisms" according to Ford's definition. It is speculated that the high frequency of these alleles in the specific populations is the result of differential selective pressure. The rare variants noted above are conversely thought to be maintained by chance mutations.

According to Melartin (1967), there are marked differences in the geographic distribution of albumin variants. Even though variants such as albumin B, albumin Reading, and albumin Gent have been designated as rare inherited traits in European populations, and in American populations of European descent, no common polymorphic variants have been reported in populations other than American Indians and Eskimos. Furthermore, the Naskapi variant is limited principally to northern North American tribes and albumin Mexico to Mexican Indians and Mestizo populations.

TOXICOLOGICAL IMPLICATIONS

It is well established that serum albumin is the principal plasma protein associated with the transport of biological substances including anions, cations, dyes, and drugs. A number of substances such as bilirubin and fatty acids are known to bind nearly exclusively to albumin. Several investigations on inherited albumin variants have demonstrated that the binding capabilities of the two fractions of the albumin in heterozygous sera may differ. For example, Sarcione and Aungst (1961) and Tarnoky and Lestas (1964) reported that only the variant band of the albumin in heterozygous sera (albumin B and albumin Reading, respectively) binds thyroxine. In Reading/A heterozygotes only albumin Reading binds bromphenol blue, while in A/B heterozygotes the dye binding capacity of the albumin fractions were similar.

Melartin (1967) has hypothesized that differential albumin variant binding capacities may affect physiological and pathological processes. However, she stated that no example of such deviations are known but there are several speculative possibilities. For example, if an albumin variant displayed greater bilirubin binding capacity than albumin A, the homozygous and heterozygous individuals would be predicted to be at a selective advantage as compared to persons with only albumin A.

The speculation of Melartin (1967) that albumin variants may differentially bind drugs was subsequently evaluated via *in vitro* experiments by Wilding et al. (1977). These researchers reported that binding experiments with albumin Al^A and Al^{me} from human plasma and isolated albumin indicated small, but statistically significant, reductions in warfarin binding of albumin Al^{me} compared to albumin Al^A. The authors speculated that such differential binding capabilities in Al^{me} variants may be able to effect altered pharmacological responses to warfarin. However, they emphasized that *in vivo* studies would be needed to assess the clinical relevance of these *in vitro* findings.

In summary, genetic variants with respect to albumin exist but in usually very low frequencies (< 1 in 1000) in the European and American Caucasian populations. However, a higher frequency of specific genetic variants has been found in North American and Mexican Indians. Limited *in vitro* experimental evidence suggests that albumin Mexico binds less avidly to the anticoagulant warfarin than albumin A. Since differential binding to albumin may markedly affect the amount of pharmacologically active drug, the occurrence of genetic albumin variants such as albumin Mexico may represent an identifiable subsegment of the population with differential responses to certain drugs.

The toxicological significance of genetic albumin variants remains to be determined. Further investigations along the lines of the Wilding et al. (1977) study for other drugs that bind to a significant extent ($\geq 90\%$) with albumin would be of value as well as using other variants. At present no research efforts have been directed toward differential pollutant binding among human albumin variants. Finally, it is unfortunately very likely that only slow progress will be made in this general area since the frequency of genetic variants is so rare (<1 in 1000) among Caucasians. This low frequency markedly reduces the likelihood of finding affected individuals, thereby reducing follow-up studies. For example, surveying 10,000 individuals and obtaining less than a dozen potential subjects for further study is a difficult endeavor, especially resource wise. Clearly, if marked progress is to be made, then efforts will need to be directed to those subgroups (e.g., Western and Southwestern Indians) (Johnson et al., 1969) within the population with a relatively high incidence of the specific genetic variants for albumin.

REFERENCES

Calabrese, E. J. (1983). *Principles of Animal Extrapolation.* Wiley, New York.

Efremov, G. and Braend, M. (1964). Serum albumin: Polymorphism in man. *Science* **146:** 1679–1680.

Johnson, F. E., Blumberg, B. S., Agarwal, S. S., Melartin, L., and Burch, T. A. (1969). Alloalbuminemia in southwestern U.S. Indians: Polymorphism of albumin Naskapi and albumin Mexico. *Hum. Biol.* **41:** 263–270.

Laurell, C. B. and Nilehn, J. E. (1966). A new type of inherited serum albumin anomaly. *J. Clin. Invest.* **45:** 1935–1945.

Melartin, L. (1967). Albumin polymorphism in man. Studies on albumin variants in North American native populations. *Acta Pathol. Microbiol. Scand. Suppl.* **191:** 9–50.

Melartin, L. and Blumberg, B. S. (1966). Albumin Naskapi, a new variant of serum albumin. *Science* **153:** 1664–1666.

Melartin, L., Blumberg, B. S., and Lisker, R. (1967). Albumin Mexico, a new variant of serum albumin. *Nature* **215:** 1288–1289.

Sarcione, E. J. and Aungst, C. W. (1961). Studies on bisalbuminemia: Binding properties of the two albumins. *Fed. Proc.* **20:** 256.

Scheurlen, P. G. (1955). Ueber Serumweissveranderungen beim Diabetes mellitus. *Klin. Wochenschr.* **33:** 198–205.

Tarnoky, A. L. and Lestas, A. N. (1964). A new type of bisalbuminemia. *Clin. Chim. Acta* **9:** 551–558.

Weitkamp, L. R. and Chagnon, N. A. (1968). Albumin Maku: A new variant of human serum albumin. *Nature* **217:** 759–760.

Weitkamp, L. R., Shreffler, D. C., Robbins, J. L., Drachmann, O., Adner, P. L., Wieme, A. J., Simon, N. M., Cooke, K. B., Sandor, G., Wuhrmann, F., Braend, M., and Tarnoky, A. L. (1967). An electrophoretic comparison of serum albumin variants from nineteen unrelated families. *Acta Genet. Stat. Med.* **17:** 399–405.

Wilding, G., Blumberg, B. S., and Vesell, E. S. (1977). Reduced warfarin binding of albumin variants. *Science* **195:** 991–994.

Wuhrmann, F. (1959). Albumindoppelzacken als vererbbare Bluteiweissanomalie. *Schweiz. Med. Wochenschr.* **89:** 150–152.

B. Pseudocholinesterase Variants

The use of insecticidal agents is an important component of modern farming practice in the United States. One type of insecticidal agent is the chemical insecticide that is designed to disrupt the normal functioning of the insect's nervous system and thus lead to its death.

In order to affect muscle or nerve innervation, the substance acetylcholine (ACH) is first released from the end of a nerve cell. The ACH then migrates to the surface of the adjacent cell in order to affect muscle or nerve innervation. Following the nerve or muscle innervation, ACH is quickly broken down by acetylcholinesterase (ACHase). The action by ACHase is necessary to protect the cell from nerve tremors. Certain insecticides including the organophosphates and carbamates are designed to inhibit ACHase and thereby lead quickly to the insect's death.

In light of the neurological activity of such insecticides as well as the remarkable similarities between human and insect neurons, the numerous reported cases of human insecticide intoxication are not unexpected. Further investigation of insecticidal action (described below) indicates that segments of the human population may be at increased risk to certain anticholinesterase insecticides.

It is important to note that there are two types of cholinesterase: ACHase and pseudocholinesterase (pACHase). As has been previously mentioned, ACHase inactivates ACH produced at the neuromuscular junction during neurotransmission. ACHase is located in numerous tissues. pACHase, although not present in red cells, is found in most tissues including the plasma. The function of pACHase is still not known; however, it has been suggested that the main role may be the hydrolysis of certain cholinesters that inhibit ACHase (Lehmann and Liddell, 1972).

Physiological studies have revealed that although most individuals have the identical pACHase, there are individuals whose pACHase has different metabolic activity than the "normal" or typical type. Biochemical studies indicate that the different pACHase have different molecular structures. The different pACHase types are referred to as variants.

The initial studies of individuals with atypical pACHase variants indicated that they were usually symptomless. However, they exhibited an extreme sensitivity to the muscle relaxant suxamethonium. Suxamethonium has been widely used by anesthetists for the past several decades since the muscular paralysis conveniently lasts for less than 5 minutes following the usual dose. The short action is caused by the hydrolysis of suxamethonium by pACHase (Evans et al., 1952; Bourne et al., 1952). However, later biochemical studies comparing the activity of the normal and atypical pACHase revealed that (1) the normal (typical) enzyme has a greater affinity for substrates than the pACHase from sensitive subjects (Davies et al., 1960) and (2) the pACHase activity of normal plasma is inhibited considerably more strongly by most of the pACHase inhibitors (e.g., dicubaine and fluoride) (see below) (Kalow and Davies, 1958). Results from such studies clearly indicated that the pACHase present in "sensitive" individuals was different from the pACHase of "normal" plasma.

Research involving screening of large numbers of humans (Kalow and Gunn, 1958; Kattamis et al., 1963) has revealed that the development of atypical or variant types of pACHase is under genetic control. The variants are usually determined by two genetic loci (Ch_1 and Ch_2). The variants at the Ch_1 locus are identified by altered sensitivity of the enzyme to inhibitors such as dicubaine, R02-0683, fluoride, and butanol. An additional rare allele (the silent gene) is also found at the Ch_1 locus; its presence results in either the complete absence (type 1) or slight levels (type 2) of cholinesterase activity (Lehmann and Liddell, 1972).

Gene frequencies have been determined for some of the variant genotypes. The most frequent "atypical" homozygous variant (the dibucaine variant) occurs with a frequency of 1 in 2800 healthy Canadians of European ancestry (Kalow and Gunn, 1958) and has been found to be extremely sensitive to R02-0683, the dimethyl carbamate of (2-hydroxy-5-phenylbenzyl) trimethyl ammonium bromide (Lehmann and Liddell, 1972). This is of particular significance in light of the widespread use of carbamate insecticides. It should be pointed out that 3–4% of the Canadian population tested were found to be heterozygote carriers (normal and dicubaine genes together) of intermediate sensitivity (Kalow and Gunn, 1958). Additionally, of the 10 recognized genotypes of pACHase at the CH_1 locus, four are known to display a marked sensitivity to suxamethonium; specifically, the homozygotes for dicubaine sensitivity and the silent gene and heterozygotes for the dicubaine–silent gene combination and the dicubaine–fluoride combination. Their combined frequency in individuals of European ancestry is 1 in 1250 (Szeinberg, 1973). Intermediate sensitivity is known to occur in individuals homozygous for the fluoride-resistant gene and those possessing fluoride-resistant silent genes. The combined frequency of these intermediately sensitive individuals is approximately 1 : 15,000 in Europeans (WHO, 1968, 1973).

In addition to genetic factors, there are several other conditions in which low pACHase activities have been found, for example, liver diseases (McArdle, 1940), malnutrition (Waterlow, 1950), infectious diseases (Hall and Lucas, 1937), and organophosphorous poisoning (Barnes and Davies, 1951).

In terms of public health, the data indicate that individuals have differential sensitivity to the activity of various neuromuscular drugs and insecticidelike chemicals. Differences in sensitivity are directly related to the occurrence of cholinesterase enzyme variants and its diminished ability to inactivate the drug or insecticide analogue. Individuals with such pACHase variants should be considered potentially at high risk to anticholinesterase insecticides (Lehmann and Ryan, 1956). It should be emphasized that not all drugs and insecticidelike compounds act with greater sensitivity in atypical pACHase variants. For example, two organophosphate insecticides, tetraethylpyrophosphonate (TEPP), and diisopropylfluorophosphonate (DFPisofluorophate), do not inhibit differentially (Kalow and Davies, 1958) with pACHase variants and would not cause a higher risk to those with an atypical variant. In a limited study of organophosphorous poisonings, Tabershaw and Cooper (1966) noted no increased susceptibility of heterozygous dicubaine carriers.

SCREENING FOR SUSCEPTIBILITY TO DRUG REACTIONS IN PATIENTS WITH CHOLINESTERASE MUTANTS

The usual technique for phenotype (variant) identification employs the benzoylcholine assay with dibucaine (Kalow and Genest, 1957) and fluoride (Harris and Whittaker, 1961). Also, the use of altered sensitivity of cholinesterase to inhibitors like R02-0683 (Liddell et al., 1963; Lehmann and Liddell, 1972) and butanol (Whittaker, 1968), also improves the identification of some genotypes.

According to Szeinberg (1973), several potential broad screening techniques have been developed that may be utilized for the detection of individuals homozygous or heterozygous for the variant genes including those homozygous for the silent gene. These screening techniques include:

1. Agar diffusion and paper spot tests (Harris and Robson, 1963).
2. Test-tube method (Morrow and Motulsky, 1968).
3. Automated methods (Boutin and Brodeur, 1969).

Simpson and Kalow (1965), using the agar diffusion technique, developed a mass screening program which tested 6500 Brazilians. They reported only 0.1% false negative determinations, but considerably more false positive results. Szeinberg et al. (1972) screened 9500 adult males with a slightly modified paper spot test and a follow-up confirmation by spectrophotometric procedures in cases where a positive determination occurred. They indicated that the spot test was easy to conduct and also inexpensive. However, the verification of positive determinations is more difficult and more costly.

Szeinberg (1973) stated that even though each case of prolonged apnea (caused by the presence of atypical pACHase in association with suxamethonium) causes a strain on the medical staff during operation procedures, no death connected with a single dose of suxamethonium has been reported. He further indicated that in well-equipped operating rooms little danger exists with regard to pACHase variant patients. Furthermore, Goedde et al. (1968) indicated that enzyme replacement has been successfully carried out by the injection of purified pACHase. Consequently, Szeinberg (1973) concluded that screening of every patient prior to suxamethonium does not appear justified.

REFERENCES

Barnes, J. M. and Davies, D. R. (1951). Blood cholinesterase levels in workers exposed to organo-phosphorous insecticides. *Br. Med. J.* **2**: 816.

Bourne, J. G., Collier, H. O., and Somers, G. F. (1952). Succinylcholine (succinoylcholine): Muscle relaxant of short action. *Lancet* **1**: 1225.

Boutin, D. and Brodeur, J. (1969). An automated method for the determination of pseudocholinesterase variants. *Clin. Biochem.* **2**: 187.

Davies, R. O., Marton, A. V., and Kalow, W. (1960). The actions of normal and atypical cholinesterase of human serum upon a series of esters of choline. *Can. J. Biochem. Physiol.* **38**: 545.

Evans, F. T., Gray, P. W. S., Lehmann, H., and Silk, E. (1952). Sensitivity to succinylcholine in relation to serum cholinesterase. *Lancet* **1**: 1229.

Giblett, E. R. (1969). *Genetic Markers in Human Blood*. Blackwell Scientific, Oxford, p. 192.

Goedde, H. W., Altland, K. J., and Schloot, W. (1968). Therapy of prolonged apnea after suxamethonium with purified pseudocholinesterase: New data on kinetics of the hydrolysis of succinyldicholine and succinylmonocholine and further data on N-acetyltransferase-polymorphism. *Ann. N.Y. Acad. Sci.* **151**: 742.

Hall, G. E. and Lucas, C. C. (1937). Choline-esterase activity of normal and pathological serum. *J. Pharmacol. Exp. Ther.* **59**: 34.

Harris, H. and Whittaker, M. (1961). Differential inhibition of human serum cholinesterase with fluoride: Recognition of two phenotypes. *Nature* **191**: 496.

Harris, H. and Robson, E. B. (1963). Screening tests for the "atypical" and "intermediate" serum cholinesterase variant. *Lancet* **ii**: 218.

Kalow, W. and Davies, R. O. (1958). The activity of various esterase inhibitors towards atypical human serum cholinesterase. *Biochem. Pharmacol.* **1**: 183.

Kalow, W. and Genest, K. (1957). A method for the detection of typical forms of human cholinesterase: Determination of dibucaine numbers. *Can. J. Biochem. Physiol.* **35**: 339.

Kalow, W. and Gunn, D. R. (1958). Some statistical data on atypical cholinesterase of human serum. *Ann. Hum. Genet.* **23**: 239.

Kattamis, C., Zannos-Mariolea, L., Franco, A. P., Liddell, J., Lehmann, H., and Davies, D. (1963). The frequency of atypical pseudocholinesterase in British Mediterranean populations. *Nature* **196**: 599.

Lehmann, H. and Liddell, J. (1964). Genetical variants of human serum pseudocholinesterase. *Prog. Med. Genet.* **3**: 75.

Lehmann, H. and Liddell, J. (1972). The cholinesterase variants. In: *The Metabolic Basis of Inherited Disease*, 3rd ed. J. B. Stanbury, J. B. Wyngaarden, and D. S. Fredrickson (eds.). McGraw-Hill, New York, p. 1730.

Lehmann, H. and Ryan, E. (1956). The familial incidence of low pseudocholinesterase level. *Lancet* **2**: 124.

Liddell, J., Lehmann, H., and Davies, D. (1963). Harris and Whittaker's pseudocholinesterase variant with increased resistance to fluoride. *Acta Genet.* **13**: 95.

McArdle, B. (1940). The serum choline esterase in jaundice and diseases of the liver. *Q. J. Med.* **9**: 875–895.

Morrow, A. C. and Motulsky, A. G. (1968). Rapid screening method for the common atypical pseudocholinesterase variant. *J. Lab. Clin. Med.* **71**: 350.

Simpson, N. E. and Kalow, W. (1965). Comparisons of two methods for typing of serum cholinesterase and prevalence of its variants in Brazilian population. *Am. J. Hum. Genet.* **17**: 156.

Table 4-3. Surveys of Chronic Obstructive Pulmonary Disease (COPD)

Location	Reference	Sample Size	Percent (MZ + FZ) COPD	Percent (MZ + FZ) Control	COPD/Control
Oslo, Norway[a]	Fagerhol and Hauge (15)	196	2.8	2.9	0.9
Northern California, US	Kueppers et al. (16)	103	24.3	14.0	1.7
Southern California, US	Mittman et al. (17)	164	11.0	2.9[b]	3.4
Southern California, US	Mittman and Lieberman (18)	240	8.3	2.9[b]	4.2
Massachusetts, US[c]	Talamo et al.	99	4.0	2.8	1.4
West Germany[c]	Kueppers and Donhardt (19)	77	5.2 ⎫	0.5	10.4 (2.6)[d]
		89	9.0 ⎭		18.0 (4.5)[d]
North Carolina, US[c]	Barnett et al. (20)	107	9.4	2.2	4.3
Toronto, Canada[c]	Present (i.e. Cox et al., 1976)	163	4.9	1.9	2.6

Source: Cox et al. (1976).

[a]Not all samples by crossed immunoelectrophoresis.
[b]Controls of Fagerhol (1968).
[c]Series used for calculations in text.
[d]Figures in brackets based on 2% MZ in control subjects.

1976), coke oven jobs (Mittman, 1978), smoking activity (Kueppers et al., 1969), as well as from living in communities with high pollutant levels (Ostrow et al., 1977). These statements are based on observations of much lower than expected prevalences of the PiMZ individual in the above situations. This "survivor effect" may result in a study population of nonrepresentative PiMZ individuals for subsequent studies.* Another nonsupportive study that showed no significant differences in pulmonary function tests between heterozygotes and controls illustrates the reasons why disagreements may arise between different studies. Smoking history was considered positive if the person smoked ⩾1 cigarette/day without any further qualification; characterization of a dusty workplace was only anecdotal; no consideration was given to any potential diurnal variation in pulmonary function nor was blindness of the technicians to the study design mentioned (Buist et al., 1979).

One of the problems with a screening program for SAT heterozygotes is that about 90% of those with that condition will not develop the disease (Mittman, 1978). In other words, the presence of SAT only explains a small percentage of the variance within the population. It is quite clear that other factors are also major contributors to the final outcome.

1. DIFFERENTIAL LEVELS OF PROTEASE

Since leukocytes contain elastolytic enzymes in their lysosomes and intratracheal instillation can lead to the development of experimental emphysema, it has been suggested by a growing number of investigators that the level of leukocyte protease may be the major variable affecting the occurrence of emphysema (Abboud et al., 1979). Thus, Galdston et al. (1973) have suggested that an individual with a PiMZ genotype but with low concentration of leukocyte proteases may be less likely to develop emphysema than individuals with the same PiMZ genotype but with normal concentrations of leukocyte proteases. Consistent with this interpretation, Lieberman (1973) proposed that individuals with normal antitrypsin may be at high risk to develop emphysema if they have a high concentration of leukocyte protease. If the levels of such proteases are under genetic control as suggested by Evans and Bognacki (1979), it may be an important variable to classify.

2. CHEMOTACTIC FACTOR INACTIVATION

It has been emphasized that leukocyte chemotactic factors are important mediators of inflammation and facilitate the marshalling of neutrophils

*Use of such PiMZ individuals would be expected to result in false negative responses.

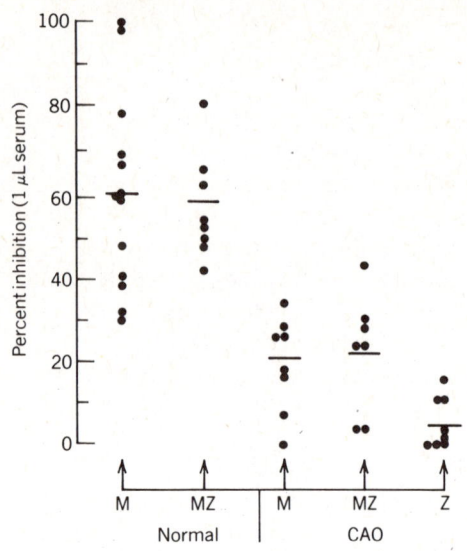

Figure 4-1. Percent inhibition of maximal chemotactic activity of leukocytes by 1.0 μL serum in the various subject groups. The horizontal bar represents the mean value of each group. *Source:* Lam et al. (1980).

in tissue (Snyderman et al., 1975). Chemotactic factor may also induce lysosomal enzyme release from leukocytes (Becker and Showell, 1974). Since an enhanced leukocyte inflammatory response may lead to more leukocyte enzyme mediated tissue damage, regulatory control over leukocyte mediators in serum is thought to be held by chemotactic factor inhibitors (CFI). Lam et al. (1980) have recently shown that the extent of CFI appeared independent of the Pi^m and PiMZ phenotypes (Figure 4-1). These results indicate that possession of both the PiMZ phenotype plus a low concentration of CFI would result in a considerably higher risk of tissue damage.

While there are other factors that may help explain the variance in emphysema prevalence, it is rapidly emerging that emphysema development is a multicausal phenomenon (Kazazian, 1976; Mittman, 1978). The data indicate that the heterozygote PiMZ phenotype is clearly a risk factor but, by itself, is not a major predictor of emphysema. It would also be important to know leukocyte protease levels and CFI serum concentrations. One could calculate the different risk potentials if these three factors segregated independently of each other and had an equal contribution to the risk. Thus, a person with a PiMZ phenotype, high leukocyte protease activity, and a low CFI should be at very high risk while the PiMZ phenotype plus a low leukocyte protease level and high CFI may be at a much lower risk. Such variability may also help to explain why various studies have not shown unanimous agreement.

Chemotactic Factor Inactivation

Table 4-4. Possible Interactions in the MZ Person[a]

White Blood Cell Elastase Activity		Does the Person Smoke?		Emphysema Risk
Normal	+	Yes	−	High
Normal	+	No	−	Intermediate
Decreased	+	Yes	−	Intermediate
Decreased	+	No	−	Low

Source: Kazazian (1978).

[a]Factors such as smoking and leukocyte elastase activity may modulate the emphysema risk to the MZ person.

In conclusion, data from animal and human studies strongly support the hypothesis that an intermediate deficiency of SAT is a contributing factor in the development of emphysema. However, it is now being recognized that emphysema is a multicausal disease and that the heterozygote state, by itself, is not a major predisposing factor in the development of the disease (Table 4-4). However, in combination with other predisposing factors such as high leukocyte protease and low CFI, the heterozygote condition is likely to be at very high risk of developing emphysema. The most reasonable position at the current time would be to try and evaluate the relative contribution of these variables in the development of emphysema. If the above associations prove true and are amenable to widespread screening, then it is likely that screening would involve an assessment of two or three important factors.

Finally, while it has been the intention of this section to assess the role of biological factors (e.g., protease and antiprotease) on the development of emphysema, it is essential to emphasize the dominant role of environmental factors in the etiology of emphysema. For example, recent studies have indicated that cigarette smoke has been found to significantly lower SAT activity in rats after only three puffs (Janoff et al., 1979), while investigations with humans have likewise found that chronic human smokers displayed a nearly twofold decrease in functional activity of SAT as compared to nonsmokers (Gadek et al., 1979). Other experimental studies have also suggested that cadmium (Chowdhury and Louria, 1976, 1977) and ambient ozone at levels approaching the current federal hourly standard of 0.12 ppm may be a contributing factor in the development of emphysema as a result of a marked ability to inhibit SAT activity* (Johnson, 1980).

---

*The hypothesized ozone action on SAT activity was based on *in vitro* inactivation of human plasma SAT activation. It is of value to see the speculative nature of this suggestion as well as the reasoning process used in such an estimate.

3. SCREENING

Several reliable tests have been developed for both the screening of large populations for SAT deficiency and for a more specific test that actually genetically types the individual deficiency. Two of the simplest are that of James et al. (1966) and Smith (1972) in which the level of antitrypsin potential is measured by determining the ability of the serum to counteract three gradations of standard tryptic activity. The procedure is able to determine relative SAT deficiency. However, a more quantitative assay involving a colorimetric determination of serum trypsin inhibition capacity is available (Briscoe et al., 1966). These methods could be inexpensively applied to large populations of any age, but they lack sensitivity and selectivity. For example, the level of antitrypsin in serum does not definitely identify an individual's phenotype. The level of SAT is known to be affected by various stimuli including infection, surgery, or estrogen (e.g., oral contraceptives) administration. Thus, while the Pi^z phenotype rarely displays a level above 30% of normal, serum concentration in individuals who are the heterozygous deficient phenotypes has been found to rise into the low normal range depending on the circumstances. Since a normal person (i.e., Pi^m) is only rarely found to have activity of < 0.80 units, between 0.40 and 0.79 units indicate an intermediate deficiency. However, this assessment does not tell the specific genotype.

Lieberman and Mittman (1972) have developed a simple screening test that detects abnormal SAT variants independent of the actual level of SAT in the serum. The procedure involves the appearance of a "doublering" antibody precipitation pattern by the SAT variants during immunodiffusion on an agarose gel containing specific antibodies against SAT. A double-ring pattern was displayed by 95–100% of the serums from subjects with PiMZ, PiMS, PiSZ, or PiMP genotypes, while only 2.1% of the normal (or Pi^m) showed this response. According to Lieberman and Mittman (1972), if these findings can be applied to large-scale field investigations, then a practical screening procedure would be available. All blood samples would be tested by the double-ring immunodiffusion and only those with the positive result would then be subjected to phenotyping electrophoresis.

REFERENCES

Abboud, R. T., Rushton, J. M., and Grzybowski, W. (1979). Interrelationships between neutrophil elastase, serum alpha$_1$-antitrypsin, lung function and chest radiography in patients with chronic air flow obstruction. *Am. Rev. Respir. Dis.* **120:** 31–40.

Barnett, T. B., Gottovi, D., and Johnson, A. (1975). Protease inhibitors in chronic obstructive pulmonary disease. *Am. Rev. Respir. Dis.* **111:** 587.

References

Becker, E. L. and Showell, H. J. (1974). The ability of chemotactic factors to induce lysosomal enzyme release. *J. Immunol.* **112**: 2055–2063.

Briscoe, W. A., Kueppers, F., Davis, A. L. and Bearn, A. G. (1966). A case of inherited deficiency of serum antitrypsin associated with pulmonary emphysema. *Ann. Rev. Respir. Dis.* **94**: 529–539.

Buist, A. S., Sexton, G. J., Azzam, A-B., and Adams, B. E. (1979). Pulmonary function in heterozygotes for alpha$_1$-antitrypsin deficiency: A case control study. *Am. Rev. Respir. Dis.* **120**: 759–766.

Chan-Yeung, M., Ashley, M. J., Corey, P., and Maledy, H. (1978). Pi phenotypes and the prevalence of chest symptoms and lung function abnormalities in workers employed in dusty industries. *Am. Rev. Respir. Dis.* **117**: 239–245.

Chowdhury, P. and Louria, D. B. (1976). Influence of cadmium and other trace metals on human α_1-antitrypsin: An *in vitro* study. *Science* **191**: 480–481.

Chowdhury, P. and Louria, D. B. (1977). A response to criticism of their original article. *Science* **19**: 57.

Cole, R. B., Nevin, N. C., Blundell, G., Merrett, J. D., McDonald, J. R., and Johnston, W. P. (1976). Relation of alpha$_1$-antitrypsin phenotype to the performance of pulmonary function tests and to the prevalence of respiratory illness in a working population. *Thorax.* **31**: 149–157.

Cox, D. W., Hoeppner, V. H., and Levison, H. (1976). Protease inhibitors in patients with chronic obstructive pulmonary disease: The alpha$_1$-antitrypsin heterozygote controversy. *Am. Rev. Respir. Dis.* **113**: 601–612.

Eriksson, S. (1965). Studies in alpha$_1$-antitrypsin deficiency. *Acta Med. Scand. (Suppl.)* **432**: 1–85.

Eriksson, S., Moestrup, T., and Hagerstrand, I. (1975). Liver, lung and malignant disease in heterozygous (PiMZ) alpha$_1$-antitrypsin deficiency. *Acta Med. Scand.* **198**: 243.

Evans, H. E. and Bognacki, N. (1979). α_1-antitrypsin deficiency and susceptibility to lung disease. *Environ. Health Perspect.* **29**: 57–61.

Gadek, J. E., Fells, G. A., and Crystal, R. G. (1979). Cigarette smoking induces functional antiprotease deficiency in the lower respiratory tract of humans. *Science* **206**: 1315–1316.

Galdston, M., Janoff, A., and Davis, A. L. (1973). Familial variation of leukocyte lysosomal protease and serum alpha$_1$-antitrypsin as determinants in chronic obstructive pulmonary disease. *Am. Rev. Respir. Dis.* **107**: 718–727.

James, K. M., Collins, M. L., and Fudenberg, H. H. (1966). A semiquantitative procedure for estimating serum antitrypsin levels. *J. Lab. Clin. Med.* **67**: 528.

Janoff, A., Carp, H., Lee, D. K., and Drew, R. T. (1979). Cigarette smoke inhalation decreases α_1-antitrypsin activity in rat lung. *Science* **206**: 1313–1314.

Johnson, D. A. (1980). Ozone inactivation of human a$_1$-proteinase inhibition. *Am. Rev. Respir. Dis.* **121**: 1031–1038.

Kazazian, H. H. (1976). A geneticist's view of lung disease. *Am. Rev. Respir. Dis.* **113**: 261–266.

Kueppers, F. and Black, L. F. (1974). Alpha$_1$-antitrypsin and its deficiency. *Am. Rev. Respir. Dis.* **110**: 176–194.

Kueppers, F. and Donhardt, A. (1974). Obstructive lung disease in heterozygotes for alpha$_1$-antitrypsin deficiency. *Ann. Intern. Med.* **80**: 209.

Kueppers, F., Fallat, R., and Larson, R. K. (1969). Obstructive lung disease and alpha$_1$-antitrypsin deficiency gene heterozygosity. *Science* **165**: 899.

Lam, S., Chan-Yeung, M., Abboud, R., and Kreutzer, D. (1980). Interrelationships between serum chemotactic factor inactivator, alpha$_1$-antitrypsin deficiency, and chronic obstructive lung disease. *Am. Rev. Respir. Dis.* **121:** 507–512.

Larsson, C., Eriksson, S., and Dirksen, H. (1977). Smoking and intermediate alpha$_1$-antitrypsin deficiency and lung function in middle-aged men. *Br. Med. J.* **2:** 922.

Laurrell, C. B. and Eriksson, S. (1963). The electrophoretic alpha$_1$-globulin pattern of serum in alpha$_1$-antitrypsin deficiency. *Scand. J. Clin. Lab. Invest.* **15:** 132–140.

Lebowitz, M. D., Knudson, R. J., Morse, R. O., and Armet, D. (1978). Closing volumes and flow volume abnormalities in alpha$_1$-antitrypsin phenotype groups in a community population. *Am. Rev. Respir. Dis.* **117:** 179–181.

Lieberman, J. (1969). Heterozygous and homozygous alpha$_1$-antitrypsin deficiency in patients with pulmonary emphysema. *N. Engl. J. Med.* **281:** 279–284.

Lieberman, J. (1973). Familial variation of leukocyte lysosomal protease and serum alpha$_1$-antitrypsin as determinants in chronic obstructive lung disease. Correspondence. *Am. Rev. Respir. Dis.* **108:** 1019–1020.

Lieberman, J. and Mittman, C. (1972). A new "double-ring" screening test for carriers of 1-antitrypsin variants. Presented at American Thoracic Society Meeting. Kansas City, May 22.

Lieberman, J., Mittman, C., and Schneider, A. S. (1969). Screening for homozygous and heterozygous alpha$_1$-antitrypsin deficiency. *J. Am. Med. Assoc.* **210:** 2055.

McDonough, D. J., Nathan, S. P., Knudson, R. J., and Lebowitz, M. D. (1979). Assessment of alpha$_1$-antitrypsin deficiency heterozygosity as a risk factor in the etiology of emphysema. *J. Clin. Invest.* **63:** 299–309.

Mittman, C. (1978). The PiMZ phenotype: Is it a significant risk factor for the development of chronic obstructive lung disease? *Am. Rev. Respir. Dis.* **118:** 649–652.

Mittman, C., Barbela, T., and Lieberman, J. (1973). Antitrypsin deficiency and abnormal protease inhibitor phenotypes. *Arch. Environ. Health* **27:** 201.

Mittman, C. and Lieberman, J. (1973). Screening for α_1-antitrypsin deficiency. In: *Genetic Polymorphisms and Diseases in Man.* B. Ramot, A. Adam, B. Bonne, R. Goodman, and A. Szeinberg (eds.). Academic, New York, pp. 185–192.

Mittman, C., Lieberman, J., and Rumsfeld, J. (1974). Prevalence of abnormal protease inhibitor phenotypes in patients with chronic obstructive lung disease. *Am. Rev. Respir. Dis.* **109:** 295.

Morse, J. O., Lebowitz, M. D., Knudson, R. J., and Burrows, B. (1977). Relation of protease inhibitor phenotypes to obstructive lung diseases in a community. *N. Engl. J. Med.* **296:** 11909–11914.

Ostrow, D. N., Manfreda, J., Tse, K., Dorman, T., and Cherniack, R. M. (1977). Intermediate alpha$_1$-antitrypsin phenotypes in a moderately polluted northern Ontario community. *Am. Rev. Respir. Dis.* **115:** 146A.

Rawlings, W., Kreiss, P., Levy D., Cohen, G., Menkes, H., Brashers, S., and Peruutt, S. (1976). Clinical, epidemiological and pulmonary function studies in alpha$_1$-antitrypsin-deficient subjects of PiZ type. *Am. Rev. Respir. Dis.* **114:** 945–953.

Richardson, R. H., Guenter, C. A., Welch, M. H., Hyde, R. M., and Hammersonsten, J. F. (1969). The pattern of inheritance of alpha$_1$-antitrypsin. *N. Engl. J. Med.* **287:** 1067.

Shigeoka, J. W., Hall, W. J., Hyde, R. W., Schwartz, R. H., Mudholkar, G. S., Speers, D. M., and Lin, C. C. (1976). The prevalence of alpha$_1$-antitrypsin heterozygotes (PiMZ) in patients with obstructive pulmonary disease. *Am. Rev. Respir. Dis.* **114:** 1077–1083.

Smith, J. M. (1972). A simple screening test for antitrypsin deficiency. *Ann. Rev. Respir. Dis.* **105:** 851.

Stokinger, H. E., Mountain, J. T., and Scheel, L. D. (1968). Pharmacogenetics in the detection of the hypersusceptible worker. *Ann. N.Y. Acad. Sci.* **151:** 968–976.

Synderman, R., Pike, M. C., and Altman, L. C. (1975). Abnormalities of leukocyte chemotaxis in human disease. *Ann. N.Y. Acad. Sci.* **256:** 368–401.

Talamo, R. C., Langley, C. E., Levine, B. W., and Kazemi, H. (1972). Genetic versus quantitative analysis of serum alpha$_1$-antitrypsin. *N. Engl. J. Med.* **287:** 1067.

D. Immunoglobulin A Deficiency

Immunoglobulin A (Ig A) is the major immunoglobulin in saliva, tears, nasal, bronchial and gastrointestinal secretions, and colostrum (Eisen, 1974; Koistinen, 1975). It is believed that the main function of Ig A is to act as a protective agent on secretory surfaces against foreign materials. In an effort to determine what, if any, clinical conditions are associated with Ig A deficiency, numerous studies have been performed on the possible role of Ig A deficiency in diseases in which the activity of the secretory surfaces may be disturbed. Also, among the various gastrointestinal disorders that have been reported in association with deficiency of serum Ig A, malabsorption syndromes and pernicious anemia are the most frequent clinical diseases (Bjernulf et al., 1971; Douglas et al., 1970; Gelzayd et al., 1971; Ginsberg and Mullinax, 1970; Hermans, 1967; Mawhinney and Tomkin, 1971; Penny et al., 1971; Savilahti, 1973; and Wagner and Grossman, 1969).

Several reports have indicated a direct relationship of patients with recurrent upper respiratory tract infections and the presence of Ig A deficiency (Ammann et al., 1970; Chia et al., 1970; Polmar et al., 1972; and Tushan et al., 1971). Infections in the lower respiratory tract together with Ig A deficiency have also been reported (Bachmann, 1965).

Reports of the frequency of selective Ig A deficiency in "normal" healthy populations varies from 1:300 to 1:3000 (Cassidy and Nordby, 1975; Collins-Williams et al., 1972; Frommel et al., 1973; Natvig et al., 1971; Pai et al., 1974; and Vyas et al., 1974). A doctoral dissertation reported an analysis of Ig A levels from all new blood donors (64,588) at the City of Helsinki Blood Donation unit (Koistinen, 1975, 1975a). The results indicated a frequency of Ig A deficiency of 1:396. Other determinations made by the author using different biochemical techniques yielded results of a somewhat less frequent occurrence (e.g., 1:507, 1:661, 1:821). Thus, assuming a frequency of 1:400 in the current U.S. population more than 500,000 people will have the Ig A deficiency (see Table 4-5).

Developmentally, cells with surface Ig A are detected during the fourth to fifth month of gestation. However, synthesis of Ig A by spleen cells does not become active until some weeks after birth. Considerable placental transfer of Ig molecules does occur with respect to Ig G, with only slight quantities of Ig M and practically no Ig A molecules being transferred (Eisen, 1974).

Table 4-5. Frequencies of Ig A Deficiency

Reference	Type of Group Tests	Number of Subjects	Frequency of Ig A deficiency
Koistinen (1975)	Blood donors	64,588	1:396
	Hospital patients	9,920	1:661
Bachmann (1965)	Healthy adults	6,995	1:700
Collins-Williams et al. (1972)	School children (5–19 years)	2,170	1:310
	Hospital patients (0–19 years)	7,261	1:458
Natvig et al. (1971)	Rheumatoid arthritis patients	3,187	1:398
	Blood donors	5,020	1:1,255
Frommel et al. (1973)	Blood donors	15,200	1:3,250

Source: Koistinen, J. (1975). Selective Ig A deficiency in blood donors. *Vox. Sang.* **29**: 1.

Serum levels of the different Ig molecules reach "adult levels" at various times. For example, adult serum levels of Ig M are reached at about 10 months of age, Ig G at about 4 years, Ig A at about 9–10 years, and Ig E at about 10–15 years (Eisen, 1974). Figure 4-2 represents the relationship of immunoglobulin levels and the age of the typical human.

A report in 1972 of children 11 and 12 years of age in the borough of the Bronx in New York City indicated that those who had lived in the heavily air polluted Bronx for 5–12 years had 20% more lower respiratory infections than those who had lived in Riverhead (also part of New York City), which is relatively free of air pollution (Fosburgh, 1974). These findings may provide a partial understanding to the generally recognized greater susceptibility of children to respiratory infection.

Also, it has recently been reported that persons with a total lack of Ig A are more susceptible to influenza than others, including those with partial deficiencies of Ig A (Koistinen, 1975, 1975a, 1975b). However, it should be pointed out that an elevation of serum Ig G and Ig M concentrations has been observed in some of the Ig A deficient blood donors. This may be an adaptational response to compensate for the Ig A deficiency. However, the role of the elevated serum Ig G is not clear. It is possible that these adaptations may contribute to the relatively good health of many of the Ig A deficient blood donors (Koistinen, 1975b; Brandtzaeg, 1971; Brandtzaeg et al., 1968).

SCREENING TESTS

A large number of different blood donors (64,588) were tested for the presence of Ig A deficiency during the years 1971–1974. Ouchterlony's

References

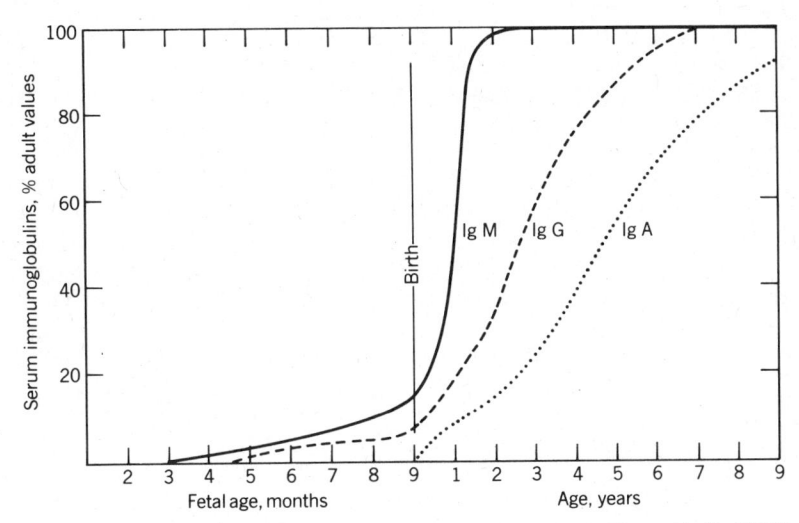

Figure 4-2. Maturation of serum Ig levels in humans. *Source:* Eisen, H. N. (1974). *Immunology.* Harper & Row, Hagerstown, Maryland, p. 505.

double diffusion test was used in the screening of Ig A deficient blood donors with confirmation by a modified racket immunoelectrophoresis technique II. Additionally, hospital patients (9,920) were also examined for Ig A deficiency by the double diffusion technique. Other Ig A deficiency technique measures (inhibition of hemagglutination and double antibody solid phase radioimmunoassay) that are more sensitive than double diffusion may be also used (Koistinen, 1975).

REFERENCES

Ammann, A. J., Roth, J., and Hong, R. (1970). Recurrent sinopulmonary infections, mental retardation and combined Ig A and Ig E deficiency. *J. Pediatr.* **77**: 802–804.

Bachmann, R. (1965). Studies on the serum gamma-A-globulin level. III. The frequency of A-gamma A-globulinemia. *Scand. J. Clin. Lab. Invest.* **17**: 316–320.

Bjernulf, A., Johnsson, S. G. O., and Parrow, A. (1971). Immunoglobulin studies in gastrointestinal dysfunction with special reference to IgA deficiency. *Acta Med. Scand.* **190**: 71–77.

Brandtzaeg, P. (1971). Human secretory immunoglobulins: II. Salivary secretion from individuals with selectively excessive or defective synthesis of serum immunoglobulins. *Clin. Exp. Immunol.* **8**: 69–85.

Brandtzaeg, P., Fjellander, I., and Gjeruldsen, S. T. (1968). Immunoglobulin M: Local synthesis and selective secretion in patients with immunoglobin A deficiency. *Science* **160**: 789–791.

Cassidy, J. T. and Nordby, G. T. (1975). Human serum immunoglobulin concentrations: prevalence of immunoglobulin deficiencies. *J. Allergy Clin. Immunol.* **55**: 35–48.

Chia, B. L., Chew, C. H., and Lee, S. (1970). Recurrent respiratory tract infection due to isolated absence of Ig A. *Med. J. Malaya* **24:** 215–217.

Collins-Williams, C., Kokubu, H. L., Lamenza, C., Nizami, R., Chiu, A. W., Lewis-McKinley, C., Comerford, T. A., and Varga, E. A. (1972). Incidence of isolated deficiency of Ig A in the serum of Canadian children. *Ann. Allergy* **30:** 11–23.

Douglas, S. D., Goldberg, L. S., Fudenberg, H. H., and Goldberg, S. B. (1970). Agammaglobulinaemia and co-existent pernicious anemia. *Clin. Exp. Med.* **127:** 181–187.

Eisen, H. (1974). *Immunology: An Introduction to Molecular and Cellular Principles of the Immune Response.* Harper & Row, New York, pp. 473–474, 504.

Fosburgh, L. (1974). Bad air's effect on young assayed. *New York Times*, Dec. 12.

Frommel, D., Moullec, J., Lambin, P., and Fine, J. M. (1973). Selective serum Ig A deficiency. Frequency among 15,200 French blood donors. *Vox Sang.* **25:** 513–518.

Gelzayd, E. A., McCleery, J. L., Melnyk, C. S., and Kraft, S. C. (1971). Intestinal malabsorption and immunoglobulin deficiency. *Arch. Intern. Med.* **127:** 141–147.

Ginsberg, A. and Mullinax, F. (1970). Pernicious anemia and monoclonal gammopathy in a patient with Ig A deficiency. *Am. J. Med.* **48:** 787–791.

Hermans, P. E. (1967). Nodular lymphoid hyperplasia of the small intestine and hypogammaglobulinemia: Theoretical and practical considerations. *Fed. Proc.* **26:** 1606–1611.

Koistinen, J. (1975). Selective Ig A deficiency in blood donors. *Vox Sang.* **29:** 1.

Koistinen, J. (1975a). Personal communication (August).

Koistinen, J. (1975b). Studies of selective deficiency of serum Ig A and its significance in blood transfusion. Doctoral Dissertation. University of Helsinki.

Mawhinney, H. and Tomkin, G. H. (1971). Gluten enteropathy associated with selective Ig A deficiency. *Lancet* **iii:** 121–124.

Natvig, J. B., Harboe, M., Fausa, O., and Tveit, A. (1971). Family studies in individuals with absence of gamma-A-globulin. *Clin. Exp. Immunol.* **8:** 229–236.

Pai, M. K. R., Davidson, M., Dedritis, I., and Zipursky, A. (1974). Selective Ig A deficiency in Rh-negative women. *Vox Sang.* **27:** 87–91.

Penny, R., Thompson, R. G., Polmar, S. H., and Schultz, R. B. (1971). Pancreatitis, malabsorption and Ig A deficiency in a child with diabetes. *J. Pediatr.* **78:** 512–516.

Polmar, S. H., Waldmann, T. A., Balestra, S. T., Jost, M. C., and Terry, W. D. (1972). Immunoglobulin E in immunologic deficiency disease. I. Relation of Ig E and Ig A to respiratory tract disease in isolated Ig E deficiency, Ig A deficiency and ataxia telangiectasis. *J. Clin. Invest.* **51:** 326–330.

Savilahti, E. (1973). Ig A deficiency in children. Immunoglobulin containing cells in the intestinal mucosa, immunoglobulins in secretions and serum Ig A levels. *Clin. Exp. Immunol.* **13:** 395–406.

Tushan, F. S., Zawadzki, Z. A., Vassallo, C. L., and Robin, E. D. (1971). Serum Ig A deficiency in a man with desquamative interstitial pneumonia. *Am. Rev. Respir. Dis.* **103:** 264–268.

Vyas, G. N., Perkins, H. A., Yang, Y. M., and Basantani, G. K. (1974). Healthy blood donors with selective absence of immunoglobulin A. Prevention of anaphylactic transfusion reaction caused by antibodies to Ig A. (Unpublished manuscript.) Cited in Koistinen, 1975.

Wagner, A. and Grossman, H. D. (1969). Malabsorptionssyndrom Bie Partiellem Immunoglobulinmangel (Ig A defekt). *Dtsche. Med. Wochenschr.* **94:** 2023–2028.

5 Cardiovascular Diseases

A. Hyperlipidemia

1. INTRODUCTION

In his review, "The Origin of Atherosclerosis," Benditt (1976) stated that this disease is "the major cause of death in the U.S. and other Western countries, not because people are living longer but because some unknown aspects of modern life are increasing [its] incidence." The pathologic lesion of atherosclerosis is a plaquelike thickening of the intima and media, the innermost and middle of the three layers of the artery wall. The thickening of the intimal and medial layers results from the accumulation of proliferating smooth muscle cells that are encompassed by interstitial substances such as collagen, elastin, glycosaminoglycans, and fibrin. However, the substance of greatest clinical importance is the cholesterol ester that is laid down in excessive amounts both in the interstitial space surrounding the smooth muscle and within the smooth muscle cells themselves.

The causes of atherosclerosis are complex and delicately interwoven. The etiology, therefore, of this disease may involve genetic predisposition, the contribution of related diseases such as diabetes mellitus, diet, high blood pressure, smoking, and possibly environmental and industrial pollutants. This section of the book is designed to evaluate to what extent genetic factors may predispose individuals to the occurrence of environmentally induced atherosclerosis. In the context of this chapter, environment is defined broadly to include both dietary factors as well as toxic substances found in the general environment and workplace.

2. CHOLESTEROL—HOW IT IS HANDLED

The much-maligned cholesterol is a fatty alcohol that plays an essential role in the normal functioning of the plasma membrane. The cholesterol of cellular membranes is usually synthesized in the liver or absorbed via the intestine and subsequently transported to the respective cells (Goldstein and Brown, 1977). As is well-known, the body does not ultimately tolerate excessive quantities of cholesterol in blood since it becomes deposited in the artery walls leading to atherosclerosis. Magnificent mechanisms have evolved for the transporting of cholesterol through the blood and to the cells while simultaneously avoiding its excessive accumulation. The cholesterol becomes incorporated into soluble protein molecules that are secreted into the blood and taken up by the target cells. In humans, the most prevalent of the cholesterol-carrying molecules is low density lipoprotein (LDL) (Kane, 1977).

It is well recognized that the cellular uptake process requires a surface receptor that binds LDL (Goldstein and Brown, 1977; Brown and Goldstein, 1974). The receptor-bound LDL molecules invaginate and pinch off to form endocytic vesicles that travel within the cytoplasm, ultimately entering the lysosomes (Anderson et al., 1977). At this time, the lipoprotein becomes hydrolyzed by proteases to amino acids, while the cholesterol esters are hydrolyzed by a lysosomal acid lipase (Goldstein et al., 1975). The cholesterol then passes across the lysosomal membrane into the cytoplasm where it may be used in membrane synthesis (Brown et al., 1975).

The released cholesterol drives a feedback control system that regulates the intracellular cholesterol concentration. This is accomplished by its suppression of the activity of 3-hydroxy-3-methylglutaryl CoA reductase (HMG CoA reductase), the rate controlling enzyme in cholesterol biosynthesis. This, in effect, can shut off intracellular cholesterol biosynthesis (Brown et al., 1974). The cholesterol can activate an esterifying enzyme (e.g., acyltransferase) such that presently unneeded cholesterol can be stored (Goldstein et al., 1974). Finally, the cholesterol shuts off the synthesis of LDL receptors, thereby protecting cells against its excessive accumulation (Brown and Goldstein, 1975). According to Goldstein and Brown (1979), "the net effect of this regulatory system is to coordinate the intracellular and extracellular source of cholesterol so as to maintain a constant level of cholesterol within the cell in the force of fluctuations in the external supply of lipoproteins."

Despite the fact that the LDL receptor helps to protect man against developing atherosclerosis, it is obvious that this disease frequently occurs. How this may happen despite the highly evolved regulatory-protective mechanism is as follows: Once saturation of the receptor takes place, LDL is taken up by a phagocytic, receptor-independent process that results in an uncontrolled accumulation of cholesteryl esters in smooth muscle cells. This pathologic sequence occurs in many individuals because the plasma LDL levels exceed that needed by the LDL receptor pathway (Goldstein and Brown, 1977).

3. GENETIC PREDISPOSITION TO HYPERCHOLESTEROLEMIA

Since the process by which cholesterol is initially synthesized, transported, taken up, utilized, or stored is very intricate and complex, there are clearly many opportunities for errors at specific points in the scheme. Figure 5-1 portrays the LDL pathway and indicates sites at which various mutational events have occurred resulting in actual defects in this process.

4. LDL-RECEPTOR DEFECT

The most common of the known defects involves a reduction in the number of LDL receptors, thereby causing LDL particles to reach high levels in the plasma. In addition, this reduced cellular intake of cholesterol results in a failure to suppress intracellular cholesterol synthesis that leads to hypercholesterolemia. Consequently, familial hypercholesterolemia caused by this genetic defect then is characterized in a clinical sense by three principal features: (1) selective elevation of plasma LDL levels, (2) deposition of LDL-derived cholesterol in abnormal sites including tendons and arteries, and (3) inheritance as an autosomal dominant (Goldstein and Brown, 1979).

While the homozygous condition is extremely rare being of the order of only one per million persons of European ancestry, heterozygotes may occur about once in every 200–500 persons (Goldstein et al., 1973; Carter et al., 1971; Leonard et al., 1976). Heterozygotes display moderate hypercholesterolemia from birth but they typically stay asymptomatic until adult life when premature coronary disease becomes evident. Their total plasma cholesterol levels are usually between 300 and 500 mg/100 mL and only infrequently exceed 600 mg/100 mL. Homozygotes are known to be more severely hypercholesterolemic (650 and 1000 mg/100 mL) than the heterozygotes with considerably earlier occurrence of clinical symptoms (Goldstein and Brown, 1979).

The typical heterozygote patient with familial hypercholesterolemia has about 50% normal and 50% abnormal receptors. At the present time, tests for the LDL-receptor defect are not yet available for general labora-

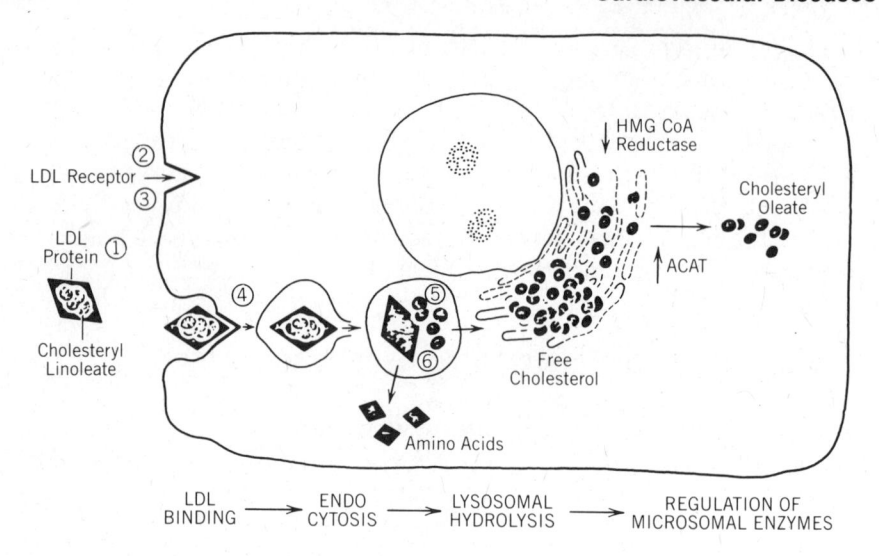

Figure 5-1. Sequential steps in the LDL pathway in cultured human fibroblasts. The numbers indicate the sites at which mutations have been identified: 1. abetalipoproteinemia; 2. familial hypercholesterolemia, receptor negative; 3. familial hypercholesterolemia, receptor defective; 4. familial hypercholesterolemia, internalization defect; 5. Wolman syndrome; and 6. cholesteryl ester storage disease. HMG-CoA reductase denotes 3-hydroxy-3-methylglutaryl coenzyme-A reductase, and ACAT denotes acyl-coenzyme A: cholesterol acyltransferase. *Source:* Goldstein and Brown (1977). Modified from Brown and Goldstein and reproduced with permission of the publisher.

tory use (Motulsky, 1976). Treatment of the heterozygote patient usually includes a reduction of dietary cholesterol and of saturated fat intake (Motulsky, 1976).

While it is known that individuals with a genetic defect in the LDL receptor are at increased risk to further cholesterol arterial deposition caused by consumption of diets high in cholesterol, it is also of concern whether exposure to environmental agents that are known to enhance the occurrence of atherosclerosis may be additional risk factors in such persons acting to accelerate the onset of this disease. Environmental agents known to affect the levels of synthesis of cholesterol, LDL, and/or related lipids are widespread and have been summarized by de Bruin (1976) and listed in Table 5-1. To what extent exposure to such toxic agents in various occupations may affect the risk of heterozygotes for the LDL receptor defect is not known, but is of public health interest. Finally, Motulsky (1976) has summarized the various genetic hyperlipidemias including several of these clinical characteristics and genetic mechanisms of inheritance and frequency in the population, and these are given in Table 5-2. Whether those with other genetic hyperlipidemia diseases are predisposed to damage from environmentally induced lipidemias is also unknown as in the case of the LDL receptor defect.

Table 5-1. Serum Lipoprotein Changes in Animal Poisoning

Compound	Animal Species	Treatment, Condition	Result
Lead acetate	Rabbit	Long-term intravenous administration	Increase of β lipoprotein, with associated hypercholesterolemia
Metallic mercury	Rat	Intraperitoneal application	Marked rise of β/α ratio; hypercholesterolemia
Mercury salt	Rabbit	Prolonged inhalation	Rise of β/α ratio; elevated β fraction
Zinc salt	Rabbit	Single dose, intravenous	Shift from α to β fraction; rise of β/α ratio
Cadmium salt	Rabbit	Acute doses	Increase of β/α ratio
Manganese-chloride	Rabbit	Chronic doses	Shift from β to α lipoprotein; marked fall in β/α ratio
	Rabbit	Low repeated doses intravenous, 3 weeks	
Hydrogensulphide		Acute and subacute poisoning	Rise of α and β fractions; decline of γ-lipoprotein component
Tetrachloromethane	Rat	Single treatment (acute)	Distinct changes in pattern, in nature dependent on time of poisoning ratio
Trichloroethylene	Rat	Vapor exposure, 4 months	Rise in β/α lipoprotein
Dichloromethane	Rat	Repeated exposure	Rise in β fraction, paralleled by fall in α lipoprotein
Carbondisulphide	Rabbit	Inhalation; 150–700 ppm	β/α rise; marked initial increase of β lipoprotein
Carbonmonoxide	Rabbit	Chronic exposure (2500 hr, 100 ppm)	Rise in β/α quotient
Benzene	Dog	Chronic poisoning	Rise in α and β fractions
Styrene	Rabbit	Intragastric administration	Rise in β/α ratio
Amylacetate	Rabbit	Chronic exposure (40 mg/L)	Shift from α to β lipoprotein
Aniline	Rabbit		Rise in α lipoprotein
N-2-fluorenyl-acetamide (2-AAF)	Rat	Prolonged feeding	Initial (preneoplastic) rise of α fraction; later increase of both α and β lipoprotein
Silica	Guinea pig	Intraperitoneal	Rise in β lipoprotein; hyperlipidemia
Quartz powder	Dog	Intrabronchially	
	Rabbit	Intravenously	Rise in β lipoprotein fraction

Source: de Bruin (1976).

Table 5-2. Genetic Hyperlipidemias

Type	Cholesterol Level	Triglyceride Level	Lipoprotein Type	Usual Appearance of Chilled Serum	Xanthomas	Premature Coronary Heart Disease	Genetic Mechanism	Frequency in Population	Frequency in Unselected Population with Mi[a] < 60 Years of Age	Comment
Familial hypercholesterolemia	Elevated	Usually normal	2a (rarely 2b)	Usually clear	Frequent tendon xanthelasma	4+	Autosomal dominant	0.1–0.5%	3–6%	LDL-receptor defects; homozygotes have coronary heart disease in adolescence
Polygenic hypercholesterolemia	Elevated	Normal	2a or 2b	Clear		2+	Polygenic	5%	Increased	
Familial hypertriglyceridemia	Normal	Elevated	4 (rarely 5)	Usually turbid (rarely creamy)		1+	Autosomal dominant	1%	5%	Probably heterogeneous
Familial combined hyperlipidemia	Elevated in 66% of cases	Elevated in 66% of cases	2a, 2b, 4 (rarely 5)	Usually turbid (rarely creamy)	Rare	3+	Autosomal dominant	1.5%	11–20%	About 33% have hypercholesterolemia; About 35% have hypertriglyceridemia; About 33% have both elevated cholesterol and triglyceride
Broad β disease	Elevated	Elevated	3	Turbid (often with creamy layer)	Frequent, tuberous, palmar creases	4+	Autosomal dominant	Rare	1%	Peripheral vascular disease common; requires tests on isolated VLDL
Familial lipoprotein lipase deficiency	Slightly elevated	Very high	1	Creamy layer on top; clear below	Eruptive	0	Autosomal recessive	Very rare		Abdominal pain common

Source: Motulsky (1976).

[a] Mi represents myocardial infarction.

REFERENCES

Anderson, R., Brown, M. S., and Goldstein, J. L. (1977). Role of the coated endocytic muscle in the uptake of receptor-bound low density lipoprotein in human fibroblasts. *Cell* **10**: 351–364.

Benditt, E. (1976). The origin of atherosclerosis. *Sci. Am.* **236**: 74–86.

Berg, K., Hamer, C., Dahlen, G., Frick, M. H., and Krisham, I. (1976). Genetic variation in serum low density lipoprotein and lipid levels in man. *Proc. Natl. Acad. Sci.* **73**(3): 937–940.

Brown, M. S., Dana, S. E., and Goldstein, J. L. (1974). Regulation of 3-hydroxy-3-methylglutaryl coenzyme A reductase activity in cultured human fibroblasts: Comparison of cells from a normal subject and from a patient with homozygous familial hypercholesterolemia. *J. Biol. Chem.* **249**: 789–796.

Brown, M. S., Faust, J. R., and Goldstein, J. L. (1975). Role of the low density lipoprotein receptor in regulating the content of free and esterified cholesterol in human fibroblasts. *J. Clin. Invest.* **55**: 783–793.

Brown, M. S. and Goldstein, J. L. (1974). Familial hypercholesterolemia: Defective binding of lipoproteins to culture fibroblasts associated with impaired regulation of 3-hydroxy-3-methylglutaryl coenzyme A reductase activity. *Proc. Natl. Acad. Sci.* **71**: 788–792.

Brown, M. S. and Goldstein, J. L. (1975). Regulation of the activity of the low density lipoprotein receptor in human fibroblasts. *Cell* **6**: 307–316.

Carter, C. O., Slack, J., and Myant, N. B. (1971). Genetics of hyperlipoproteinaemias. *Lancet* **1**: 400–401.

De Bruin, A. (1976). *Biochemistry and Toxicology of Environmental Agents*. Elsevier, Amsterdam, pp. 565–603.

Goldstein, J. L. and Brown, M. S. (1977). The low-density lipoprotein pathway and its relation to atherosclerosis. *Ann. Rev. Biochem.* **46**: 897–930.

Goldstein, J. L. and Brown, M. S. (1979). The LDL receptor locus and the genetics of familial hypercholesterolemia. *Ann. Rev. Genet.* **13**: 259–289.

Goldstein, J. L., Dana, S. E., and Brown, M. S. (1974). Esterification of low density lipoprotein cholesterol in human fibroblasts and its absence in homozygous familial hypercholesterolemia. *Proc. Natl. Acad. Sci.* **71**: 4288–4292.

Goldstein, J. L., Dana, S. E., Faust, J. R., Beaudet, A. L., and Brown, M. S. (1975). Role of lysosomal and lipase in the metabolism of plasma LDL: Observations in cultured fibroblasts from a patient with cholesteryl ester storage disease. *J. Biol. Chem.* **250**: 8487–8495.

Goldstein, J. L., Schrott, H. C., Hazzard, N. R., Bierman, E. L., and Motulsky, A. G. (1973). Hyperlipidemia in coronary heart disease. II. Genetic analysis of lipid levels in 176 families and delineation of a new inherited disorder, combined hyperlipidemia. *J. Clin. Invest.* **52**: 1544–1568.

Kane, J. P. (1977). Plasma lipoproteins: Structure and metabolism. In: *Lipid Metabolism in Mammals*, Vol. 1. F. Synder (ed.). Plenum, New York, pp. 209–257.

Leonard, J. V., Fosbrooke, A. S., Lloyd, J. K., and Wolff, O. H. (1976). Screening for familial hyper-B-lipoproteinaemia in children in hospital. *Arch. Dis. Child,* **51**: 842–847.

Motulsky, A. G. (1976). Current concepts in genetics. The genetic hyperlipidemias. *N. Engl. J. Med.* **294**: 823–827.

Norum, K. R. (1978). Genetic and non-genetic hyperlipidemia and western diets. In: *Human Genetic Variations in Response to Medical and Environmental Agents: Pharmacogenetics and Ecogenetics*, pp. 125–129. *Human Genet.*, Supp. 1.

B. Homocystinuria

Arteriosclerosis is the most common cause of death in the United States. Considerably large efforts have been made to identify and quantify the risk factors associated with this condition. In essence, many of these research efforts have resulted in a greater understanding of the role of high-cholesterol diets, hypertension, smoking, and a lack of exercise in the pathogenesis of arteriosclerosis. However, over the past decade a novel hypothesis by McCully (1969, 1972) has attempted to explain many of the anomalies of arteriosclerosis. He proposed that genetic and dietary factors that led to significantly increased concentrations of plasma homocystine may lead to the development of arteriosclerosis.

The initial formulation of this theory emerged from the study of several patients with an inherited disease that was characterized in part by the presence of large amounts of homocystine in their urine. Homocystine is not normally found in urine and the disease was called homocystinuria (Carson et al., 1963). Upon the death of these patients at the age of 13 or less, autopsies indicated extensive vascular disease and associated thromboses (McCully, 1969).

These findings led McCully (1969) to hypothesize that elevated concentrations of homocystine in the tissue could cause arterial damage leading to arteriosclerosis. Later experimentation with rabbits that were injected with homocystine developed fibrous intimal and medial plaques with an identical distribution and histopathology as those children with hereditary homocystinuria and in young adults with early fibrous arteriosclerotic plaques (McCully and Ragsdale, 1970).

Homocystinuria, a genetically recessive trait, is a very rare risk factor in arteriosclerosis since it occurs in but one of 80,000 people. While this would generally be considered an insignificant percentage of the population, that is not the case for the heterozygote which is estimated in various populations to have frequencies of 0.5–1.5% (Mudd et al., 1981).

From a biochemical perspective, those with the inherited condition of homocystinuria have a cystathionine synthase deficiency. Homocysteine is normally produced from the amino acid methionine, which is supplied in the protein. Homocysteine is considered a toxic intermediate and is rapidly converted to cystathionine, which is used in other biochemical processes. Individuals with homocystinuria cannot efficiently convert homocysteine to cystathionine thereby leading to increased levels of homocysteine. In blood vessels the homocysteine is thought to affect abnormal cellular proliferation and cause the normally thin cellular linings of the wall to become very porous. Such alterations in cellular functioning are thought to lay the foundation for subsequent arteriosclerotic changes (Gruberg and Raymond, 1979). Heterozygotes possess about half the normal cystathionine synthase activity in liver biopsy specimens (Finkelstein et al., 1964). That heterozygotes may be at increased risk was suggested by several research groups (Sardharwalla et al., 1974;

Kang et al., 1979) who indicated that heterozygotes may accumulate abnormally high plasma concentrations of homocysteine or its derivative (i.e., homocystine). These suggestions were consistent with the finding of an Australian study linking mild homocysteinemia with the pathogenesis of coronary artery disease in patients under 50 years old with angiographic evidence of ischemic heart disease (Wilcken and Wilchen, 1976). The authors reported that after a methionine load, the mixed disulfide of homocysteine and cysteine was detected in 17 of 25 patients but in only 5 of 22 control subjects free of known ischemic disease. In seven of the ischemic heart disease patients but in only one of the controls (Wilcken and Wilcken, 1976), the concentrations of homocysteine-cysteine disulfide were similar to those achieved following a similar loading of obligate heterozygotes for cystathionine synthase deficiency (Sardharwalla et al., 1974).

These findings have led to the suggestion that heterozygotes for cystathionine synthase deficiency may be at a significantly enhanced risk of developing coronary artery disease by age 50. Mudd et al. (1981) hypothesized that "if these heterozygotes are at serious risk of developing early coronary artery disease, an explanation would then be provided for a substantial portion of such disease currently unassociated with known risk factors." An attempt to assess the risk of cardiovascular disease in heterozygotes was published in 1981 by Mudd et al. No statistically significant increases in the incidence of heart attacks or strokes were consistently detected. It should be pointed out that this investigation has a number of methodological limitations that mitigate the strength of any negative findings. No information was included on smoking history, drinking water quality, life-style, occupational history, diet, and other factors of recognized importance in determining risk to cardiovascular disease.

In addition to the medical concern over the possible risk of homocystinuria heterozygotes to developing cardiovascular disease, it has also been speculated that diets inadequate in vitamin B_6 may be a risk factor. This vitamin presumably enhances the conversion of homocysteine to cystathionine. Consequently, it has been shown that monkeys reared on diets deficient in vitamin B_6 displayed a markedly enhanced risk of developing arteriosclerosis. In addition, humans maintained on a low vitamin B_6 diet for 3 weeks began excreting homocysteine (Gruberg and Raymond, 1979). Of further interest is that diets associated with an increased risk of cardiovascular disease such as those with meats and dairy products have considerably higher methionine to vitamin B_6 ratios than from fruits and vegetables (Gruberg and Raymond, 1979).

1. IDENTIFICATION OF THE HETEROZYGOTE

Finkelstein et al. (1964) reported that heterozygotes for homocystinuria display approximately half the normal cystathionine synthase activity in liver biopsy samples as individuals normal for that trait. This differential level of enzyme activity suggested that a functional test could be developed that could effectively identify those in the heterozygous condition.

The general thrust of some of the initial metabolic efforts to identify the carrier state centered on the idea that carriers should metabolize L-methionine more slowly than normal individuals. For example, Brenton et al. (1965) loaded two parents (obligate heterozygotes), affected children (i.e., homozygous recessives), and three normal adult male controls with L-methionine (0.1 g/kg body weight). However, no significant differences between the normal and carrier in the level of sulfur amino acids in the blood and urine occurred. Comparable findings were also presented by Kennedy et al. (1965) and White et al. (1964). Other research by Laster et al. (1965) who even viewed the dose between 0.1 and 0.5 mg of L-methionine/kg body weight could not differentiate between controls and carriers with respect to the levels of inorganic sulfate excretion during the postload period. In contrast to the initial failures to distinguish between normals and carriers, Dunn et al. (1966) showed no overlap between the two groups at 4 hr after an L-methionine load (0.1 g/kg body weight) after a 14-hr fast. Unfortunately, the difference between the two groups was insufficient to make a reliable differentiation for screening/identification purposes. Chase et al. (1967) likewise were able to differentiate between the normal and carrier state in a study that involved the determination of plasma methionine concentration 2 hr after an oral load of L-methionine in four carriers and eight normal controls. The value of these findings though were questioned since the number of heterozygotes was small and the value of the plasma methionine parameter was doubtful. In contrast, Sardharwalla et al. (1974) have reported on a reliable L-methionine loading procedure whereby heterozygotes may be identified.

Their investigation has revealed several parameters which indicate that methionine metabolism is impaired in heterozygotes for homocystinuria, the number of subjects studied being large enough for conclusions to be made with confidence:

1. Plasma concentration of homocystine. The plasma of each heterozygote contained homocystine at 6 and 9 hours after the L-methionine load. No homocystine was detected in the plasma of normal subjects at any time.
2. Plasma concentrations of cysteine-homocysteine disulfide. Concentrations of this amino acid were significantly higher in heterozygotes than normals during the post-load period. Clear distinction of the two groups was observed at 6, 9, and 12 hours post-load.
3. Urine concentration of homocystine and homocystine:cystine ratios. These parameters gave a clear distinction of the two groups in the 16 to 24 hour and 0 to 24 hour post-load periods.
4. Urine cysteine-homocystein disulfide:cystine ratios. There was no overlap of the normal and heterozygote ranges of this parameter in the 8 to 16 hour and 0 to 24 hour post-load periods.

5. Homocysteic acid:cysteic acid ratios in oxidized urine. Differences in these parameters were highly significant and there was no overlap of the two groups at 16 to 24 hours post-load.

Although any single parameter considered in isolation may not provide an unequivocal distinction, the summation of the evidence affords a conclusive distinction between those heterozygotes for homocystinuria and the normal subjects investigated and underlines the value of this method when heterozygote detection is required for genetic counselling in affected families.

2. ENVIRONMENTAL IMPLICATIONS

Are heterozygotes at increased risk to environmental agents that affect the pathogenesis of arteriosclerosis? This question has never been assessed, yet it has been suggested by Gruberg and Raymond (1979) that consumption of contraceptive pills which significantly lower blood levels of vitamin B_6 may enhance vascular pathology. Pollutants such as carbon monoxide and carbon disulfide, which create vitamin B_6 deficiency states, are known to lead to the development of arteriosclerotic changes (deBruin, 1977). It is reasonable to hypothesize that persons heterozygous for homocystinuria may be at increased risk to vascular pathology from exposure to carbon disulfide and carbon monoxide.

REFERENCES

Brenton, D. P., Cusworth, D. C., and Gaull, G. E. (1965). Homocystinuria: Metabolic studies on 3 patients. *J. Pediatr.* **67:** 58.

Carson, N. A., Cusworth, D. S., Dent, C. E., Field, C. M. B., Neil, D. W., and Westall, R. G. (1963). Homocystinuria: A new inborn error of metabolism associated with mental deficiency. *Arch. Dis. Child.* **38:** 425.

Chase, H. P., Goodman, S. I., and O'Brien, D. (1967). Treatment of homocystinuria. *Arch. Dis. Child.* **42:** 514.

deBruin, A. (1977). *Biochemical Toxicology of Environmental Agents*. Elsevier, The Netherlands.

Dunn, H. G., Perry, T. L., and Dolman, C. L. (1966). Homocystinuria, a recently discovered cause of mental defects and cerebrovascular thrombosis. *Neurology* **16:** 407.

Finkelstein, J. D., Mudd, S. H., Irreverre, F., and Laster, L. (1964). Homocystinuria due to cystathionine synthetase deficiency; the mode of inheritance. *Science* **146:** 785.

Fowler, B. and Robbins, A. J. (1972). Methods for the quantitative analysis of sulphur-containing compounds in physiological fluids. *J. Chromatogr.* **72:** 105.

Fowler, B., Sardharwalla, I. B., and Robins, A. J. (1971). The detection of heterozygotes for homocystinuria by oral loading with L-methionine. *Biochem. J.* **122:** 23.

Gruberg, E. R. and Raymond, S. A. (1979). Beyond cholesterol. *Atlantic Monthly*, May, 59–65.

Kang, S-S., Wong, P. K. W., and Becker, N. (1979). Protein-bound homocysteine in normal subjects and in patients with homocysteine. *Pediatr. Res.* **13:** 1141–1143.

Kennedy, C., Shih, V. E., and Rowland, L. P. (1965). Homocystinuria: A report of two siblings. *Pediatrics* **36:** 736.

Laster, L., Mudd, S. H., Finkelstein, J. D., and Irreverre, F. (1965). Homocystinuria due to cystathionine synthase deficiency; the metabolism of L-methionine. *J. Clin. Invest.* **44:** 1708.

McCully, K. S. (1969). Vascular pathology of homocysteinemia: Implications for the pathogenesis of arteriosclerosis. *Am. J. Pathol.* **56:** 111–122.

McCully, K. S. (1972). Homocysteinemia and arteriosclerosis. *Am. Heart J.* **83**(4): 571–573.

McCully, K. S. and Ragsdale, B. D. (1970). Production of arteriosclerosis by homocysteinemia. *Am. J. Pathol.* **61**(1): 1–8.

Mudd, S. H., Havlik, R., Levy, H. L., McKusick, V. A., and Feinleib, M. (1981). A study of cardiovascular risk in heterozygotes for homocystinuria. *Am. J. Hum. Genet.* **33:** 883–893.

Sardharwalla, I. B., Fowler, B., Robbins, A. J., and Komrower, G. M. (1974). Detection of heterozygotes for homocystinuria. *Arch. Dis. Child.* **49:** 554–559.

White, H. H., Araki, S., Thompson, H. L., Rowland, L. P., and Cowen, D. (1964). Homocystinuria. *Trans. Am. Neurol. Assoc.* **89:** 24.

Wilcken, D. E. L. and Wilchen, B. (1976). The pathogenesis of coronary disease—a possible role for methionine metabolism. *J. Clin. Invest.* **58:** 1079–1082.

C. Animal Models for Cardiovascular Disease

The preponderance of literature on the causes of cardiovascular diseases in humans has dealt with dietary, genetic, and socioeconomic factors. More recently there has emerged the recognition that exposure to environmental contaminants may also contribute to the occurrence and severity of cardiovascular diseases as well. For example, environmental exposures to arsenic, barium, cadmium, lead, and sodium as well as noise have been either suspected or found to be associated with the occurrence of various aspects of cardiovascular disease based on investigations with animal models and/or humans.

Since cardiovascular diseases are so prevalent in western society accounting for approximately 40% of yearly human deaths and extensive disability, it is essential to develop predictive models that can provide a better understanding of the causes and mechanisms of the various types of cardiovascular diseases. This need has of course been long recognized and extensive efforts have been made to develop animal models that could provide predictive insights into the pathogenesis of human cardiovascular disease.

1. ATHEROSCLEROSIS

When discussing animal extrapolation with respect to human cardiovascular disease, it is necessary to first differentiate between the various types of cardiovascular disease of importance to humans such as athero-

Atherosclerosis

sclerosis, hypertension, and myocardial injury. The pathogenesis of these diseases in man occurs over many years and involves a multistage process. Consequently, the complexity of the pathological process accompanied by the multiple contributory causal agents poses great difficulties in attempts to provide qualitative and especially quantitative extrapolation. To enhance the accuracy of the extrapolation process, investigators have formulated the desired characteristics of the ideal animal model with respect to the specific type of cardiovascular disease. An example of this type of guidance was provided by McCauley and Bull (1980) who defined some of the qualities of an ideal laboratory model for the study of atherosclerosis:

1. It should have dietary characteristics similar to humans, since diet and metabolism of diet in humans has been established as an arteriosclerotic risk factor.
2. It should develop arteriosclerosis in vital arteries of the heart and brain as well as in the aorta.
3. It should develop complications of atherosclerosis such as stroke, hypertension, and myocardial infarction.
4. It should be small, cheap, and easy to handle and house for studies involving large numbers of animals and it should be large enough to facilitate morphologic observation of arteries and simple surgery.
5. It should have functional, structural, and biochemical similarities in arteries and heart to those in humans.

These authors were quick to point out that the idealized animal model is really only in the mind of the investigator, for no one animal model fulfills all of the above criteria. Therefore, each investigator will be challenged to select the most appropriate model or complementary models with respect to the specific outcomes to be measured.

Historically, the use of animal models for research on atherosclerosis involved either rabbits or chickens. Rabbits were essentially first employed by Anitschkow in 1914 who induced atherosclerosis by feeding fats and egg yolks. As a result of consumption of this diet, yellow-appearing lesions in the aortic intima developed that were comparable to those found in humans but distributed slightly differently, being more in the proximal aorta than the distal. Since this time many investigators have used this dietary approach to induce atherosclerosis in the rabbit. However, Clarkson (1972) has concluded that despite the apparent similarity between the appearance of rabbit and human atherosclerosis, the total syndrome in the rabbit only superficially resembles human atherosclerosis with the most impressive difference being the marked hypercholesterolemia (about 10–40 times greater than normal) and the presence of lipid in almost all of the organs and tissues of the rabbit's body (see Prior et al., 1961), thereby appearing as that of a fat storage disease. As for the chicken, it develops spontaneous aortic atherosclerosis

(Dauber, 1944; Paterson et al., 1948; Katz and Stamler, 1953). This model has proved helpful in predicting the capacity of estrogens to reduce coronary artery disease, but according to Clarkson (1972), this model has not proved to be of particular predictive value in studies dealing with the effects of ACTH, cortisone, and in pancreatectomized cholesterol fed chickens.

Despite their extensive utilization in numerous and diverse types of toxicological studies, mice and rats are not extensively used in the study of atherosclerosis, the prime difficulty with the rat being that most strains do not develop this disease naturally and in fact are quite resistant to its development (Clarkson, 1972). Despite their relatively minor use in the study of arteriosclerosis, the development of new rodent strains with more appropriate predictive capabilities is a distinct possibility (see Adel et al., 1969). For example, genetic mutant models such as the Zuker-Koletsky rat (Zucker and Zucker, 1961; Koletsky, 1973) for the study of hypercholesterolemia and atherosclerosis are now available. Another model that has been used widely in toxicological assays but has not been of notable value here is the dog. This model does not usually develop atherosclerosis unless thyroid activity is depressed and very large quantities of cholesterol are added to the diet (Schenk et al., 1965).

In contrast to dogs, the domestic cat has been a useful model in studying atherosclerosis because of the anatomic location of the coronary artery lesion induced by dietary manipulation with relevance to the human disease (Clarkson, 1972; Manning and Clarkson, 1970).

The pig has been used extensively as a model in the study of atherosclerosis principally because it develops this disease naturally and it is susceptible to dietary modification (French et al., 1963; Rowsell et al., 1965). Furthermore, the dietary habits of the pig resemble those of humans more than any other species. Other similarities to man include the structure of the aorta and coronary arteries as well as the development of atherosclerosis beginning early in life and requiring years to develop (e.g., 3–12 years in the swine). In addition, Moreland et al. (1963) have reported that, like men, when pigs are given diets high in fat and cholesterol, alterations in serum lipid and blood-clotting patterns occur along with an increase in the severity of atherosclerotic lesions.

Several avian models such as the turkey and especially the pigeon have been widely used in the study of atherosclerosis. The pigeon has been widely used in the area of cardiovascular research since it develops a naturally occurring atherosclerosis similar to man's and because the pattern of pathogenesis of this disease varies according to the specific breed of pigeon. Of particular importance with respect to the pigeon model is that it develops lesions in predictable sections of the aorta after short-term cholesterol feeding. For these reasons, the pigeon is quite useful in assessing the initial metabolic alterations associated with the development of lesions. Pigeons are also of further value as a model since only small quantities of cholesterol are needed to increase the level of serum lipids and aggravate arterial lesions (Clarkson and Lofland, 1961). In addition, myocardial infarction is also known to occur in pigeons as a result of obstructive coronary lesions (Prichard et al., 1963). Of particular comparable value is the finding of resistant substrains of pigeon that may be helpful in attempting to elucidate the reasons for differential susceptibility in men to atherosclerosis (McCauley and Bull, 1980). Finally, Reevis et al. (1980) have attempted to utilize the pigeon model in the study of heavy metals (i.e., cadmium and lead) induced atherosclerotic lesions.

Considerable evidence has accumulated over the past two decades on the pathogenesis of atherosclerosis in nonhuman primates (Clarkson et al., 1976; Lugenbuhl et al., 1977; Wissler and Vesselinovitch, 1977; Vesselinovitch, 1979). Because of their phylogenetic proximity to man, it has been widely thought that the findings obtained from these models would be of greater relevance to the human situation than the results of an evolutionarily more distant species. Experimentally induced atherosclerosis has been produced in a wide range of both Old and New World monkeys via alteration of the diet (Clarkson, 1972; Gresham, 1973).

In general, the nonhuman primates have been found to vary considerably in their sensitivity to the development of atherosclerotic lesions as well as the location of lesions within the vasculature. However, this variability can be be used to a great advantage if properly directed toward specific aspects of the human disease process. For example:

1. Chimpanzees develop diet induced lesions that are very much like those of human atherosclerosis especially in the cerebral arteries (Clarkson, 1972).

2. While baboons did not usually develop coronary artery atherosclerosis and are not likely to be an important model for studying atherosclerosis lesions, they do have an experimental value for the study of whole-body cholesterol metabolism because they absorb cholesterol with a comparable degree of efficiency as man and excrete it in the form of neutral steroids as has been found in some men (Eggen and Strong, 1969).

3. Rhesus monkeys have been extensively used because they are very sensitive to the development of extensive atherosclerosis following consumption of diets high in cholesterol. According to Clarkson (1972), the "extent and complexity of diet-induced atherosclerosis in this species exceeds that of any other species . . . we have studied; additionally, it is one of few species in which myocardial infarction occurs with reasonable frequency. They not only develop extensive aortic atherosclerosis, but extensive coronary, cerebral, renal, and femoral artery atherosclerosis as well. The anatomic

distribution of the coronary artery lesions in these amounts is another advantage; the lesions are most significant in the epicardially distributed proximal branches of the coronary arteries."

This brief summary of but three types of nonhuman primates reveals the flexibility that investigators have in studying the causes and pathogenesis of atherosclerosis. Thus, nonhuman primate models are now available to study the occurrence of specific facets of the disease including cholesterol absorption, cholesterol excretion, aortic atherosclerosis, coronary artery atherosclerosis, cerebral artery atherosclerosis, or peripheral arterial disease.

McCauley and Bull (1980) have recently offered a suggestion of how these various types of models could be used as part of an overall methodology for assessing the influence of environmental agents on the pathogenesis of atherosclerosis. These authors state that any such methodology needs to be cost-effective thereby taking into account time and spacing factors. Accordingly, they proposed that

a reasonable approach would be to utilize smaller species in which the atherosclerosis process might be accelerated. For this purpose studies in the pigeon, chicken, or even the rabbit might be preferred. However, such systems may give rise to artifactual results for various reasons some of which have been mentioned above. Consequently, confirmatory experiments in selected incidences should be conducted in mammalian species which develop spontaneous atherosclerosis without artifactual manipulations. For these purposes the swine might be preferred followed by nonhuman primates.

Finally, Table 5-3 offers a summarization of the strengths and weaknesses of commonly used animal models for the study of arteriosclerosis.

2. HYPERTENSION

According to McCauley and Bull (1980) the best animal model to study hypertension is the young rat. Among the compelling reasons for using the rat is that it is very resistant to spontaneous arteriosclerosis. Thus, use of this model would permit the investigator to differentiate those substances that are purely hypertensive. These authors stressed the complementarity of different models. For instance, the rat could be used along with a spontaneously atherosclerotic species like the pigeon, as a way of differentiating primary hypertensive agents from chemicals causing hypertension secondary to atherosclerosis.

A wide selection of rat models exist that are used to study hypertension. McCauley and Bull (1980) suggest that these different hypertensive rat models may be used to simulate sensitive human population sub-

Table 5-3. Some Species Used in Laboratory Studies of Atherosclerosis

Human	*Advantages:* Obvious *Disadvantages:* Severely limited *in vivo* experimentation; difficulty of acquiring tissue for *in vitro* analysis
Rabbit	*Advantages:* Well characterized physiology and pathology, ease of handling, low expense *Disadvantages:* Low rate of and usual atypical spontaneous lesion develop; herbivorous; manipulations needed to develop atherosclerotic plaque differ from other species; hyperlipidemic models demonstrate unusual composition of lipoproteins
Rat	*Advantages:* Small, well characterized; many different genetic hypercholesterolemic strains *Disadvantages:* Usual lab strains do not develop spontaneous atherosclerosis; difficult to induce by diet; major structural differences in aorta; (1) no vasa vasorum; (2) extremely thin intimal layer; extremely hypolipidemic; ratio of cholesterol in LDL to HDL very low.
Chicken	*Advantages:* Spontaneous atherosclerosis; similar serum cholesterol levels to humans; easily induceable lesion and hyperlipidemia; genetic hyperlipidemic models available; atherosclerosis may develop into arterial stenosis; omnivorous *Disadvantages:* Limited distribution of lesions; considerable small artery involvement; avian
Pigeon	*Advantages:* Spontaneous atherosclerosis resembling that of humans in both morphology and distribution; atherosclerotic resistant and prone breeds; genetic hypercholesteremic variants; atherosclerotic plaques do eventually obstruct lumen; myocardial infarcts have been reported *Disadvantages:* Foam cell derived from smooth muscle only; herbivorous; avian
Swine	*Advantages:* Spontaneous atherosclerosis; omnivorous with similar aorta and coronary artery structure; spontaneous atherosclerosis of similar type and distribution as in humans; atherosclerotic resistant genetic variant (von Willebrands disease) known *Disadvantages:* Large size and difficulty of handling; myocardial infarction and sudden death only reported after severe physical coronary artery injury; 6–12 years necessary to develop atherosclerotic lesions but may be sped up with diet or diet plus artery injury
Nonhuman primates	*Advantages:* Omnivorous in most cases; many species and genetic variants available with a variety of characteristics pertinent to atherosclerosis; closest species phylogenetically to humans *Disadvantages:* Difficult to handle; prone to stress related complications; may acquire or transmit human diseases; difficult to procure and house in large numbers necessary for large studies

Source: McCauley and Bull (1980).

groups. For example, (1) the hypertension of the Wistar Kyoto spontaneously hypertensive rat may simulate the occurrence of essential hypertension in humans (Yamori and Okamoto, 1973; Yamori, 1977), (2) the Dahl rat displays an enhanced susceptibility to sodium induced hypertension (Dahl et al., 1962), and (3) another rat (spontaneous, renal hypertensive rat, Milan strain) developed nonrenin mediated renal hypertension (Bianchi et al., 1974, 1975). Other rat models of hypertension exist such as the adrenal hypertensive rat (Selye et al., 1943) and the glomerulonephritis model (McCauley and Bull, 1980). This great diversity of models offers considerable experimental advantages since the causes of hypertension are so variable that any one strain would not be very likely to simulate adequately more than one type of hypertension pathology.

REFERENCES

Adel, H. N., Deming, Q. B., and Brun, L. (1969). Genetic hypercholesterolemia in rats. *Circulation Suppl. 3* **39:** 1

Anitschkow, N. (1914). Über die atherosclerose der aorta beim kaninchen und über derin entschungsbedingungen. *Beitr. Pathol. Anat. Allg. Pathol.* **59:** 308–48.

Bianchi, G., Baer, P. G., Fox, V., Duzzi, L., Pagetti, D., and Giovannetti, A. M. (1975). Changes in renin, water balance and sodium balance during development of high blood pressure in genetically hypertensive rats. *Cir. Res. Suppl. 1,* **36** and **37,** I-153–I-161.

Bianchi, G., Fox, V., Difrancesco, G. F., Giovannetti, A. M., and Pagetti, D. (1974). Blood pressure changes produced by kidney cross-transplantation between spontaneously hypertensive rats and normotensive rats. *Clin. Sci. Mol. Med.* **47:** 435–448.

Clarkson, T. B. (1972). Animal models of arteriosclerosis. *Ad. Vet. Sci. Comp. Med.* **16:** 151–173.

Clarkson, T. B. and Lofland, H. B. (1961). Effects of cholesterol–fat diets on pigeons susceptible and resistant to atherosclerosis. *Cir. Res.* **9:** 106–109.

Clarkson, T. B., Prichard, R. W., Bullock, B. C., St. Clair, R. W., Lehner, N. D. M., Jones, D. C., Wagner, W. B., and Rudel, L. L. (1976). Pathogenesis of atherosclerosis; some advances from using animal models. *Exp. Mol. Pathol.* **24:** 264–286.

Clarkson, T. B., Prichard, R. W., Wetsky, M. G., and Lofland, H. B. (1959). Atherosclerosis in pigeons. Its spontaneous occurrence and resemblance in human atherosclerosis. *Arch. Pathol.* **68:** 143–147.

Dahl, L. K., Heine, M., and Tassinari, C. (1962). Role of genetic factors in susceptibility to experimental hypertension due to chronic excess salt ingestion. *Nature* **194:** 480–482.

Dauber, D. V. (1944). Spontaneous arteriosclerosis in chickens. *Arch. Pathol.* **88:** 46.

Day, C. E., Phillips, W. A., and Schun, P. E. (1979). Animal models for an integrated approach to the pharmacologic control of atherosclerosis. *Artery* **5:** 90–109.

DePalma, R. G., Insull, W., Bellon, E. M., Roth, W. T., and Robinson, A. V. (1972). Animal models for the study of progression and regression of atherosclerosis. *Surgery* **72:** 268–278.

Dieterich, R. A. and Preston, D. J. (1979). Atherosclerosis in lemmings and voles fed a high fat, high cholesterol diet. *Atherosclerosis* **33:** 181–189.

Eggen, D. A. and Strong, J. P. (1969). Cholesterol metabolism in baboon, rhesus, and squirrel monkeys on an atherogenic diet. *Cir. Res.* **40:** III–7.

References

Florentein, R. A., Nam, S. C., Daoud, A. S., Jones, R., Scott, R. F., Morrison, F. S., Kim, D. N., Lee, K. T., Thomas, W. A., Dodds, W. J., and Miller, K. D. (1968). Dietary-induced atherosclerosis in miniature swine. *Exp. Mol. Pathol.* **8:** 263–301.

French, J. E., Jennings, M. A., Poole, J. C. F., Robertson, D. S., and Florey, H. (1963). Intimal changes in the arteries of aging pigs. *Proc. R. Soc. London Ser. B* **158:** 24–42.

Fuster, V., Bowie, E. J. W., and Brown, A. L. (1975). Spontaneous arterial lesions in normal pigs and pigs with Von Willebrands Disease. In: *Atherosclerosis, Metabolic, Morphologic and Clinical Aspects.* G. W. Manning and M. D. Haust (eds.). Plenum, New York, pp. 315–317.

Gresham, G. A. (1973). The use of primates in cardiovascular research. In: *Nonhuman Primates and Medical Research.* G. H. Bourne (ed.). Academic, New York, pp. 225–244.

Koletsky, S. (1973). Obese spontaneously hypertensive rats—a model for the study of atherosclerosis. *Exp. Mol. Pathol.* **19:** 53–60.

Katz, L. N. and Stamler, J. (1953). *Experimental Atherosclerosis.* Thomas, Springfield, Illinois, p. 375.

Lugenbuhl, H., Rossi, G. L., Ratcliffe, H. L., and Muller, R. (1977). Comparative atherosclerosis. In: *Advances in Veterinary Science and Comparative Medicine.* Academic, New York, pp. 421–448.

Manning, P. J. and Clarkson, T. B. (1970). Diet-induced atherosclerosis of the cat. *Arch. Pathol.* **89:** 271–278.

McCauley, P. T. and Bull, R. J. (1980). Experimental approaches to evaluating the role of environmental factors in the development of cardiovascular disease. *J. Environ. Pathol. Toxicol.* **4:** 27–50.

Moreland, A. F., Clarkson, T. B., and Lofland, H. B. (1963). Atherosclerosis in "miniature" swine. *Arch. Pathol.* **76:** 209–215.

Paterson, J. C., Slinger, S. J., and Gartley, K. M. (1948). Experimental coronary sclerosis. *Arch. Pathol.* **45:** 306.

Prichard, R. W., Clarkson, T. B., and Lofland, H. B. (1963). Myocardial infarcts in pigeons. *Am. J. Pathol.* **43:** 651–659.

Prior, J. T., Kurtz, D. M., and Ziegler, D. D. (1961). The hypercholesterolemic rabbit. *Arch. Pathol.* **71:** 672–684.

Reevis, N. W., Major, T. C., and Horton, C. Y. (1980). The effects of calcium, magnesium, lead, or cadmium on lipoprotein metabolism and atherosclerosis in the pigeon. *J. Environ. Pathol. Toxicol.* **4:** 293–304.

Rowsell, H. C., Mustard, J. E., and Downie, H. G. (1965). Experimental atherosclerosis in swine. *Ann. N.Y. Acad. Sci.* **127:** 743–762.

Schenk, E. A., Penn, L., and Schwartz, S. (1965). Experimental atherosclerosis in the dog. *Arch. Pathol.* **80:** 102–109.

Selye, H., Hall, C. E., and Rowley, E. M. (1943). Malignant hypertension produced by treatment with desoxycorticosterone acetate and sodium chloride. *Can. Med. Assoc. J.* **49:** 88–92.

Vesselinovitch, D. (1979). Animal models of atherosclerosis, their contribution and pitfalls. *Artery* **5:** 193–208.

Vesselinovitch, D. and Wissler, R. W. (1975). Comparison of primates and rabbit as animal models in experimental atherosclerosis. In: *Atherosclerosis, Metabolic, Morphologic and Clinical Aspects.* G. W. Manning and M. D. Hurst (eds.). Plenum, New York, pp. 614–622.

Wissler, R. W. and Vesselinovitch, D. (1977). Atherosclerosis in non-human primates. In: *Advances in Veterinary Science and Comparative Medicine, Vol. 21.* C. A. Bradley, C. E. Cornelius, and C. F. Simpson (eds.). Academic, New York, pp. 351–420.

sues, is known to catalyze the first step in the metabolism of benzo(a)pyrene (BP) and other PAHs. Being an inducible enzyme, AHH displays increased activity following administration of a number of agents such as PAHs, various drugs, steroids, and insecticides among others (Conney, 1967).

A large number of studies have indicated that the mutagenic and carcinogenic action of BP is contingent on its enzymatic conversion to active forms (Gelboin et al., 1969, 1972; Grover et al., 1971, 1972; Daly et al., 1972). AHH is thought to play a key role in the bioactivation of PAHs such as BP by metabolizing them to epoxides (Figure 6-1). The epoxide may bind to DNA and other macromolecules or it may be converted to other products such as phenols and dihydrodiols or be conjugated with GSH. Sophisticated studies have revealed diol-epoxides as the principal mutagenic metabolite of PAHs such as 7,8-diol-9,10 epoxide as in the case with BP. Diol epoxide binding to DNA and other macromolecules is thought to be the initial cause of malignant transformation in cells. Consequently, PAH metabolism via AHH results both in detoxification for excretion and in activation to more highly mutagenic and carcinogenic intermediates.

A number of investigators have shown that there is considerable variation in the extent of AHH induction in cultured leukocytes and that it is tightly under genetic control (Kellermann et al., 1973). According to Kellermann et al. (1973a), the normal Caucasian population can be divided into three distinct groups with low, intermediate, and high degrees of inducibility. This variation was hypothesized to result from two alleles at a single locus with the three groups representing homozygous low and high alleles and the intermediate heterozygote. The distribution of the gene is thought to follow the Hardy-Weinberg equilibrium with the gene frequencies of the low and high alleles within the study population being 0.717 and 0.283, respectively. This translates into phenotypic frequencies of 53%, 37%, and 10%.

Since the inducibility of AHH was found to be under genetic control and exhibited wide variation in the population, Kellermann et al. (1973a) sought to evaluate whether it may help to explain differential susceptibility to lung cancer presumably caused by PAHs that may have been bioactivated to carcinogenic products. They selected 50 patients with bronchogenic carcinoma (i.e., 58% squamous cell carcinoma, 24% adenocarcinoma, 18% oat cell carcinoma) as well as 85 healthy controls and 46 tumor controls,[*] and compared their capacity to have 3-methylcholanthrene (MCA) induce the synthesis of AHH in leukocytes. They reported that the healthy and tumor control groups did not differ between each other and displayed phenotypic frequencies nearly identical with that

---

[*]Tumor controls were selected to show that low or high inducibility was not simply the result of having a tumor.

AHH Inducibility and Lung Cancer

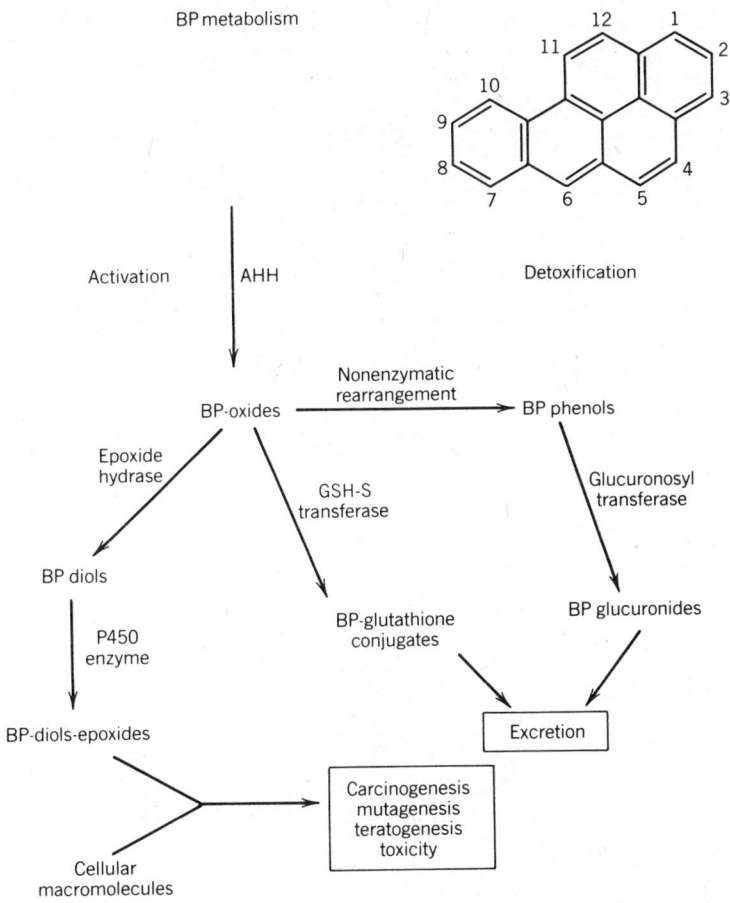

Figure 6-1. Metabolism of BP leading to activation and detoxification. *Source:* Arnott et al. (1979).

noted above. However, the lung cancer patients displayed a marked shift to the right in that only 4% were low inducers, 66% were moderate, and 30% were high (Table 6-1). The authors concluded that the risk of lung cancer for the groups with intermediate and high inducibility was 16 and 36 times, respectively, greater than that of the low-inducibility group.

As striking as this initial report of Kellermann et al. (1973a) was, their study displayed several important limitations that markedly reduced the strength of proposed conclusions. For example, in the selection of control groups there was no matching for sex, age, smoking history, or other relevant variables. No indication was provided by the researchers as to whether measurements of AHH were made blindly (i.e., without a knowledge of which study group the sample was derived) thereby leading to a potential for observational bias (Epidemiology Resources, 1982). This is an inherent weakness in all subsequently cited studies with AHH. In

Table 6-1. Comparison of Healthy Control Group, Tumor Control Group, and Lung-Cancer Group for AHH Inducibility

Group	Number in Group	AHH Inducibility (%) AA[b]	AB[c]	BB[d]	Gene Frequencies[a] AHH	AHH
Healthy control	85	44.7	45.9	9.4	0.676	0.324
Tumor control	46	43.5	45.6	10.9	0.663	0.337
Lung cancer	50	4.0	66.0	30.0	0.370	0.630

Source: Kellermann et al. (1973).
[a] Superscripts refer to the two alleles.
[b] Low.
[c] Intermediate.
[d] High.

addition, approximately ¼ of the 50 cases had adenocarcinomas, which are not very closely associated with smoking and therefore presumably with AHH formed ultimate carcinogens.

In the intervening years, a variety of research teams have sought to replicate and extend the initial findings of Kellermann et al. (1973a) because of their massive public health implications. On the positive side, seven studies have supported the initial findings (Guirgis et al., 1976; Trell et al., 1976; Emery et al., 1978; Brandenburg and Kellermann, 1978; Prasad et al., 1979; Gahmberg et al., 1979; Jett et al., 1979). For the most part, these studies have shown that persons with lung but also laryngeal cancer displayed significantly greater lymphocyte AHH inducibility than controls. With some exception,* these studies were better designed than the original Kellermann et al. (1973a) report, in that a better attempt at appropriately matching controls with cases occurred and the tumors were usually more specifically related to those potentially caused by PAHs.

In addition, Kellermann et al. (1973a) determined AHH activity fluorometrically by measuring the amount of only one of the many possible metabolites produced from MCA via the action of AHH. However, Guirgis et al. (1976) developed an improved assay that measured the production of all water soluble products. This technique is considered superior since it determines the rate of production of labeled water soluble metabolites, not the induction ratio of Kellermann et al. (1973a). Gahmberg et al. (1979) later used the assay of Guirgis et al. (1976).

One area that the original and subsequent positive or negative studies have not incorporated has been a consideration of nutritional factors that

---

*Trell et al. (1976) did not employ any concurrent control groups but used the Kellermann control group phenotypic frequencies even though the Kellermann et al. (1973a) data were from persons living in Houston and the Trell et al. (1976) data were from Sweden.

may have differed between subjects. Since it is known that a wide variety of nutrients can markedly affect the functioning of mixed function oxidase enzymes such as AHH, a lack of consideration of this variable may contribute to lack of agreement between studies. This may be particularly important when only one metabolite is used in determining the extent of AHH induction (see Kellermann et al., 1973a; Emery et al., 1978; and the negative studies of Paigen et al., 1977). Presumably, an altered or different nutritional status could shift the pattern of metabolites produced (Calabrese, 1980).

Not all the reports, however, supported the original association between AHH inducibility and susceptibility to bronchogenic carcinoma. For example, Paigen et al. (1977) reported that they were unable to measure AHH induction in 6 of 12 lung cancer patients* who had been operated on but not given subsequent chemoradiation therapy. Because of this lack of response, Paigen et al. (1977a) compared AHH inducibility in the progeny of lung cancer patients with controls that were the spouses of each case. This was justified on the basis of similarity of age, ethnic background, and environment such as diet and neighborhood, and since AHH inducibility had been shown not be affected by sex. The results indicated that the progeny of the patients did not differ from the matched control population in induced AHH activity levels or in basal AHH levels.†

The question, of course, emerges as to why the findings of Paigen et al. (1977, 1977a) contradict those of Kellermann et al. (1973a). Paigen et al. (1977) feel that the higher AHH inducibility by patients with lung cancer is probably caused by the presence of the disease itself and is not representative of the AHH inducibility level that existed prior to the carcinogenesis.‡

---

*The lymphocytes of these lung cancer patients displayed poor survival, reduced protein synthesis and reduced mitogen response when grown in culture.

†However, a more recent progeny study in Finland by Gahmberg et al. (1979) did show a genetic relationship in absolute induced levels of AHH activity thereby supporting the theory of Kellermann et al.

‡According to Kellermann et al. (1980), the principal reason for such conflicting findings is likely to be the use of mitogen-stimulated lymphocytes that gave highly reproducible AHH results in some individuals but quite variable in others. Furthermore, the lymphocytes of certain cancer patients as in the Paigen et al. (1977) study display a decreased response to the stimulation by phytohemagglutinin and pokeweed, which makes this test system unsuitable for the assessment of individual AHH values in cancer patients. Other researchers (Ducos et al., 1970; Garrioch et al., 1970; Han and Takita, 1972; Rao, 1974) have reported that lymphocytes from cancer patients do not respond well to mitogen stimulation. Other methodological issues that may lead to difficulties in reproducing the work of others is the seasonal variation in AHH levels, which means that population distributions of AHH activity cannot be collected over prolonged periods of time thereby often limiting data collection to but a few months of the year. Paigen et al. (1977) also have claimed that the lymphocyte AHH inducibility assay is difficult to standardize necessitating pretesting lots of media and fetal calf serum that may contain inducers of AHH that vary according to the

To make their case even stronger Paigen and associates (see Ward et al., 1978) evaluated 59 patients (32 lung and 27 laryngeal cancers) after they had been operated on and were free from disease for a period of time. No differences in AHH inducibility between the cancer groups and controls were noted. This study allowed the authors to avoid the criticism that any altered enzyme measurements found in diseased persons result from the disease itself. Other nonsupportive studies were also published which indicated that AHH inducibility and baseline levels were not enhanced in the lymphocytes or macrophages of individuals with lung cancer (McLemore et al., 1977, 1978).

The real issue is not actually how does lymphocyte AHH inducibility relate to lung cancer but does the response of the lymphocyte predict the activity or biochemical responses of lung tissue. In 1978, McLemore et al. addressed this issue in a limited study involving 14 smokers; 7 with lung cancer and 7 without. They found that noncancerous individuals displayed inducible levels of AHH in lymphocytes that were statistically significant and highly correlated (0.97) with lung tissue AHH levels. In contrast, the cancer patients did not show any statistically significant association in this regard. Such findings led these authors to conclude that

the use of a single tissue such as the peripheral blood lymphocyte for determination of AHH in patients with lung cancer is questionable. Although previous studies have shown a significant increase in the number of lung cancer patients (compared to noncancer patients) who have high AHH inducibility, the validity of these studies is subject to question since the levels of AHH present in lymphocytes from individual lung cancer patients do not necessarily reflect the levels of the enzyme in freshly excised lung tissue.

The evidence of whether there is a genetic basis affecting susceptibility to environmentally induced lung cancer is overwhelmingly documented in animal studies (Nebert, 1977, 1979), and supported via human epidemiological investigations (Tokuhata, 1964). However, the identification of a precise and reliable marker or predictor of risk to lung cancer such as AHH inducibility remains unresolved. More recent attempts by Kellermann et al. (1976, 1978, 1980) have tried to circumvent the reproducibility problems with the lymphocyte inducibility assay by measuring the T1/2 in plasma and saliva of antipyrine, which is thought to be metabolized via AHH in similar fashion to BP. This line of research was supported by several studies which showed that the plasma half-lifes of

---

lot (see Atlas et al., 1976). They also pointed out that AHH activity levels in the same person have often been found to vary on different days separated by only a week or two. As a result of such limitations, Paigen et al. (1977) asserted that a significant improvement in the cell culture procedure or a completely different way of measuring the genetic trait is necessary before large-scale population studies can be undertaken.

AHH Inducibility and Lung Cancer

Table 6-2. Correlation Between the Induction of Benzo[a]pyrene Hydroxylase in Mitogen-Stimulated Lymphocytes and the Plasma Disappearance Rates of Antipyrine, Aminopyrine, Phenylbutazone, Theophylline, Phenacetin, Acetanilide, Phenobarbital, Mephobarbital, Hexobarbital, and Warfarin

Pairs			r	P
Benzo[a]pyrene	vs.	antipyrine	−0.95	<0.01
Benzo[a]pyrene	vs.	aminopyrine	−0.84	<0.01
Benzo[a]pyrene	vs.	phenylbutazone	−0.86	<0.01
Benzo[a]pyrene	vs.	theophylline	−0.85	<0.01
Benzo[a]pyrene	vs.	phenacetin	−0.79	<0.01
Benzo[a]pyrene	vs.	acetanilide	−0.72	<0.01
Benzo[a]pyrene	vs.	phenobarbital	−0.88	<0.01
Benzo[a]pyrene	vs.	mephobarbital	−0.80	<0.01
Benzo[a]pyrene	vs.	hexobarbital	−0.70	<0.01
Benzo[a]pyrene	vs.	warfarin	−0.57	<0.01

Source: Kellermann et al. (1978).

antipyrine and nine other drugs are correlated with AHH inducibility in cultured lymphocytes of the same person (Kellermann et al., 1978) (Table 6-2), and that the metabolism of BP correlated with that of five drugs in the livers of humans at autopsy (Kapitulnik et al., 1977). In fact, Kellermann et al. (1976, 1978, 1980) have been able to differentiate healthy controls from lung cancer patients since the patients have a significantly reduced saliva half-life of antipyrine thereby implying higher AHH activity (Figure 6-2). This different approach to the problem still requires some type of accountability to lymphocyte AHH inducibility while not addressing the relationship of T1/2 and lung AHH inducibility.

The theory of Kellermann and associates that susceptibility to PAH induced lung cancer is in part a function of the ability to induce AHH remains to be unequivocally established. The negative reports of Paigen et al. (1977, 1977a) and others have clearly revealed that current methodologies used to establish the above relationship are of limited value. However, the fact that at least nine different studies involving patients in different parts of the United States and in Europe have supported the hypothesis that lymphocyte AHH inducibility is associated with enhanced risk to PAH induced lung and/or laryngeal cancer, along with the impressive support from animal studies* (Nebert, 1977; Nebert

---

*Cytochrome P-450 is induced to a considerable extent by PAH compounds in the B6 (the inbred C57BL/6 mouse strain) and in other responsive mouse strains. In contrast, the induction of this form of cytochrome by PAHs does not occur in liver and is significantly reduced in other organs such as the lung, bowel, kidney, and skin in the D2 and other nonresponsive mouse strains. The responsiveness to aromatic hydrocarbons was called the Ah locus. The Ah regulatory locus in mice is a cytosolic receptor which may bind to certain PAHs leading to the activation of structural genes and increases in enzymes that

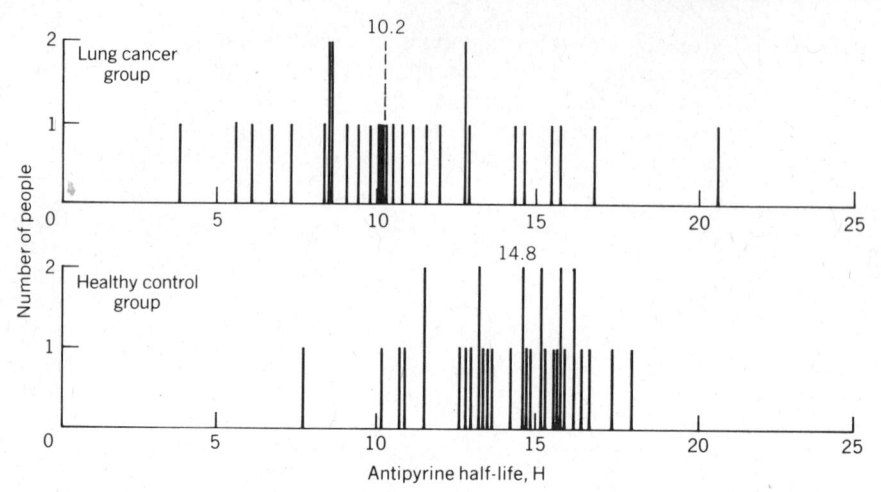

Figure 6-2. Distribution of antipyrine saliva half-lives (h) in 31 male patients with bronchogenic carcinoma and in 31 healthy matched control subjects. *Source:* Kellermann et al. (1978).

et al., 1977, 1979), strongly supports the validity that genetic determinants of PAH metabolism are somehow linked to susceptibility of lung and laryngeal cancer in humans.

This theory of Kellermann et al. therefore remains very viable and of overwhelming public health concern. To date the collective number of cancer patients who have been studied in the testing of this hypothesis is under 1000. When one considered that in 1975 the number of deaths from lung cancer in the United States alone was estimated to be more than 80,000 (Harris, 1977), the need to evaluate this hypothesis is obvious.

AHH, INDUCIBILITY, LUNG CANCER, AND RECOGNITION OF CONFOUNDING VARIABLES

Determination of the causes of differential susceptibility to PAH induced lung cancer is of extraordinary public health concern. One indirect approach toward gaining a greater understanding of the extent of interindividual differences of susceptibility to PAH induced lung cancer can be

---

metabolize the inducers and other PAHs, leading to a wide range of substances including mutagens, teratogens, carcinogens, and detoxified products (Nebert et al., 1979). Fibrosarcomas initiated by SC administered methylcholanthrene have been associated with genetically determined PAH responsiveness among 14 inbred strains of mice (Figure 6-3). Table 6-3 indicates that the carcinogenesis index for SC methylcholanthrene in the progeny from crosses involving the B6 and D2 inbred parental strains is related to the Ah response allele. The carcinogenesis index was 43 for the responsive phenotype groups while being less than 11 for all nonresponsive phenotype groups (Nebert et al., 1979).

AHH, Inducibility, Lung Cancer, and Confounding Variables

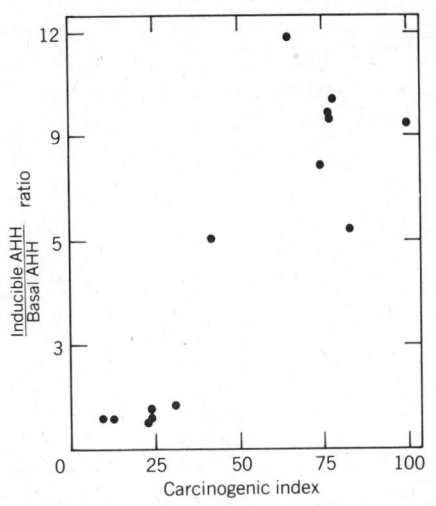

Figure 6-3. Relationship between the carcinogenesis index [defined by Ibal (1939)] for subcutaneous MC and the genetically mediated induction of AHH activity by MC for each of 14 inbred strains; the correlation coefficient r is 0.90 ($p < 0.001$) (Nebert et al. 1974). Each solid circle represents the average result from a group of 30 inbred mice of a certain strain. The carcinogenesis index was evaluated after a subcutaneous dose of 150 μg of MC had been given to a minimum of 30 weanling mice of each strain. The "inducible AHH/basal AHH ratio" reflects the mean of hepatic AHH activity in MC-treated mice divided by the mean hepatic enzyme activity in control mice ($N \geq 5$ for each of the two groups). Whether the MC-inducible AHH activity in the nonhepatic tissues appears to segregate as a single gene with the inducible hepatic AHH activity has not been examined for many of these strains. *Source:* Nebert et al. (1979). Reproduced from Nebert et al. 1974, with permission of Marcel-Dekker, Inc.

ascertained by viewing the binding of activated forms of PAH carcinogens to cellular DNA. Studies along these lines have assessed the variation in both the formation of carcinogen-DNA adducts among individuals and within each individual with respect to various target organs. Figure 6-4 reveals that variation among individuals in the binding of BP to DNA is 50–100-fold. Harris (1980), commenting on Thorgeirsson (1980), implied that the principal cause of these interindividual variations may be of genetic origin since the extent of this variation is comparable to that known to occur in pharmacogenetic investigations of drug metabolism in humans, citing the example of desmethylimipramine (Figure 6-5). While genetic findings may play a critical role in affecting the occurrence of PAH induced bronchial cancer, it is important to recognize that other factors may also contribute in a significant manner to the outcome. For example, BP binding to DNA may be markedly affected by the vitamin A nutritional status of the individual (Genta et al., 1974), as well as the intake of synthetic antioxidants such as BHT (Wattenberg et al., 1972). To what extent the individuals involved in the above study differed in these variables (i.e., vitamin A and BHT/BHA) is unknown. However,

Table 6-3. Relationship Between Aromatic Hydrocarbon Responsiveness and Susceptibility to Subcutaneous MC- and BP-Initiated Tumors Among Offspring From Appropriate Crosses Involving the B6, C3, and D2 Strains of Mice[a]

Strain or Offspring	Expression at Ah Locus[b]	Carcinogenesis Index for MC	for BP
B6	+ +	61	
D2	0	11	
B6D2F$_1$	+ +	43	
F$_1$ × B6	+ +	58	
F$_1$ × D2	+ +	54	
	0	8	
F$_2$	+ +	63	
	0	6	
C3	+ +	73	56
D2	0	10	4
C3D2F$_1$	+	37	19
F$_1$ × C3	+ +	74	27
	+	60	24
F$_1$ × D2	+	46	1
	0	9	1
F$_2$	+ +	69	31
	+	61	7
	0	17	2

Source: Nebert et al. (1979).

[a] Animals received, as weanlings, 150 µg of MC or BP in trioctanoin subcutaneously, and the carcinogenesis index was determined over an eight-month period (Kouri and Nebert 1977). The carcinogenesis index is defined by Iball (1939) as the percent incidence of subcutaneous fibrosarcomas divided by the average latency in days times 100. Further details are described elsewhere (Kouri et al. 1974, Kouri 1976).

[b] The phenotypic expression at the Ah locus is ranked as: + + = fully responsive, 0 = nonresponsive, + = intermediate responsive. (Reproduced from Kouri and Nebert 1977, with permission of Cold Spring Harbor Laboratory.)

when one attempts to evaluate the effects of genetic factors such as differential AHH inducibility on the occurrence of BP-DNA adduct formation on lung cancer, we must carefully consider the myriad of potential confounding factors as alluded to here. Not only may a lack of consideration of these factors lead to incorrect assessments of the role of genetic factors in the above outcomes, but it may also lead to disagreements between studies as is not uncommon in the literature associated with AHH inducibility and the susceptibility of humans to PAH induced lung cancer.

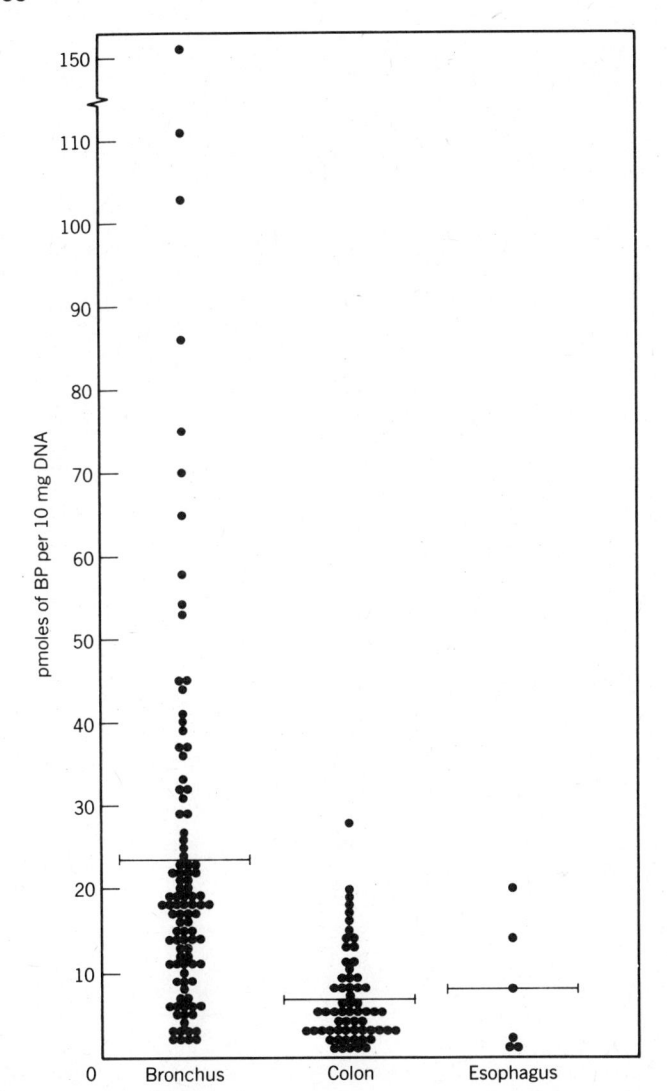

Figure 6-4. A wide interindividual variation for binding levels of BP to DNA in cultured human tissues. Each solid circle represents a value from an individual. *Source:* Harris (1980).

REFERENCES

Arnott, M. S., Yamauchi, T., and Johnston, D. A. (1979). Aryl hydrocarbon hydroxylase in normal and cancer populations. In: *Carcinogens: Identifications and Mechanisms of Action.* A. C. Griffin and C. R. Shaw (eds.). Raven, New York, p. 147.

Atlas, S. A., Vesell, E. S., and Nebert, D. W. (1976). Genetic control of interindividual variations in the inducibility of aryl hydrocarbon hydroxylase in cultured human lymphocytes. *Cancer Res.* **36**: 4619–4630.

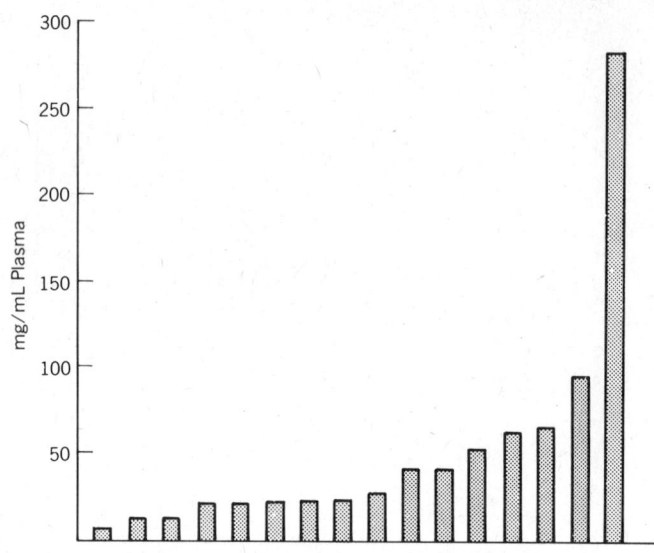

Figure 6-5. Steady-state plasma levels of desmethylimipramine during daily oral dosage of 3–25 mg. Each bar represents a different patient. *Source:* Hammer and Sjoqvist (1967).

Brandenburg, J. H. and Kellermann, G. (1978). Aryl hydrocarbon hydroxylase inducibility in laryngeal carcinoma. *Arch. Otolaryngol.* **104:** 151–152.

Calabrese, E. J. (1980). *Nutrition and Environmental Health. Vol. 1. The Vitamins.* Wiley, New York.

Conney, A. H. (1967). Pharmacological implications of microsomal enzyme induction. *Pharmacol. Rev.* **19:** 317–366.

Daly, J. W., Jerina, D. M., and Witkop, B. (1972). Arene oxides and the NIH shift: The metabolism, toxicity and carcinogenicity of aromatic compounds. *Experientia* **28:** 1129–1149.

Ducos, J., Migueres, J., Colombier, P., Kessous, A., and Poujoulet, N. (1970). Lymphocyte response to PHA in patients with lung cancer. *Lancet* **1:** 1111–1112.

Emery, A. E. H., Danford, N., Anand, R., Duncan, W., and Patton, L. (1978). Arylhydrocarbon-hydroxylase inducibility patients with cancer. *Lancet* **1:** 470–472.

Epidemiology Resources Inc. (1982). Report for the U.S. Office of Technology Assessment. Chestnut Hill, Massachusetts.

Gahmberg, C. G., Sekki, A., Kosuhen, T. U., Holsti, L. R., and Makela, O. (1979). Induction of aryl hydrocarbon hydroxylase activity and pulmonary carcinoma. *Int. J. Cancer* **23:** 302–305.

Garrioch, D. B., Good, R. A., and Gatti, R. A. (1970). Lymphocyte response to P.H.A. in patients with non-lymphoid tumors. *Lancet* **1:** 618.

Gelboin, H. V., Huberman, E., and Sachs, L. (1969). Enzymatic hydroxylation of benzopyrene and its relationship to cytotoxicity. *Proc. Nat. Acad. Sci.* **64:** 1188–1194.

Gelboin, H. V., Konoshita, N., and Wiebel, F. J. (1972). Microsomal hydroxylases: Induction and role in polycyclic hydrocarbon carcinogenesis and toxicity. *Fed. Proc.* **31:** 1298–1309.

Genta, V. M., Kaufman, D. G., Harris, C. C., Smith, J. M., Sporn, M. B., and Saffiotti, B.

References

(1974). Vitamin A deficiency enhances binding of benzo(a)pyrene to tracheal epithelial DNA. *Nature* **247**: 48–49.

Grover, P. L., Hewer, A., and Sims, P. (1971). Epoxides as microsomal metabolites of polycyclic hydrocarbons. *FEBS Lett.* **18**: 76–80.

Grover, P. L. and Sims, P. (1972). K-region epoxides of polycyclic hydrocarbons: Reactions with nucleic acids and polyribonucleotides. *Biochem. Pharmacol.* **22**: 661–666.

Guirgis, H. A., Lynch, H. T., Mate, T., Harris, R. E., Wells, I., Caha, L., Anderson, J., Maloney, K., and Rankin, L. (1976). Aryl-hydrocarbon hydroxylase activity in lymphocytes from lung cancer patients and normal controls. *Oncology* **33**: 105–109.

Han, T. and Takita, H. (1972). Immunologic impairment in bronchogenic carcinoma: A study of lymphocyte response to phytohaemagglutinin. *Cancer* **30**: 616–620.

Harris, C. (1977). Respiratory carcinogenesis. In: *Lung Cancer: Clinical Prognosis and Treatment*. M. J. Stravus (ed.). Grune and Stratton, London, pp. 1–17.

Harris, C. (1980). Individual differences in cancer susceptibility. *Ann. Intern. Med.* **92**: 809–825.

Jett, J. R., Branum, E. L., Fontana, R. S., Taylor, W. F., and Moses, H. L. (1979). Macromolecular binding of ^{3}H-benzo(a)pyrene metabolites and lymphocytes transformation in patients with lung cancer, and in smoking and nonsmoking control subjects. *Am. Rev. Respir. Dis.* **120**: 369–375.

Kapitulnik, J., Popper, P. J., and Conney, A. H. (1977). Comparative metabolism of benzo(a)pyrene and drugs in human liver. *Clin. Pharmacol. Ther.* **21**: 166.

Kellermann, G., Jett, J. R., Luyten-Kellermann, M., Moses, H. L., and Fontana, R. S. (1980). Variations of microsomal mixed function oxidases and human lung cancer. *Cancer* **45**: 1438–1442.

Kellermann, G., Luyten-Kellermann, M., Horning, M. G., and Stafford, M. (1976). Elimination of antipyrine and benzo(a)pyrene metabolism in cultured human lymphocytes. *Clin. Pharmacol. Ther.* **20**(1): 72–80.

Kellermann, G., Luyten-Kellermann, M., Jett, J. R., Moses, H. L., and Fontana, R. S. (1978). Aryl hydrocarbon hydroxylase in man and lung cancer. *Hum. Genet. Suppl. 1*, pp. 161–168.

Kellermann, G., Luyten-Kellermann, M., and Shaw, C. R. (1973). Genetic variation of aryl hydrocarbon hydroxylase in human lymphocytes. *Am. J. Hum. Genet.* **25**: 327–331.

Kellermann, G., Shaw, C. R., and Luyten-Kellermann, M. (1973a). Aryl hydrocarbon hydroxylase inducibility and bronchogenic carcinoma. *N. Engl. J. Med.* **280**: 934–937.

Kouri, R. E. and Nebert, D. W. (1977). Genetic regulation of susceptibility to polycyclic hydrocarbon induced tumors in the mouse. In: *Origins of Human Cancer*. H. H. Hiatt, J. D. Watson, and J. A. Winston (eds.). Cold Spring Harbor Laboratory, Cold Spring Harbor, New York, pp. 811–815.

McLemore, T. L., Martin, R. R., Busbee, D. L., Richie, R. C., Springer, R. R., Toppell, K. L., and Cantrell, E. T. (1977). Aryl hydrocarbon hydroxylase activity in pulmonary macrophages and lymphocytes from lung cancer and noncancer patients. *Cancer Res.* **37**: 1175–1181.

McLemore, T. L., Martin, R. R., Wray, N. P., Cantrell, E. T., and Busbee, D. L. (1978). Dissociation between aryl hydrocarbon hydroxylase activity in cultured pulmonary macrophages and blood lymphocytes from lung cancer patients. *Cancer Res.* **38**: 3805–3811.

Nebert, D. W. (1977). Genetic control of drug metabolism. In: *Pharmacological Intervention in the Aging Process*, Vol. 97. J. Roberto, R. Adelman, and V. Cristofalo (eds.). *Advances in Experimental Medicine and Biology*. Plenum, New York, pp. 55–83.

Nebert, D. W., Levitt, R. C., Orlando, M. M., and Felton, J. S. (1977). Effects of environmental chemicals on the genetic regulation of microsomal enzyme systems. *Clin. Pharmacol. Ther.* **22:** 640–658.

Nebert, D. W., Levitt, R. C., and Pelkonen, O. (1979). Genetic variation in metabolism of chemical carcinogens associated with susceptibility to tumorigenesis. In: *Carcinogens: Identification and Mechanisms of Action.* A. C. Griffin and C. R. Shaw (eds.). Raven, New York, pp. 157–185.

Paigen, B., Gurtoo, H. L., Minowada, J., Houten, L., Vincent, R., Paigen, K., Parker, N. B., Ward, E., and Hayner, N. T. (1977). Questionable relationship of aryl hydrocarbon hydroxylase to lung cancer risk. *N. Engl. J. Med.* **297:** 346–350.

Paigen, B., Minowada, J., Gurtoo, H. L., Paigen, K., Parker, N. B., Ward, E., Hayner, N. T., Bross, I. D. J., Boch, F., and Vincent, R. (1977a). Distribution of aryl hydrocarbon hydroxylase inducibility in cultured human lymphocytes. *Cancer Res.* **37:** 1829–1837.

Prasad, R., Prasad, N., Harrell, J. E., Thornby, J., Liem, J. H., Hodgins, P. T., and Tsuang, J. (1979). Aryl hydrocarbon hydroxylase inducibility and lymphoblast formation in lung cancer patients. *Int. J. Cancer* **23:** 316–320.

Rao, L. G. S. (1974). AHH inducibility and lung cancer. *Lancet* **1:** 1228.

Speier, J. L. and Wattenburg, L. W. (1975). Alterations on microsomal metabolism of benzo(a)pyrene in mice fed butylated hydroxyanisole. *J. Natl. Cancer Inst.* **55**(2): 470.

Tokuhata, G. K. (1964). Familial factors in human lung cancer and smoking. *Am. J. Public Health* **54:** 24–32.

Trell, E., Korsgaard, R., Hood, B., Kitzing, P., Norden, G., and Simonsson, B. G. (1976). Aryl hydrocarbon hydroxylase inducibility and laryngeal carcinomas. *Lancet* **2:** 140.

Ward, E., Paigen, B., Steenland, K., Vincent, R., Minowada, J., Gurtoo, H. L., Sartori, P., and Havens, M. B. (1978). Aryl hydrocarbon hydroxylase in persons with lung or laryngeal cancer. *Int. J. Cancer* **22:** 384–389.

Wattenberg, L. W. (1972). Inhibition of carcinogenic and toxic effects of polycyclic hydrocarbons by phenolic antioxidants and ethoxyguin. *J. Natl. Cancer Instit.* **48:** 1425–1430.

B. Cystic Fibrosis

Cystic fibrosis (CF) of the pancreas is a fairly common genetic disease of children and young adults in Caucasians of European origin. The clinical features result from a defect in exocrine gland function. Mucous secretions are often tenacious and thick with extremely high levels of sodium and chloride characteristically present. The increased viscosity of mucus is an important cause of obstructive pulmonary disease in afflicted people (Lobeck, 1972).

The most important clinical features are both pulmonary and gastrointestinal. Frequent pulmonary observations include progressive bronchiolus obstruction with complicating pulmonary infection, hyperaeration, and cor pulmonale. There is also a substance in plasma that causes ciliary dyskinesia (Spock et al., 1967).

CF is seen clinically only in the homozygous state. Based on population surveys, it is estimated that 1 in 2000 Caucasians have CF. The incidence of heterozygotes in the Caucasian population is approximately 4%. Based on observations that parents of CF offspring do not exhibit an increased incidence of chronic pulmonary or gastrointestinal disease (Batten et al., 1963; Anderson et al., 1962; Hallet et al., 1965; Orzales et al., 1963), Lobeck (1972) concluded that heterozygotes for CF do not have clinical findings of CF. However, the literature contains various reports indicating that adult heterozygotes have exhibited "adult mucoviscidosis" expressed as pulmonary disease associated with somewhat elevated levels of sodium or chloride in sweat. It is important to note that chronic bronchitis may be associated with a tendency to higher sweat concentrations of sodium and chloride (Lobeck, 1972). Additionally, the presence of the ciliary dyskinesis factor in the serum of heterozygotes (Spock et al., 1967) further suggests the possibility of a predisposition to pulmonary disease in these individuals. Thus, it is expected that individuals homozygous and possibly those heterozygous for the CF trait will be at increased risk to respiratory tract irritants such as ozone, sulfur dioxide, particulate sulfates, and numerous heavy metals.

SCREENING TESTS

The increased concentrations of sodium and chloride in endocrine sweat is the most constant and best defined abnormality that has been found in CF patients. A wide variety of screening techniques have been developed and are evaluated by Lobeck (1972).

The detection of heterozygotes is difficult because of occasional overlapping values in controls with regard to sodium and chloride concentrations in sweat. However, the presence of the cilia dyskinetic factor in sera and the observation of metachromasia in cultured fibroblasts of heterozygotes serves as an aid in their identification (Lobeck, 1972).

REFERENCES

Anderson, C. M., Freeman, M., Allan, J., and Hubbard, L. (1962). Observations on (i) sweat sodium levels in relation to chronic respiratory disease in adults and (ii) the incidence of respiratory and other disease in parents and siblings of patients with fibrocystic disease of the pancreas. *Med. J. Aust.* **1**: 965.

Batten, J., Muir, D., Simon, G., and Carter, C. (1963). The prevalence of respiratory disease in heterozygotes for the gene for fibrocystic disease of the pancreas. *Lancet* **1**: 1348.

Hallet, W. Y., Knudson, A. G., Jr., and Massey, F. J., Jr. (1965). Absence of detrimental effect of the carrier state for the cystic fibrosis gene. *Am. Rev. Respir. Dis.* **92**: 714.

Lobeck, C. C. (1972). Cystic fibrosis. In: *The Metabolic Basis of Inherited Disease*. J. B. Stanbury, J. B. Wyngaarden, and D. S. Fredrickson (eds.). McGraw-Hill, New York, pp. 1605–1626.

Orzales, M. M., Kohner, D., Cook, C. D., and Shwachman, H. (1963). Anamnesis, sweat electrolyte and pulmonary function studies in parents of patients with cystic fibrosis of the pancreas. *Acta Paediatr. Scand.* **52**: 267.

Spock, A., Heick, H. M. C., Cress, H., and Logan, W. S. (1967). Abnormal serum factor in patients with cystic fibrosis of the pancreas. *Pediatr. Res.* **1:** 173.

C. EMPHYSEMA

a. Genetic Predisposition: SAT Deficiency

Individuals with an inherited deficiency of SAT have been found to be predisposed to alveolar destruction even in the absence of chronic bronchitis. This was first reported by Laurell and Eriksson (1963) and subsequently verified in numerous reports (Tarkoff et al., 1968; Talamo et al., 1966, 1971, 1973; Briscoe et al., 1966). Other findings have lent support to the postulate that an inherited deficiency of SAT predisposes an individual to the development of pulmonary emphysema by allowing digestion of human lung tissue by proteases (proteolytic enzymes) from leukocytes or macrophages (Lieberman and Mohamed, 1971; U.S. Dept. HEW, 1971). These proteolytic enzymes include trypsin, chymotrypsin, elastase, collagenase, thrombin, and plasmin (Vogel et al., 1968; Eisen et al., 1970). Normal human serum is capable of inhibiting these enzymes, thereby protecting the individual from pulmonary emphysema. Research has strongly suggested that the lung is protected from proteolytic enzymes by a series of serum proteins, of which SAT is one. Most of the human serum antitryptic activity (85–90%) is associated with an α_1 protein, which is SAT (Jacobsson, 1955). The release of proteolytic enzymes occurs from cells (leukocytes and macrophages) that are responsible for cleaning the surfaces of the alveoli and bronchioles and removing foreign material and bacteria from these regions (Lieberman, 1972; Janoff, 1972; Cohn and Wiener, 1964). A wide variety of different challenges, both particulate and gaseous, and particularly both of these together, leads to an increase of these two types of cells in the lung as a more or less inevitable consequence either of direct irritation or of deposition in the alveoli of foreign material (Vogel et al., 1968; Bates, 1972; Kilburn, 1973; Corrin and King, 1970; Wood, 1951; Morrow, 1967; Coffin et al., 1968). A study performed on rabbits exposed to ozone showed an influx of heterophilic leukocytes and a diminution of pulmonary alveolar macrophages obtained by pulmonary lavage. This phenomenon was shown to be dose related. Additionally, a decrease in the ability of the alveolar macrophages to engulf streptococci was demonstrated. The data suggest that ozone either destroys the alveolar macrophage *in situ* or renders them sensitive to lysis during the process of lavage (Coffin et al., 1968).

A significant number of cases of chronic obstructive lung disease are associated with SAT deficiency in its severe (homozygous) and intermediate (heterozygous) forms (Lieberman, 1969). Blood serum of patients (suffering from pulmonary emphysema) homozygous for the SAT deficiency was found to contain less than 10% of the SAT of that of "normal" individuals. It has also been shown that these homozygous individuals (SAT-deficient) tended to develop emphysema in the third or fourth decade of life without symptoms of bronchitis. Additionally, heterozygous individuals were found to have a concentration of SAT between 50% and 60% of "normal" (Kueppers et al., 1964). However, not all carriers of the deficiency develop overt lung disease (Mittman et al., 1971).

Although patients with the homozygous condition are quite rare,* nearly 10% in certain segments of the population have intermediate levels of SAT (see Table 4-2 on page 177) (Mittman and Lieberman, 1973). Medical surveys have indicated the intermediate deficiency occurs more frequently than statistically expected in patients with obstructive lung disease, thus implying that the intermediate deficiency also predisposes the development of obstructive lung disease† (Lieberman et al., 1969; Kueppers et al., 1969).

Some individuals with severe SAT deficiency and most with the intermediate deficiency never develop lung disease. However, physiological studies in asymptomatic carriers of SAT deficiency imply that many have the lesions of anatomic emphysema (Mittman et al., 1972). Nonsmoking volunteers with an intermediate deficiency show little evidence of obstructive disease. However, a significantly greater number of apparently healthy individuals with the intermediate deficiency who smoke have early but definite airway obstruction (Mittman et al., 1972). Mittman et al. (1971) have reported that most patients with obstructive lung disease and the intermediate deficiency have smoked and their backgrounds suggest that the disease developed after less exposure to cigarettes than typical patients with normal SAT levels.

A factor which can complicate the interpretation of SAT screening tests is that in certain individuals the levels of leukocyte lysosomal protease-antiprotease are imbalanced (Galdston et al., 1972). They found some individuals with severe SAT deficiency but without the expected clinical phenotype of chronic obstructive pulmonary disease. The finding of low levels of leukolytic protease together with the SAT deficiency in such individuals may offer a possible explanation for the absence of chronic obstructive pulmonary disease in people with severe SAT deficiency.

Chowdhury and Louria (1976) have recently investigated the influence of Cd and other trace metals on human SAT. Cd was the only trace metal

---

*An incidence of approximately 0.8 per 1000 was found in a population of 16,190 individuals from Norway and Sweden (Fagerhol and Laurell, 1970).
†It is important to note that emphysema has been observed less commonly in blacks than whites and appears infrequently in American Indians and Orientals. Variations in the occurrence of SAT deficiency in different ethnic groups might account for these observations (Mittman and Lieberman, 1973).

that reduced the concentration of SAT and depressed the trypsin inhibitory capacity. Divalent Pb, Hg, Ni, and Zn ions showed no such effects. The combination of Cd-Ni, Cd-Pb, and Cd-Fe did not produce greater decreases of SAT activity than did Cd at the same concentration. However, the combination of Cd-Hg appeared to be more effective in decreasing the SAT activity than Cd alone. These results are of significance since emphysema can be caused by prolonged exposure to Cd (Lane and Campbell, 1954; Holden, 1965). The concentrations of Cd used in the Chowdhury and Louria (1976) study are far greater than concentrations of Cd found in the blood of normal adults (Kubota et al., 1968), but they are similar to blood concentrations in industrially poisoned workers (Friberg et al., 1972). Additionally, in those workers dying with Cd-induced emphysema, concentrations of 50–600 mg/g of lung tissue have been noted (Lane and Campbell, 1954; Flick et al., 1971). These were considerably above the concentration (1–5 mg/100 mL) used in the Chowdhury and Louria study. Their study may indicate a possible explanation for the emphysema that occurs in industrial workers exposed to Cd. These results imply that individuals either homozygous or heterozygous for the SAT deficiency may be considered at high risk with regard to both Cd and Cd-Hg exposure [see Glaser et al. (1977) for a critique of the Chowdhury and Louria (1976) study].

b. Animal Models to Study SAT Deficiency

As a result of the association between a SAT deficiency in humans and a predisposition to developing emphysema, attention has focused on the extent to which serum antitrypsins may vary in laboratory animals. For as Ihrig et al. (1971) notes, if the serum levels of antitrypsin can be modified—up and down—in the model, this may be of considerable value in attempting to assess the impact of trypsin inhibitors on a number of disease states including emphysema.

In an initial survey of the total serum antitrypsin activity levels of many common laboratory animals including man, wide interspecies variations were noted with the hamster and guinea pig displaying the lowest and highest respective values (Figure 6-6). While no apparent phylogenetic pattern occurred with regard to the total amount of antitrypsin activity, an evolutionary pattern did develop for the specific antitrypsin components. More specifically, while rodents (hamster, guinea pig) and lagomorphs (rabbit) displayed at least two peaks with less than $\frac{1}{2}$ the total activity in the α_1 position, primates including man had just one major active fraction, the α_1 antitrypsin. The goat displayed an intermediate pattern with the major activity found in the α_1 region.

Functional similarity tests by Ihrig et al. (1971) revealed that injection of turpentine and inhalation of NO_2 by the guinea pig produced an in-

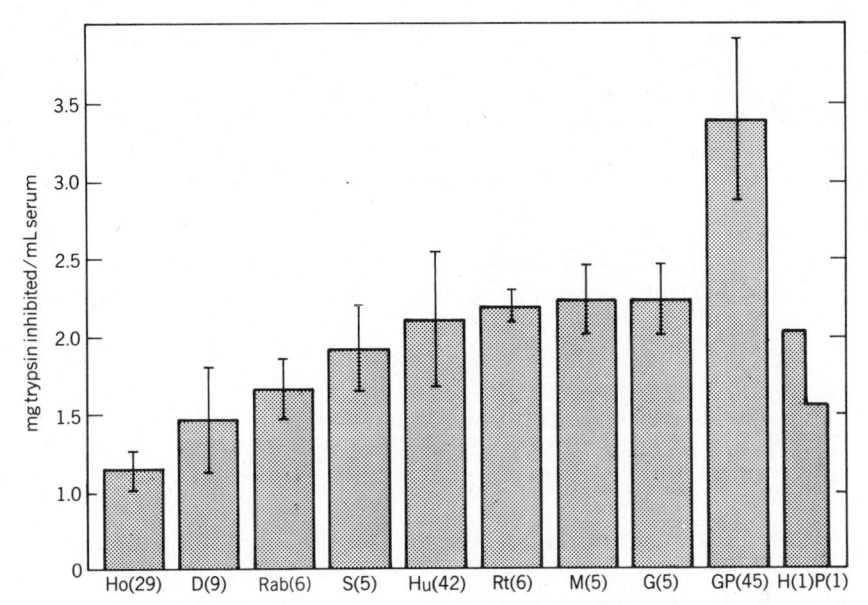

Figure 6-6. Antitrypsin activity (milligrams trypsin inhibited per milliliter of serum) in hamster (Ha), dog (D), rabbit (Rab), sheep (S), human (Hu), rat (Rt), monkey (M), goat (G), guinea pig (GP), horse (H), and pony (P). The height of each bar is the mean value, and ±SD is indicated by the vertical line. Number of animals in sample is in parameters under each bar. *Source:* Ihrig et al. (1971).

crease in serum antitrypsin activity (actually the α_1 component) as well as did spontaneous pulmonary infections. These findings suggest that guinea pigs acted in a comparable way to those stimuli as did humans.

In concluding their research, Ihrig et al. (1971) stated that in assessing the use of these laboratory animals (i.e., guinea pigs), certain points should be considered. They are analogous to humans in their response to stimuli such as tissue injury in increasing their total serum antitrypsin considerations. In choosing an animal model, however, consideration should be given to the specific pattern of antitrypsin substances in the particular species and the reactions of the several components to various stimuli should be understood. Depression of one antitrypsin substance might not necessarily have as severe an effect as in humans because the other active substances might be active as buffers. On the other hand, even stimuli that would increase only one substance might have a profound effect in the model similar to that in humans. It is important to recognize . . . that animals with spontaneous disease will have increased concentrations of antitrypsin. If such disease is subclinical, falsely elevated "control" values might be observed. Finally, some components of the antitrypsin substance might deteriorate during handling or storage.

Studies of experimental emphysema have used proteolytic enzymes such as papain for the production of lesions (Gross et al., 1965; Paleck et al., 1967; Goldring et al., 1968). Another approach might be the manipulation of native inhibitors of proteolytic enzymes alone or in association with other agents. An antibody to one or more of the major components might be a means of experimentally lowering the activities of these substances, as injury is a means of increasing them. The antibodies produced to date, however, are specific but low in titer and do not appear to be inhibitory *in vitro*, although they might be effective *in vivo* . . . . If it is true that antitrypsins protect against excessive tissue breakdown in the lung by inhibiting the proteases released from damaged cells or leukocytes, then experimentally varying the concentrations of inhibitor substances could alter the susceptibility of the laboratory animal to such changes, and provide further evidence of the role of antitrypsin in diseases such as emphysema.

It may be asked if there is any potential animal model to study humans with a SAT deficiency. According to Cheville (1980), a SAT deficiency is known to occur in turkeys but apparently is not related to a predisposition for this species to develop emphysema.

c. Broader Consideration of Animal Models for Emphysema

In a more sweeping assessment of the use of animal models in the study of emphysema, Karlinsky and Snider (1978) developed both anatomic and physiologic criteria for emphysematous animal models. In essence, their criteria are contingent on the truism of Reid (1980) that "an animal model is only as good as our understanding of the human disease."

In general, there are four major types of emphysema as determined by the primary anatomic location of lesions. The centrilobular type principally involves the respiratory bronchioles. The panlobular type affects an overall involvement of respiratory bronchioles, alveolar ducts and alveoli. Puraseptal emphysema is characterized by the destruction of respiratory epithelium along lobular septae. The final form results in destruction of lobular material.

Emphysema-like conditions have been reported spontaneously to occur in a wide variety of animals including cattle (Tucker and Maki, 1962; Maki and Tucker, 1962; Moulton et al., 1961, 1963), horses (Alexander, 1959; Nowell et al., 1971; McLaughlin and Edwards, 1966; Foley and Lowell, 1966; Gillepsie et al., 1964, 1966; Eberly et al., 1966; Cook and Rossdale, 1963), rabbits (Strawbridge, 1960; Boatman and Martin, 1965), dogs (Hernandez et al., 1966), and rats (Palecek and Holusa, 1971), with various attempts to relate the animal conditions to specific forms of the human disease. In addition, Fisk and Kuhn (1976) reported the occurrence of a "blotchy mouse" displaying a genetically controlled progressively developing panlobular emphysema caused by a defect in the inter- and intramolecular cross-linking of elastin. The lungs of these animals exhibited enlarged terminal airspaces, attenuated alveolar walls and functional symptoms typical of emphysema. However, this condition is not age related and is found in immature animals. According to Karlinsky and Snider (1978), this model is of at least qualitative significance since it nicely illustrates the role of alterations in connective tissue as an immediate cause of panlobular emphysema.

According to Slauson and Hahn (1980) there is often insufficient information available to critically assess the precise anatomic and/or physiologic similarities of the respective animal models to the human condition. These investigators emphasized the critical need to be aware of interspecies differences in subgross anatomy of mammalian lungs when assessing any particular animal model for emphysema. The following assessment by these researchers underscores the type of analysis needed:

In human lungs, the divisions distal to the terminal bronchiole are the gas-exchanging units that comprise a primary lobule. Several primary lobules are grouped together within connective tissue septa to form secondary lobules. A paucity of interlobular connective tissue prevents the definition of lobules in lungs of rodents, lagamorphs, carnivores, and nonhuman primates. Human lungs have several orders of branching respiratory bronchioles, as do lungs of dogs, cats, and monkeys. In contrast, respiratory bronchioles are rarely found in lungs of rats, hamsters, and mice; instead, the terminal bronchioles lead directly into alveolar ducts. Differences in structure of the terminal bronchioles of monkeys have also been reported. *Macaca radiata* (bonnet) have long terminal bronchioles as found in man, while *M. mulatta* (rhesus) and *M. arctoides* (stumptail) have shorter terminal bronchioles. The terminal bronchioles of *M. mulatta* also have been noted to have "transitional" areas with mixed epithelial types characteristic of both terminal and respiratory bronchioles and occasional alveoli. In the strictest sense, therefore, the term "centrilobular" would not be correctly applied to emphysematous lesions in lungs of animals (rodents and lagamorphs) not having respiratory bronchioles as found in man; thus, models for human centrilobular emphysema could only be developed in those species with well-defined respiratory bronchioles. A similar problem exists in using the term "paraseptal" in species without well-defined lobular septae. Panlobular and paracicatricial emphysema could be observed in all of the species mentioned.

REFERENCES

Alexander, A. F. (1959). Chronic alveolar emphysema in the horse. *Am. Rev. Respir. Dis.* **80:** 141.

Baskerville, A., Cox, C. S., and Stirrup, J. A. (1978). Quantitative studies on the development of the bronchi and bronchial glands of the pig from birth to maturity. *Acta Anat.* **100:** 386–390.

Bates, D. V. (1972). Air pollutants and the human lung. *Am. Rev. Respir. Dis.* **105:** 1–13.

Boatman, E. S. and Martin, H. B. (1965). Electron microscopy in pulmonary emphysema of rabbits. *Am. Rev. Respir. Dis.* **91:** 197.

Briscoe, W. A., Kueppers, F., Davis, A. L., and Bearn, A. G. (1966). A case of inherited deficiency of serum alpha$_1$-antitrypsin associated with pulmonary emphysema. *Am. Rev. Respir. Dis.* **94:** 529–539.

Cheville, N. (1980). Discussion comments. *Am. J. Pathol.* **101:** S127.

Coffin, D. L., Gardner, D. E., Holzman, R. S., and Wolcocks, F. J. (1968). Influence of ozone on pulmonary cells. *Arch. Environ. Health* **16:** 633–636.

Cohn, Z. A. and Wiener, E. (1964). The particulate hydrolases of macrophages: II. Biochemical and morphological response to particle ingestion. *J. Exp. Med.* **118:** 1009–1023.

Cook, W. R. and Rossdale, P. D. (1963). The syndrome of "broken wind" in the horse. *Proc. R. Soc. Med.* **56:** 972.

Chowdhury, P. and Louria, D. B. (1976). Influence of cadmium and other trace metals on human a_1-antitrypsins: An *in vitro* study. *Science* **191:** 480.

Committee on Diagnostic Standards for Nontuberculous Respiratory Diseases (1962). Chronic bronchitis, asthma and pulmonary emphysema. *Am. Rev. Respir. Dis.* **85:** 762–778.

Corrin, B. and King, E. (1970). Pathogenesis of experimental pulmonary alveolar proteinosis. *Thorax* **25:** 230–236.

Eberly, V. E., Tyler, W. S., and Gillespie, J. R. (1966). Cardiovascular parameters in emphysematous and control horses. *J. Appl. Physiol.* **21:** 883.

Eisen, A. Z., Block, K. J., and Sakai, T. (1970). Inhibition of human skin collagenase by human serum. *J. Lab. Clin. Med.* **75:** 258.

Eisen, H. (1974). *Immunology: An Introduction to Molecular and Cellular Principles of the Immune Response.* Harper & Row, New York, pp. 473–74, 504.

Fagerhol, M. K. (1973). Recent findings and ideas concerning the Pi polymorphism and diseases associated with alpha$_1$-antitrypsin deficiency. In: *Fundamental Problems of Cystic Fibrosis and Related Diseases.* Intercontinental, New York, P. 402.

Fagerhol, M. K. and Laurell, C. B. (1970). The Pi system: inherited variants of serum alpha$_1$-antitrypsin. In: *Progress in Medical Genetics,* Vol. 7. A. G. Steinberg and A. G. Bearn (eds.). Grune and Stratton, New York, p. 96.

Fisk, D. E. and Kuhn, C. (1976). Emphysema-like changes in the lungs of the blotchy mouse. *Am. Rev. Respir. Dis.* **113:** 787–797.

Flick, D. F., Kraybill, H. F., and Dimitroff, S. M. (1971). Toxic effects of calcium: a review. *Environ. Res.* **4:** 71–85.

Foley, F. D. and Lowell, F. C. (1966). Equine centrilobular emphysema with further observations on the pathology of heaves. *Am. Rev. Respir. Dis.* **93:** 17.

Friberg, L. T., Piscator, M., and Nordberg, G. F. (1972). *Cadmium in the Environment.* Chemical Rubber, Cleveland, Ohio.

Galdston, M., Gottwelt, C., and Davies, A. L. (1972). Leucocyte lysosomal enzymes and a_1-antitrypsin: genetic determinants in chronic obstructive pulmonary disease. *Fed. Proc.* (abstract).

Gillespie, J. R. and Tyler, W. S. (1969). Chronic alveolar emphysema in the horse. *Adv. Vet. Sci. Comp. Med.* **13:** 59–99.

Gillespie, J. R., Tyler, W. S., and Eberly, V. E. (1966). Pulmonary ventilation and resistance in emphysematous and control horses. *J. Appl. Physiol.* **21:** 416.

Gillespie, J. R., Tyler, W. S., and Eberly, V. E. (1964). Blood pH, O_2, and CO_2 tensions in exercised control and emphysematous horses. *Am. J. Physiol.* **207:** 1067.

Glaser, L. B., Karic, L., Huffaker, T., and Fallat, R. J. (1977). Influence of cadmium on human alpha-1-antitrypsin: a reexamination. *Science* **196:** 556–557.

Goldring, I. P., Greenburg, L., and Ratner, I. M. (1968). On the production of emphysema in Syrian hamsters by aerosol inhalation of papain. *Arch. Environ. Health* **16:** 59.

Gross, P., Pfitzer, E. A., Tolker, E., Babyak, M. A., and Kaschak, M. (1965). Experimental emphysema: its production with papain in normal and silicotic rats. *Arch. Environ. Health* **11:** 50.

References

Hernandez, J. A., Anderson, A. F., Jr., Holmes, W. L., and Foraker, A. G. (1966). Pulmonary parenchymal defects in dogs following prolonged cigarette smoke exposure. *Am. Rev. Respir. Dis.* **93**: 78.

Holden, H. (1965). Cadmium fume. *Ann. Occup. Hyg.* **8**: 51–54.

Ihrig, J., Kleinerman, J., and Rynbrandt, D. J. (1971). Serum antitrypsins in animals: studies of species variations, components, and the influence of certain irritants. *Am. Rev. Respir. Dis.* **103**: 377.

Jacobsson, K. (1955). Studies on the trypsin and plasmin inhibitors in human blood serum. *Scand. J. Clin. Lab. Invest., Suppl. 14.* **7**: 55–102.

Janoff, A. (1972). Elastase-like proteases of human granulocytes and alveolar macrophages. In: *Pulmonary Emphysema and Proteolysis.* C. Mittman (ed.). Academic, New York, p. 562.

Karlinsky, J. B. and Snider, G. L. (1978). Animal models of emphysema. *Am. Rev. Respir. Dis.* **117**: 1109–1133.

Kennedy, A. R. and Little, J. B. (1979). Respiratory system differences relevant to lung carcinogenesis between Syrian hamsters and other species. *Prog. Exp. Tumor Res.* **24**: 302–314.

Kilburn, K. H. (1973). Biological effects of cigarette smoking in the pathogenesis of pulmonary disease. *J. Occup. Med.* **15**(3): 198–201.

Kubota, J., Lazar, V. A., and Losee, F. (1968). Copper, zinc, cadmium and lead in human blood from 19 locations in the United States. *Arch. Environ. Health* **16**: 788–793.

Kueppers, F., Briscoe, W. A., and Bearn, A. G. (1964). Heriditary deficiency of serum α_1-antitrypsin. *Science* **146**: 1678–1679.

Kueppers, F., Fallat, R. and Larson, R. K. (1969). Obstructive lung disease and alpha$_1$-antitrypsin deficiency gene heterozygosity. *Science* **165**: 899.

Lane, R. E. and Campbell, A. C. P. (1954). Fatal emphysema in two men making a copper cadmium alloy. *Br. J. Ind. Med.* **11**: 118–122.

Laurell, C. B. and Eriksson, S. (1963). Electrophoretic a$_1$-globulin pattern of serum in α_1-antitrypsin deficiency. *Scand. J. Clin. Lab. Invest.* **15**: 132.

Lieberman, J. (1969). Heterozygous and homozygous alpha$_1$ antitrypsin deficiency in patients with pulmonary emphysema. *N. Engl. J. Med.* **281**: 279–284.

Lieberman, J. (1972). Digestion of antitrypsin deficient lung by leukoproteases. In: *Pulmonary Emphysema and Proteolysis.* C. Mittman (ed.). Academic, New York, p. 562.

Lieberman, J. and Mittman, C. (May 22, 1972). A new "double-ring" screening test for carriers of α_1-antitrypsin variants. Presented at American Thoracic Society Meeting, Kansas City, Missouri.

Lieberman, J., Mittman, C., and Schneider, A. S. (1969). Screening for homozygous and heterozygous alpha$_1$-antitrypsin deficiency. *J. Am. Med. Assoc.* **210**: 2055.

Lieberman, J. and Mohamed, A. G. (1971). Inhibitors and activators of leukocytic proteases in purulent sputum. *J. Lab. Clin. Med.* **77**: 713–727.

Maki, L. R. and Tucker, J. O. (1962). Acute pulmonary emphysema in cattle. II. Etiology. *Am. J. Vet. Res.* **23**: 824.

Martorana, P. A. and Share, N. N. (1976). Effect of human α-antitrypsin in papain induced emphysema in the hamster. *Am. Rev. Respir. Dis.* **113**: 607–612.

McLaughlin, R. F., Jr. and Edwards, D. W. (1966). Naturally occurring emphysema, the fine gross and histopathologic counterpart of human emphysema. *Am. Rev. Respir. Dis.* **93**: 22.

McLaughlin, R. F., Tyler, W. S., and Canada, R. O. (1961). A study of the subgross pulmonary anatomy in various mammals. *Am. J. Anat.* **108**: 149–165.

McLaughlin, R. F., Tyler, W. S., and Canada, R. O. (1966). Subgross pulmonary anatomy of the rabbit, rat, and guinea pig, with additional notes on the human lung. *Am. Rev. Respir. Dis.* **94**: 380–387.

Mittman, C. and Lieberman, J. (1973). Screening for a_1-antitrypsin deficiency. In: *Genetic Polymorphisms and Diseases in Man.* B. Ramot, A. Adam, B. Bonne, R. Goodman, and A. Szeinberg (eds.). Academic, New York, pp. 185–192.

Mittman, C., Lieberman, J., Marasso, F., and Miranda, A. (1971). Smoking and chronic obstructive lung disease in alpha$_1$-antitrypsin deficiency. *Chest* **60**: 214.

Mittman, C., Lieberman, J., Miranda, A., and Marasso, F. (1972). Pulmonary disease in intermediate alpha$_1$-antitrypsin deficiency. In: *Pulmonary Emphysema and Proteolysis.* C. Mittman (ed.). Academic, New York, p. 33.

Mohr, U. and Ketkar, M. B. (1980). Animal model of human disease: bronchogenic carcinoma of the lung. *Am. J. Pathol.* **99**: 521–524.

Morrow, P. E. (1967). Adaptations of the respiratory tract to air pollutants. *Arch. Environ. Health* **14**: 127–136.

Moulton, J. E., Cornelius, C. F., and Osburn, B. L. (1963). Acute pulmonary emphysema in cattle. *J. Am. Vet. Med. Assoc.* **142**: 133.

Moulton, J. E., Harrold, J. B., and Horning, M. A. (1961). Pulmonary emphysema in cattle. *J. Am. Vet. Med. Assoc.* **139**: 669.

Negus, V. (1965). *The Biology of Respiration.* E&S Livingstone, Edinburgh, Scotland. 228 pp.

Nowell, J. A., Gillespie, J. R., and Tyler, W. S. (1971). Scanning electron microscopy of chronic pulmonary emphysema. A study of the equine model. Scanning Electron Microscopy/1971 Proceedings of the 4th Annual Scanning Electron Microscope Symposium. I. Johann (ed.). I.I.T. Research Institute, Chicago, pp. 105–113.

Palecek, F. and Holusa, R. (1971). Spontaneous occurrence of lung emphysema in laboratory rats. *Physiol. Bohemoslov.* **20**: 335–344.

Paleck, F., Palecekova, M., and Aviado, D. M. (1967). Emphysema in immature rats: condition produced by tracheal constriction and papain. *Arch. Environ. Health* **15**: 332.

Petty, T. L. and Nett, L. M. (1967). *For Those Who Live and Breathe with Emphysema and Chronic Bronchitis.* Thomas, Springfield, Illinois, 108 pp.

Phalen, R. F. (1974). Respiratory tract morphology: summary of a conference. *Bioscience* **24**: 612–614.

Pirie, H. M. and Wheeldon, E. B. (1976). Chronic bronchitis in the dog. *Adv. Vet. Sci. Comp. Med.* **20**: 253–276.

Reid, L. (1954). Pathology of chronic bronchitis. *Lancet* **1**: 275–278.

Reid, L. (1967). *The Pathology of Emphysema.* Lloyd-Luke, London, 372 pp.

Reid, L. (1980). Needs for animal models of human diseases of the respiratory system. *Am. J. Pathol.* **101**: S89–S101.

Schmidt-Nielsen, K. (1972). *How Animals Work.* Cambridge University, Cambridge, 114 pp.

Slauson, D. O. and Hahn, F. F. (1980). Criteria for development of animal models of diseases of the respiratory system. *Am. J. Pathol.* **101**: S103–S129.

Strawbridge, H. T. G. (1960). Chronic pulmonary emphysema. II. Spontaneous pulmonary emphysema in rabbits. *Am. J. Pathol.* **37**: 309.

Takizawa, T. and Thurlbeck, W. M. (1971). A comparative study of four methods of assessing the morphologic changes in chronic bronchitis. *Am. Rev. Respir. Dis.* **103**: 774–781.

Talamo, R. C., Blennerhassett, J. B., and Austen, K. F. (1966). Familial emphysema and alpha$_1$ antitrypsin deficiency. *N. Engl. J. Med.* **275**: 1301–1304.

References

Talamo, R. C., Langley, C. E., Reed, C. E., and Makino, S. (1973). Alpha$_1$-antitrypsin deficiency: a variant with no detectable alpha$_1$-antitrypsin. *Science* **181:** 70–71.

Talamo, R. C., Levison, H., Lynch, M. J., Hercy, A., Hyslop, N. E., Jr., and Bain, H. W. (1971). Symptomatic pulmonary emphysema in childhood associated with hereditary alpha$_1$-antitrypsin and elastase inhibitor deficiency. *J. Pediatr.* **79:** 20–26.

Tarkoff, M. P., Kueppers, F. and Miller, W. F. (1968). Pulmonary emphysema and alpha$_1$-antitrypsin deficiency. *Am. J. Med.* **45:** 220–228.

Thurlbeck, W. M. and Lowell, F. C. (1964). Heaves in horses. *Am. Rev. Respir. Dis.* **89:** 82.

Tucker, J. O. and Maki, L. R. (1962). Acute pulmonary emphysema in cattle. I. Experimental production. *Am. J. Vet. Res.* **23:** 821.

U.S. Dept. HEW (1971). The Health Consequences of Smoking: A Report to the Surgeon General—1971. Public Health Service and Mental Health Administration, Washington, D.C.

Vogel, R., Trautschold, I., and Werle, E. (1968). *Natural Proteinase Inhibitors*. Academic, New York.

Wheeldon, E. B. and Pirie, H. M. (1974). Measurement of bronchial wall components of young dogs, adult normal dogs and adult dogs with chronic bronchitis. *Am. Rev. Respir. Dis.* **110:** 609–615.

Wood, W. B., Jr. (1951). White blood cells v. bacteria. *Sci. Am.* **184:** 48–52.

7 Renal Disorders

A. Diabetes with Respect to Renal Nephropathy

1. INTRODUCTION

It is generally accepted that diabetes is not a single disease type but a heterogeneous group of syndromes that result from a variety of genetic and environmental factors. Consequently, diabetes will be broadly defined according to the suggestion of Grodsky et al. (1982) as a "demonstrable abnormality in metabolism resulting in hyperglycemia." These authors have summarized some of the causes and amplifying factors of diabetes in Figure 7-1. In a general sense, insulin-dependent diabetes mellitus is characterized by an absolute deficiency of insulin along with considerable β-cell lesions and necroses. The causes of the β-cell lesions are uncertain but may be caused by viral agents and chemical substances, as well as autoimmune processes. Susceptibility of the β-cells is very dependent on the genetic and metabolic status of the individual. Noninsulin dependent diabetes (i.e., adult onset) is characterized by significant insulin production ranging from less than normal to above normal but invariably in quantities insufficient to sustain glucose hemeostasis. As in the case with insulin-dependent diabetes, these types can be amplified by one's genetic status, diet, and obesity, as well as endocrine and autonomic nervous system variables (Grodsky et al., 1982).

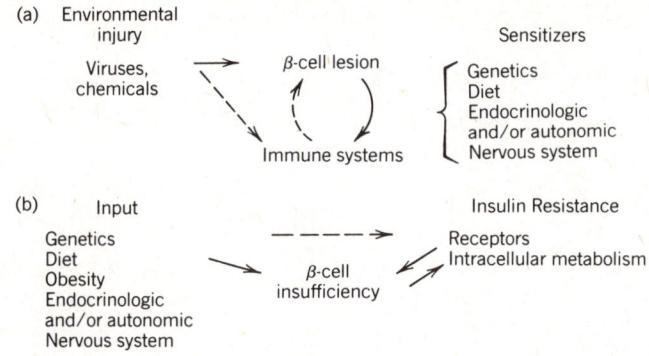

Figure 7-1. Hypothetical scheme for pathogenesis of (a) IDDM and (b) NIDDM. *Source:* Grodsky et al. (1982).

This section will address the issue of whether diabetics may be at increased risk to environmental pollutants. More specifically, the assessment will evaluate: (1) the prevalence of diabetes in the population; (2) the effects of arsenic, a known renotoxin, on a chemically induced diabetic model; (3) the risk of those with diabetes insipidus to developing dental fluorosis; (4) whether diabetics are at increased risk to acute X-ray induced toxicity; (5) the interaction of smoking and diabetes in humans; (6) the effect of the diabetic state on xenobiotic metabolism; (7) whether the diabetic state enhances the hepatotoxic effects of several organic solvents; and (8) the occurrence of environmental diabetogenic agents.

2. PREVALENCE OF DIABETES

Renold et al. (1978) have indicated that the prevalence of diabetes is hard to determine because of a lack of precision in how diabetes is defined as well as methodological problems in many potentially applicable studies. They assert that blood glucose determination is the only practical way to make such estimations. In general, the blood glucose levels display a unimodal population distribution, thereby making any differentiation between normality and disease arbitrary. Nevertheless, sufficient evidence has accumulated so as to promote a reasonable estimate of prevalence. The 1973 Health Interview Survey of the National Commission on Diabetes estimated that 2% of the U.S. population (i.e., greater than 4 million persons) are diabetics. A similar percentage of the population are thought to have undiagnosed diabetes. The National Diabetes Commission further determined that there are about 1.25 million Americans who take insulin, 1.25 million take oral antidiabetic agents, and between 2.0 and 3.5 million diabetics are treated with diet alone with between 1.5 and 4 million cases undiagnosed, giving an overall estimate of about 5% of the population affected. Juvenile onset diabetes is much more accurately quantified. For example, 1 in about 500–600 school age students have been proved to have diabetes in Minnesota and Michigan (Kyllo and Nuttall, 1976; Gorwitz et al., 1976).

3. ARSENIC-ALLOXAN-INDUCED DIABETES IN THE RAT

In 1981, Schiller et al. reported on the results of a study concerned with assessing the effect of a previous exposure to arsenate on urinary enzyme levels in rats developing alloxan-induced diabetes. The rationale for such a study was based on the well recognized phenomena that the kidney is a target organ for arsenic toxicity and that persons with diabetes mellitus characteristically display general kidney dysfunction along with microvascular changes and abnormal urinary enzyme patterns. The fact that both arsenic toxicity and diabetes mellitus adversely affect the kidney suggested that the interaction of the two may be worthy of study since it may be hypothesized that the presence of the one condition may enhance the toxicity of the other.

Schiller et al. (1981) reported some of the findings of this study in Table 7-1. Alkaline phosphatase was the only enzyme that was increased significantly in normal rats treated with arsenic. The alloxan diabetes treatment resulted in a statistically significant effect on the excretion of acid phosphatase, a phenomenon enhanced further by the arsenate. β-glucuronidase excretion was increased significantly by these combined treatments. In contrast, each of the other enzymes evaluated (i.e., N-acetyl-β-D-glucosaminidase, lactate dehydrogenase, and L-glutamate-oxaloacetate transminase) were considerably increased only in the alloxan-diabetic rats that were exposed previously to oral arsenic.

The significance of these findings for the human condition is unclear since the exposure that was via drinking water reflected a level some 800 times the current EPA primary drinking water standard. Nevertheless, these findings are of considerable interest since they represent the first experimental study of a heavy metal-diabetes interaction.

4. DIABETES AND FLUORIDE

Among those considered at potentially increased risk to fluoride are persons with a compromised kidney function. Since renal nephropathy is a common occurrence in diabetics, it follows that persons with this disease may be at increased risk to fluoride. While studies concerning the effects of fluoride on diabetics are few, the studies that do exist suggest that the above hypothesis may be correct. For instance, a 1974 report by Greenberg et al. noted the occurrence of systemic fluorosis in two patients with a variety of renal complications including nephrogenic diabetes in-

Table 7-1. Activities of Selected Enzymes in Urines From Normal and Alloxan-Diabetic Rats Exposed to Arsenic in the Drinking Water for 3 Weeks

Specific Gravity[a]

	1 Lactate Dehydrogenase	2 L-Glutamate-Oxalo-Acetate Transaminase	3 Alkaline Phosphatase	4 Acid Phosphatase	5 β-Glucuronidase	6 N-Acetyl-β-D-Glucosaminidase
			Normal			
Control	159 ± 38	156 ± 31	544 ± 114	245 ± 38	65.2 ± 33.5	369 ± 79
Arsenic	145 ± 50	138 ± 25	787 ± 107[b]	306 ± 92	38.1 ± 21.3	381 ± 110
			Alloxan-Diabetic			
Control	197 ± 26	184 ± 53	541 ± 136	602 ± 126[b]	38.7 ± 14.4	275 ± 65
Arsenic	412 ± 17[b,c]	685 ± 104[b,c]	823 ± 234	969 ± 105[b,c]	154 ± 79[b,c]	714 ± 62[b,c]

Source: Schiller et al. (1981).

[a]Values are specific activities given as the mean ±1 S.E.M. for duplicate determinations on three to six separate urine samples. Enzyme specific activities are expressed as nmoles per minute per milligram of urinary protein.
[b]Statistically significant difference from control (two $P < 0.10$) based on the two-sided U test [25].
[c]Statistically significant difference from normal arsenic (two $P < 0.10$) based on the two-sided U test [25].

sipidus. The levels of fluoride in both incisor and molar teeth were 3–6 times greater than in the controls. Both patients were boys, aged 10 and 11, who had lived their entire lives in communities where the drinking water did not contain more than 1 ppm fluoride. Greenberg et al. (1974) estimated their daily water intake to be between 1250 and 3000 mL, which is equivalent to a fluoride intake of $1\frac{1}{4}$–3 mg from the water. A second clinical report has also associated fluorosis with diabetes insipidus (Klein, 1975). In this case, four children with diabetes insipidus who had spent their entire lives where the drinking water had a fluoride content of about 0.5 ppm developed fluorosis. The author stated that the daily intake of water by these children was 4–5 L thereby leading to 2–2.5 mg fluoride/day from drinking water. In addition, individuals with diabetes mellitus who live in communities with fluoridated drinking water were also found to have their renal clearance of fluoride significantly impaired. However, the renal clearance of fluoride by individuals with diabetes mellitus was not significantly affected in a nonfluoridated community (Hanhijarvi, 1974). Based on these studies, Marier (1977) suggested that more cases of fluorosis with underlying nephropathic diabetes may soon be recognized given the 6% increase in diabetes incidence over the decade from 1965 to 1975 and the widespread occurrence of fluoridation programs.

Finally, Hanhijarvi (1974) reported that patients with severe renal insufficiency have significantly higher levels of free ionized plasma fluoride as compared to unaffected individuals, when such individuals are exposed to levels of fluoridated water containing either 1.0 or 0.2 ppm. According to Guy and Taves (1973) similar plasma inorganic fluoride levels are found in people who drink water containing about 2–3 ppm fluoride. Further studies revealed that the renal fluoride clearance activity is significantly lower in patients with renal insufficiency in both fluoridated and nonfluoridated areas. Similar results were found for individuals with nephritis (Hanhijarvi, 1974).

5. DIABETES AND SUSCEPTIBILITY TO X RAY

Exposure to levels of radiation in the acute LD_{50} range results in gross disturbances in the metabolism of fats, carbohydrates, and proteins, changes suggestive of severe diabetes (Bacq and Alexander, 1961).

In addition to their comparable metabolic similarities, both diabetes and severe radiation intoxication accelerate chronic degenerative disease. Based on these observations, Cember and Thorson (1978) hypothesized that diabetic individuals may be more susceptible to a single large accidental overexposure to radiation than normal persons. This hypothesis was tested by exposing female Holtzman S-D rats made diabetic by an intravenous injection of streptozotocin to the lethal effects of

X rays. The results indicated that the LD_{50} exposures for the diabetic rats and the control rats were 436 R and 617 R, respectively. While these values appear markedly different they are not significant at the 0.1 level due to the large variability in individual responses. Consequently, the conclusions of the authors that their study shows diabetic rats to be more susceptible to radiation death than nondiabetic rats is premature. Nevertheless, this apparent interaction between X irradiation and the chronically induced diabetic state is worthy of further evaluation.

6. DIABETES AND SMOKING

While investigations concerning the interaction of diabetes and pollutants are few, studies with animal models in which diabetes was chemically induced suggest that the diabetic state may enhance the toxicity of X rays, arsenic, and several organic solvents. The one attempt to investigate whether diabetes in humans as opposed to only animal models enhance pollutant toxicity is seen in the study of smoking and diabetic microangiopathy (i.e., widespread lesions in the small vessels and capillaries).

The initial report associating cigarette smoking and microangiopathy was made in 1977 by Paetkau et al. in the journal *Diabetes*. The investigators concluded that cigarette smoking diabetics displayed an enhanced risk of retinopathy. However, this initial report by Paetkau et al. (1977) was soon criticized in a letter to the editor of *Lancet* in early 1978 (March) since the smoking and nonsmoking diabetics were not sufficiently matched for the type and duration of diabetes (Christiansen and Nerup, 1978). Nevertheless, these two critics reported that smoking may be a risk factor in diabetic nephropathy, which not only is a manifestation of diabetic microangiopathy, but also the most frequent cause of death in young diabetics. They presented evidence in their letter to the editor and in a subsequent paper (Christiansen, 1978) that the prevalence of diabetic nephropathy in several hundred insulin-dependent juvenile-onset diabetics was significantly related to current and past smoking activity and to the duration of the disease. The findings of Christiansen and Nerup (1978) were partially supported by Nielsen and Hjollund (1978). In a study of 246 juvenile-onset diabetic subjects, they found no relationship between smoking and retinopathy or nephropathy in women but men displayed an excess of both lesions.

As a result of the previous reports that indicated associations between smoking and the frequency of both diabetic nephropathy (Christiansen and Nerup, 1978; Christiansen 1978; Nielsen and Hjollund, 1978) and retinopathy (Nielsen and Hjollund, 1978; Paetkau et al., 1977), Madsbad et al. (1980) evaluated whether there may be any metabolic differences between smoking and nonsmoking diabetics. In a cross-sectional study of

163 adult insulin treated patients, they noted that smokers (≥ 10 cigarettes/day) displayed a 15–20% higher insulin requirement ($P < 0.001$) and significantly higher serum triglyceride levels ($P < 0.05$), which reached a 30% increase in heavy smokers (20 cigarettes/day). However, the degree of retinopathy did not differ on the basis of smoking. In addition to previous studies, West et al. (1980) concluded a study that showed no association between cigarette smoking and the risk of glomerulosclerosis and retinopathy in a much larger population (973 patients).

Why these studies differ is of considerable concern. First, the West et al. (1980) report involved only adult onset diabetics in contrast to research by Christiansen (1978) and Nielsen and Hjollund (1978), which involved only juvenile-onset diabetics. However, West et al. (1980) did not think that this was a determining factor and claimed that "the most likely reason for these results is that smoking does not increase risk of microangiopathy." They felt their study was more valid than the earlier reports because their sample size was much larger: 973 vs. about 180 (Christiansen, 1978) and was less affected by chance and unmeasured confounding variables. Nevertheless, it is clear that the West et al. (1980) study did not test the same hypothesis as that of Christiansen (1978). Further research investigating the relationship between smoking and diabetic microangiopathy, especially among those with juvenile onset diabetes is needed.

Another smoking–diabetic interaction was reported by McNair et al. (1980), which indicated that smoking aggravates bone loss in insulin treated diabetes. In this cross-sectional epidemiologic study, the average bone loss as measured by photon absorptiometry in both forearms was twice as high in the smoking as compared to the nonsmoking diabetics. In addition, the amount of mineral reduced was directly related to the number of cigarettes smoked per day. This smoking-induced bone loss appears to be additive to the already known reduction in bone mineral content in insulin treated diabetics. The mechanism by which bone mineral content is reduced by smoking tobacco is thought to involve a suppression of gonadal function.

7. EFFECT OF THE DIABETIC STATE ON XENOBIOTIC METABOLISM

In 1961 Dixon et al. reported the first study on the effects of alloxan-induced diabetes on drug metabolism in male rats. They noted that the drug-metabolizing enzymes of the liver microsomes from such animal models displayed a reduced capacity to metabolize hexobarbital, chlorpromazine, and codeine. In addition, the *in vivo* hexobarbital induced sleeping time was prolonged after alloxan pretreatment. Subsequent investigations have tended to support these findings but they also revealed that the activities of drug metabolizing liver enzymes from alloxan-diabetic rats were increased toward several type II spectral binding substrates such as aniline (Dixon et al., 1963; Kato and Gillette, 1965; Kato et al., 1970), but this effect was not observed in streptozotocin diabetes (Ackerman and Liebman, 1977).

In partial contrast to studies in male rats, Kato et al. have reported increases over control values for the microsomal metabolism of all substrates studied in the alloxan-diabetic female rat (Kato and Gillette, 1965; Kato et al., 1968, 1970; Kato and Takahash, 1969). These researchers have hypothesized that the diabetic state caused these sex dependent metabolic alterations by interfering with androgen-dependent microsomal pathways in the male by possibly affecting a decreased substrate binding to cytochrome P-450 (Kato and Gillette, 1965; Kato et al., 1970, 1970b).

Hepatic microsomal cytochrome P-450 content has been found to be unaffected in diabetic male rats (Kato et al., 1970a; Ackerman and Liebman, 1977) but increased over control values in diabetic female rats (Kato et al., 1970). Insulin treatment of diabetic male rats corrects most abnormalities of mixed function oxidation with the exception of aminopyrene and aniline metabolism (Dixon et al., 1961, 1963; Weiner et al., 1972; Dajani et al., 1974).

Metabolic studies in streptozotocin-induced diabetic rats have been reported by Reinke et al. (1978, 1979) and Ackerman and Leibman (1977). Reinke et al. (1978, 1979) reported a sex-dependent effect of streptozotocin-induced diabetes on aminopyrene N-demethylase activity and aryl hydrocarbon hydroxylase in Sprague-Dawley rats. In diabetic male rats, the activities in these enzymes were inhibited; in contrast, they were enhanced in diabetic female rats. Aniline hydroxylase activity and cytochrome P-450 content were enhanced as compared to controls in both sexes of diabetic rats. Of importance was that insulin treatment of both male and female diabetic rats eliminated all physical and biochemical abnormalities of the diabetic condition. These findings are in apparent conflict with those of Dixon et al. (1963) who reported that the decreased aminopyrene N-demethylase activity of hepatic microsomes from diabetic male rats was not affected by insulin treatment. However, Reinke et al. (1978) pointed out that the respective studies used different insulin treatment methodologies that may have contributed to the variation between studies.

How the insulin affects liver microsomal metabolism is uncertain. However, Ackerman and Leibman (1977) demonstrated that infusion of glucose did not affect drug metabolism and they concluded that hyperglycemia is not the explanation for changes in drug-metabolizing activities in diabetic animals.

Not only does the diabetic condition affect the metabolism of xenobiotics in the liver, it also affects the extra hepatic metabolism of BP. In studies with streptozotocin-induced diabetes in female Sprague-Dawley rats BP mono-oxygenase activity and cytochrome P-450 in hepatic microsomes was 75% greater than in the controls. Insulin treatment totally reversed these alterations. Similarly, intestinal BP mono-oxygenase activity in the diabetic female rat was increased to 2.7 times that of the controls. Once again the insulin treatment reversed this change. In contrast to the increased activity of the liver and intestine, the BP mono-oxygenase activity of lung microsomes in the diabetic rats was markedly reduced and once again insulin reversed this effect. No changes in the BP mono-oxygenase activity were found in tissue homogenates of kidney and adrenals.*

While the above research was conducted in animal models, it is important to note that Dajani et al. (1974) found that human diabetics from whom insulin had been withheld for 48 hr excreted more of a dose of phenacetin as the unchanged drug and less as the 0-de-ethylated metabolite and its conjugates than after their diabetic state had been stabilized by insulin treatment and as compared to nondiabetic control subjects.

8. DIABETES AND THE HEPATOTOXIC EFFECTS OF SEVERAL ORGANIC SOLVENTS

That the diabetic state may affect the hepatotoxicity of organic solvents was first examined in 1975 by Hanasono et al. These researchers decided to test this idea based on previous results by Traiger and Plaa (1972, 1973) and Traiger and Robert (1973) who showed that pretreatment of rats with isopropanol, acetone, 2-butanol, or 2-butanone markedly enhanced CCl_4 mediated liver damage. They suspected that metabolic disease states that cause the generation of abnormal amounts of ketones may affect CCl_4 hepatotoxicity. On the way to evaluating that notion it was first necessary to determine whether diabetes induced by alloxan or streptozotocin could modify the toxic response of the liver to CCl_4.

In their study, male Sprague-Dawley rats were pretreated with single intravenous injections of alloxan (40 or 80 mg/kg) or streptozotocin (65 mg/kg) and then given a dose of CCl_4 (0.1 mL/kg IP) 4 days after alloxan pretreatment or 5 days following streptozotocin pretreatment. The animals were sacrificed 24 hr after the CCl_4 exposure. Subsequent analyses revealed that the pretreatment with either agent significantly increased CCl_4-induced hepatotoxicity as determined by serum glutamine pyruvic transaminase (SGPT) activity. Hepatic triglyceride levels in the diabetic

---

*It is thought that the female rat may be a more appropriate model than the male for extrapolation purposes since both sexes of alloxan-diabetic mice and rats showed the same effects on microsomal drug metabolism as the streptozotocin-induced diabetic female rat (Kato et al., 1970).

rats were similarly markedly increased above control values after CCl_4 exposure. By themselves, either diabetogenic agent did not enhance the occurrence of liver damage. Interestingly, insulin treatment of rats administered alloxan (80 mg/kg) significantly protected against CCl_4-induced hepatotoxicity.

The mechanism by which the diabetic state enhanced CCl_4 toxicity is not well understood. Alterations in food intake of the diabetic rats could not explain the enhanced susceptibility nor could the occurrence of ketone bodies since streptozotocin treatment does not effectively lead to their development in contrast to alloxan. Metabolic enhancement of CCl_4 toxicity via bioactivation does not seem a likely candidate either since the action of alloxan, in general, leads to reduced activity of microsomal drug metabolizing pathways (Dixon et al., 1961; Kato and Gillette, 1965; Kato et al., 1970).

Later in 1975, Hanasona et al. (1975a) reported on a follow-up study in which they assessed whether the alloxan induced diabetic state in male rats could enhance the action of a variety of other hepatotoxic agents (i.e., chloroform, 1,1,2-trichloroethane, trichloroethylene, and 1,1,1-trichloroethane) each displaying a lower hepatotoxic capability than CCl_4. In addition, they assessed the effects of alloxan on the hepatotoxic responses of chemical agents whose toxic effects markedly differ from those resulting from the above noted solvents. These additional hepatotoxins were galactosamine (focal necrosis), beryllium (midzonal necrosis), and α-naphthylisothiocyanate (cholestasis). The findings indicated that alloxan diabetes enhanced the hepatotoxicity (i.e., as indicated by SGPT values and hepatic triglyceride levels) to chloroform and 1,1,2-trichloroethane but not to trichloroethylene or 1,1,1-trichloroethane. As in the case with CCl_4, insulin treatment partially protected against the occurrence of alloxan enhanced chloroform hepatotoxicity. The alloxan induced diabetes increased the hepatotoxicity of galactosamine but not of beryllium nor α-naphthylisothiocyanate.

This follow-up study unequivocally demonstrated that alloxan pretreatment can potentiate the hepatotoxicity of several agents and is not just specific for CCl_4. The data indicated that the alloxan-induced diabetes enhanced the toxicity of agents with known high hepatotoxic capabilities but not those agents usually viewed as milder liver toxins (e.g., trichloroethylene, 1,1,1-trichloroethane).

9. DIABETOGENIC AGENTS OF ENVIRONMENTAL CONCERN

A variety of experimental agents have been proved to be diabetogenic although they may act via different mechanisms. For example, alloxan and streptozotocin induce diabetes by causing β-cell destruction and glucose intolerance with streptozotocin having a higher specificity as a β-cytotoxic agent and a lower toxicity in rats than alloxan (Ackerman and Leibman, 1977). 6-aminonicotinamide, an antimetabolite of $NADP^+$ synthesis, is thought to produce diabetes by blocking insulin release (Ammon and Steinke, 1972). N-methylacetamide causes a diabetic state by making the organism insulin resistant (Peters et al., 1966). Neither insulin synthesis nor release are affected, but the insulin that is released is not effective in lowering the blood glucose level, even though the plasma insulin level rises along with that of the glucose. The administration of insulin to animals previously treated with alloxan, streptozotocin, or 6-aminonicotinamide reverses the diabetic effects caused by such compounds but lacks any effect on those treated with N-methylacetamide (Peters et al., 1966).

a. The Rodenticide PNU

While the above-mentioned diabetogenic agents are not of environmental concern, there is a growing amount of evidence which indicates that a rodenticide and a nitrosamine can induce diabetes in humans. The rodenticide, N-3-pyridylmethyl N'-p-nitrophenyl urea (PNU) was introduced into the United States in July of 1975. Within 2 years of its introduction, the manufacturer became aware of 15 cases of acute ingestion by human adults—14 intentional and one accidental. Four of the subjects died while all 11 survivors displayed insulin-dependent diabetes mellitus (Prosser and Karam, 1978). Other clinical reports associating the ingestion of this rodenticide with sustained insulin dependent diabetes mellitus have been reported (Pont et al., 1979; Miller et al., 1978; Anonymous, 1975). While the mechanism of action is unknown, it is most likely by β-cell destruction. Circumstantial evidence supporting this hypothesis is found in the fact that the chemical structures of streptozotocin, alloxan, and PNU (Figure 7-2) are very similar in containing a urea group.

b. Nitrosamine

In 1974, Berne et al. presented evidence that the nitrosamine N-nitroso-N-methylurea was diabetogenic in both mice and Chinese hamsters causing destruction of the β cells. These findings were confirmed in a 1975 report by Wilander and Gunnarsson. Based on their findings these authors concluded that nitrosamines be studied not only for their carcinogenic and mutagenic potential, but also for their capacity to be diabetogenic. A 1980 clinical report indicated that humans were also sensitive to the diabetogenic effects of a nitrosamine. In this instance, a 44-year-old woman developed diabetes mellitus following a repeated attempt at criminal poisoning with the agent N-nitrosodimethylamine (Fussgaenger and

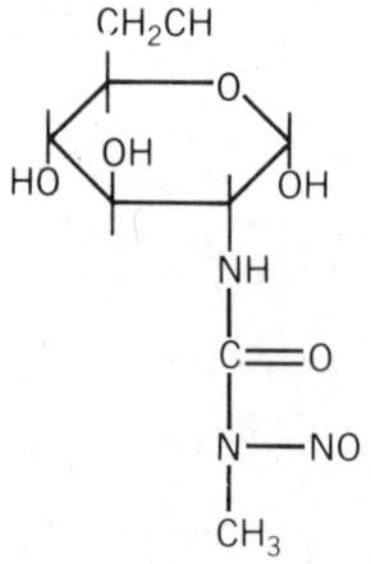

Streptozocin

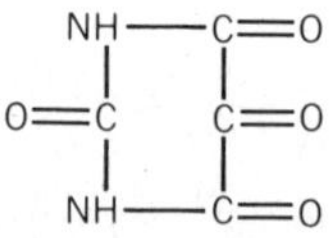

Alloxan

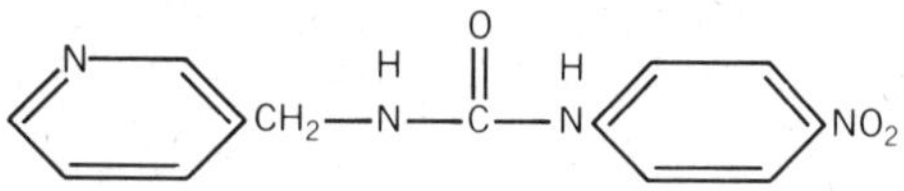

N-3 pyridylmethyl *N'*-0-nitrophenyl urea

Figure 7-2. Chemical structures of rodenticide (PNU) and two known pancreatic β-cell toxins showing common urea component. *Source:* Prosser and Koram (1978).

Ditschuneit, 1980). Before her death, she had also developed other indications of nitrosamine intoxication such as liver cirrhosis.

c. Carbon Disulfide

There are a number of occupational studies that have associated exposure to carbon disulfide with decreased glucose tolerance (i.e., latent diabetes) (Franco et al., 1978; Candura et al., 1979). The mechanism by which carbon disulfide may alter glucose tolerance is thought to involve a disorder in tryptophan degradation resulting from altered pyridoxine metabolism. The tryptophan is thought to form a complex with insulin, a complex that has less biological activity than insulin itself (Kujalova et al., 1979).

10. SUMMARY

The effects of pollutants on diabetics is an emerging area of environmental toxicology. At present, only one study with heavy metals (i.e., arsenic) has been published. This study clearly revealed that a previous exposure to arsenic enhanced the occurrence of renal abnormalities in alloxan-induced rats. Continued research with arsenic, as well as other renal toxins such as cadmium, mercury, and lead, would be of considerable biomedical interest. The fact that kidney complications such as glomerulosclerosis are so common and serious in diabetics clearly underlines the need to assess the extent to which persons with such diabetes-related pathology may be at increased risk to renal toxins. In addition to the well-known renal toxins, subdiabetogenic doses of multiple agents (i.e., several viruses and streptozotocin) have induced diabetes in rodents and may be of possible concern to high-risk individuals (Toniolo et al., 1980).

Limited research has revealed that alloxan and streptozotocin-induced diabetes in the rodent markedly altered the metabolism of a variety of mixed function oxidize enzymes. With some exceptions, the induced diabetic state slowed the rate of metabolism. Injection of insulin, however, tended to reverse metabolic activity to control levels.

REFERENCES

Ackerman, D. M. and Leibman, K. C. (1977). Effects of experimental diabetes on drug metabolism. *Intern. J. Neuropharmacol.* **7**: 405–410.

Ammon, H. P. T. and Steinke, J. (1972). *Diabetes* **21**: 143.

Anon. (1975). Rodenticide blamed for Korean deaths. *Chem. Eng. News* **53**: 8.

Bacq, Z. M. and Alexander, P. (1961). *Fundamentals of Radiobiology*, 2nd ed. Pergamon, Oxford.

Berne, C., Gunnarsson, R., Hellerstrom, C., and Wilander, E. (1974). Diabetogenic nitrosamines. *Lancet* **1**: 173–194.

Candura, F., Franco, G., Malamani, T., and Piazza, A. (1979). Altered glucose tolerance in CS_2 exposed workers. *Acta Diabetol. Lat.* **16**: 259–263.

Cember, H. and Thorson, T. M., Jr. (1978). X-ray lethality in diabetic rats. *Am. Ind. Hyg. Assoc. J.* **39**(4): 331–333.

Christiansen, J. S. (1978). Cigarette smoking and prevalence of microangiopathy in juvenile-onset insulin dependent diabetes mellitus. *Diabetes Care* **1**(3): 146–149.

Christiansen, J. S. and Nerup, J. (1978). Smoking and diabetic nephropathy. *Lancet* **1**: 605.

Dajani, R. M., Kaygali, S., Saheb, S. E., and Birbari, A. (1974). *Comp. Gen. Pharmacol.* **5**: 1.

Dixon, R. L., Hart, L. G., and Fouts, J. R. (1961). The metabolism of drugs by liver microsomes from alloxan-diabetic rats. *J. Pharmacol. Exp. Ther.* **133**: 7–11.

Dixon, R. L., Hart, L. G., Rogers, L. A., and Fouts, J. R. (1963). The metabolism of drugs by liver microsomes from alloxan-diabetic rats: Long term diabetes. *J. Pharmacol. Exp. Ther.* **142**: 312–317.

Franco, G., Malamani, T., and Piazzi, A. (1978). Glucose tolerance and occupational exposure to carbon disulphide. *Lancet* **2**: 1208.

Fussgaenger, R. D. and Ditschuneit, H. (1980). Lethal exitus of a patient with N-nitrosodimethylamine poisoning, 2.5 years following the first ingestion and signs of intoxication. *Oncology* **37**: 273–277.

Gorwitz, K., Howen, G. G., and Thompson, T. (1976). Prevalence of diabetes in Michigan school age children. *Diabetes* **25**: 122.

Greenberg, L. W., Nelsen, C. E., and Kramer, N. (1974). *Pediatrics* **54**: 320.

Grodsky, G. M., Anderson, C. E., Coleman, D. L., Craighead, J. E., Gerritsen, G. C., Hansen,

C. T., Herberg, L., Howard, C. F., Jr., Lernmark, A., Matschinsky, F. M., Rayfield, E., Riley, W. J., and Rossini, A. A. (1982). Metabolic and underlying causes of diabetes mellitus. *Diabetes Suppl. 31* **45:** 53.

Guy, W. S. and Taves, D. R. (1973). *Int. Assoc. Dent. Res. Abstract.* 718. (Cited in Marier, 1977.)

Hanasono, G. K., Witschi, H., and Plaa, G. (1975). Potentiation of the hepatotoxic responses to chemicals in alloxan-diabetic rats. *Proc. Soc. Exp. Biol. Med.* **149:** 903–907.

Hanasono, G. K., Cote, M. G., and Plaa, G. L. (1975). Potentiation of CCl_4-induced hepatotoxicity in alloxan or streptozotocin-diabetic rats. *J. Pharmacol. Exp. Ther.* **192:** 592–604.

Hanhijarvi, H. (1974). Comparison of free ionized fluoride concentrations of plasma and renal clearance in patients of artificially fluoridated and non-fluoridated drinking water areas. *Proc. Finn. Dent. Soc. Suppl. III* **70:** 12.

Karam, J. H., Prosser, P. R., LeWitt, P. A. (1978). Islet-cell surface antibodies in a patient with diabetes mellitus after rodenticide ingestion. *N. Engl. J. Med.* **299:** 1191.

Kato, R. and Gillette, J. R. (1965). Sex differences in the effects of abnormal physiological states on the metabolism of drugs by rat liver microsomes. *J. Pharmacol. Exp. Ther.* **150:** 285–294.

Kato, R., Onoda, K., and Takanaka, A. (1970). Species difference in drug metabolism by liver microsomes in alloxan diabetic or fasted animals. (1) The activity of drug-metabolizing enzymes and electron transport system. *Jpn. J. Pharmacol.* **20:** 546–553.

Klein, H. (1975). *Oral Surg. Oral Med. Oral Pathol.* **40:** 736.

Kujalova, V., Lukas, E., Sperlingova, I., and Frantik, E. (1979). Glucose tolerance and occupational exposure to carbon disulphide. *Lancet* **1:** 664.

Kyllo, C., and Nuttall, F. Q. (1976). Prevalence of diabetes mellitus among public school children in Minnesota. *Diabetes* **25,** Suppl, Abstract 205.

Madsbad, S., McNair, P., Christiansen, M. S., Christiansen, C., Faber, O. K., Binder, C., and Transbol, I. (1980). Influence of smoking on insulin requirement and metabolic status in diabetes mellitus. *Diabetes Care* **3:** 41–43.

Mansford, K. R. L. and Opie, L. (1968). Comparison of metabolic abnormalities in diabetes mellitus induced by streptozotocin or by alloxan. *Lancet* **1:** 670–671.

Marier, J. P. (1977). Some current aspects of environmental fluoride. *Sci. Total Environ.* **8:** 253–265.

McNair, P., Christiansen, M. S., Madsbad, S., Christiansen, C., Binder, C., and Trensbol, I. (1980). Bone loss in patients with diabetes mellitus. *Miner. Electrolyte Metabolism* **3:** 94–97.

Miller, L. V., Stokes, J. D., and Silpipat, C. (1978). Diabetes mellitus and autonomic dysfunction after rat or rodenticide ingestion. *Diabetes Care* **1:** 73–76.

National Commission on Diabetes (Report to the Congress of the United States) (1973). The Long Range Plan to Combat Diabetes. U.S. DHEW No. 76-1018, NIH.

Nielsen, M. M. and Hjollund, E. (1978). Smoking and diabetic microangiopathy. *Lancet* **2:** 533–534.

Paetkau, M. E. (1978). Diabetic retinopathy and smoking. *Lancet* **2:** 1098–1099.

Paetkau, M. E., Boyd, T. A. S., Winship, B., and Grace, M. (1977). Cigarette smoking and diabetic retinopathy. *Diabetes* **26:** 46–48.

Peters, G., Guidox, R., and Grassi, L. (1966). *Naunyn-Schmiedebergs Arch. Exp. Pathol. Pharmakol.* **255:** 58.

Pont, A., Rubino, J. M., Bishop, D., and Peal, R. (1979). Diabetes mellitus and neuropathy following vacor ingestion in man. *Arch. Intern. Med.* **139:** 185–187.

Prosser, P. R. and Karam, J. H. (1978). Diabetes mellitus following rodenticide ingestion in man. *J. Am. Med. Assoc.* **239:** 1148–1150.

Reinke, L. A., Stohs, S. J., and Rosenberg, H. (1978). Altered activity of hepatic mixed-function monooxygenases enzymes in streptozotocin–induced diabetic rats. *Xenobiotica,* **8**(10): 611–619.

Reinke, L. A., Stohs, S. J. and Rosenberg, H. (1979). Increased aryl hydrocarbon hydroxylase activity in hepatic microsomes from streptozotocin–diabetic female rats. *Xenobiotica,* **8**(12): 769–778.

Renold, A. E., Mintz, D. H., Muller, W. A., Cahill, G. E., Jr. (1978). Diabetes mellitus. In: *The Metabolic Basis of Inherited Disease.* J. B. Stanbury, J. B. Wyngaardon, and D. S. Frederickson (eds.). McGraw-Hill, New York, p. 80.

Rerup, C. C. (1970). Drugs producing diabetes through damage of insulin-secreting cells. *Pharmacol. Rev.* **22:** 485–520.

Schiller, C. M., Walden, R., and Fowler, B. A. (1981). Interaction between arsenic and alloxan-induced diabetes: effects on rat urinary enzyme levels. *Biochem. Pharmacol.* **30:** 168–170.

Toniolo, A., Onodera, T., Yoon, J-W., and Notkins, A. B. (1980). Induction of diabetes by cumulative environmental insults from viruses and chemicals. *Nature* **288:** 383–385.

Traiger, G. J. and Plaa, G. L. (1971). Differences in the potentiation of carbon tetrachloride in rats by ethanol and isopropanol pretreatment. *Toxicol. Appl. Pharmacol.* **20:** 105–112.

Traiger, G. J. and Plaa, G. L. (1972). Relationship of alcohol metabolism to the potentiation of CCl_4 hepatotoxicity induced by aliphatic alcohols. *J. Pharmacol. Exp. Ther.* **183:** 481–488.

Traiger, G. J. and Plaa, G. L. (1973). Effect of aminotriazole on isopropanol and acetone-induced potentiation of CCl_4 hepatotoxicity. *Can. J. Physiol. Pharmacol.* **51:** 291–296.

Traiger, G. and Robert, T. (1973). Relationship of alcohol metabolism to the potentiation of CCl_4 hepatotoxicity by 2-butanol. *Pharmacologist* **15:** 260.

Weiner, M., Buterbaugh, G. G., and Blake, D. A. (1972). *Res. Commun. Chem. Pathol. Pharmacol.* **4:** 37.

West, K. M., Erdreich, L. S., and Stober, J. A. (1980). Absence of a relationship between smoking and diabetic microangiopathy. *Diabetes Care* **3:** 250–252.

Wilander, E. and Gunnarsson, R. (1975). Diabetogenic effects of *N*-nitrosomethylurea in the Chinese hamster. *Acta Pathol. Microbiol. Scand. Sect. A* **83:** 206–212.

B. Animal Models for Diabetes Study

The studies that have investigated the effects of pollutants on the diabetic state are few in number. Schiller et al. (1981) evaluated how a previous exposure to arsenic affected the renal function of alloxan-induced diabetes in rats. Earlier, Traiger and his associates (1971, 1972, 1973) studied the influence of chemically-induced (i.e., alloxan and/or streptozotocin) diabetes on the capacity to metabolize several organic solvents and how this could be subsequently affected by insulin treatment. Consequently, whether the diabetic state is a predisposing condition enhancing the toxic effects of specific environmental pollutants is for the most part unknown. Nevertheless, an extraordinary range of animal models to study the diabetic state exists and has been exploited by the biomedical community to study the pathogenesis and pathophysiology of this disease.

Mordes and Rossini (1981) have recently articulated the important role that studies with animal models of diabetes may have in our understanding of this disease as well as in the development of therapeutic treatments. Among those advantages articulated were: (1) the facilitation of genetic studies with many generations over a relatively short period; (2) the ability to isolate specific genetic, dietary, or environmental variables for study; and (3) the chance to assess the developmental pathophysiology of the disease.

The occurrence of hyperglycemia is widespread in the animal kingdom having been found in cats, dogs, cattle, sheep, horses, pigs, shrews, monkeys, hippopotamuses, and dolphins. However, the disease has been best characterized in rodents primarily because they have been studied most as a result of their availability and low cost.

Animal models for the diabetic state were classified by Mordes and Rossini (1981) into several broad categories: (1) spontaneous diabetes in lean animals, (2) spontaneous diabetes in obese animals, (3) animals with diabetes related to dietary stress, (4) experimentally induced diabetes, and (5) diabetes induced by toxic agents. Table 7-2 summarizes the characteristics of spontaneously diabetic animals including body shape, level of plasma insulin, degree of ketosis, and β-cell pathology, while Table 7-3 provides information on genetics of some of the spontaneously hyperglycemic models.

The experimentally induced examples of diabetes involve the use of: (1) insulin antagonists, such as epinephrine, glucocorticoids, and growth hormone; (2) agents such as gold thioglucose that cause hypothalamic lesions leading to obesity, hyperglycemia, hyperinsulinemia, and insulin resistance; (3) virus induction of diabetes such as with rubella, encephalitis, and foot and mouth disease. The only experimental models that have been used in pollutant related studies have been the toxic agent induced diabetes using alloxan and streptozotocin, both β-cell destroyers.

Since human diabetes affects a variety of systems including the renal (e.g., nephropathy), cardiovascular (e.g., myocardial infarction), and nervous systems (i.e., retinopathy), it is often best to select a model of diabetes related to the disease process of interest.

1. EXAMPLE OF ANIMAL MODELS OF DIABETIC NEPHROPATHY

According to Brown et al. (1982) the initial requirement of animal models of this disease entity is that the structural changes are comparable to those occurring in humans. These authors have indicated that extraordi-

Table 7-2. Characteristics of Some Spontaneously Diabetic Animals

Name	Body Habitus	Plasma Insulin	Ketosis	β-Cell Pathology
BB rat	Lean	Very low	++++	Insulitis, β-cell necrosis
Chinese hamster (*Cricetulus griseus*)	Lean	Normal or low	+	Degranulation, decreased β-cell numbers; with ketosis; necrosis
South African hamster (*Mystromys albicaudatus*)	Lean	?	++	Most show hyperplasia and glycogen infiltration with ketosis; necrosis
C57BL/6J *ob/ob* mouse	Very obese	Early, very high; late, normal		Early: degranulation; later: hyperplasia and regranulation
C57BL/KsJ *db/db* mouse	Very obese	Early, high; late, low	Rarely	Early: degranulation; later: β-cell necrosis
Yellow mouse	Obese	High		Hypertrophy and hyperplasia
KK mouse	Obese	High		Hypertrophy and hyperplasia
Yellow-KK mouse	Obese	High		Hypertrophy and hyperplasia
Sand rat (*Psammomys obesus*)	Obese[a]	High	Rarely	Islet enlargement; β-cell degranulation; glycogen deposits; with ketosis: necrosis
Spiny mouse (*Acomys cahirinus*)	Obese[a]	High	Rarely	Massive hypertrophy and hyperplasia
Wellesley hybrid mouse (C3hf X I)	Obese	High		β-cell hyperplasia, hypertrophy, and degranulation
Celebese black ape (*Macaca nigra*)	Lean	Normal		Amyloid
Djungarian hamster (*Phodopus sungorus*)	Obese	High	+	Hypertrophy and hyperplasia
Tuco-tuco (*Ctenomys talarum*)	Very obese	?	+++	Hypertrophy
Keeshond dog	Lean	Low		Absence of islet β cells; solitary pancreatic β cells

Source: Mordes and Rossini (1981).
[a] In the wild, animals are lean.

Table 7-3. Genetics of Spontaneously Hyperglycemic Animals

Single gene mutants
   Obese mouse (ob/ob)   (autosomal recessive)
   Diabetic mouse (db/db)   (autosomal recessive)
   Obese yellow mouse (A^{vy})   (autosomal dominant)
Polygenic inheritance
   Chinese hamster
   NZO mouse
   KK mouse (one component dominant, penetrance influenced by recessive modifiers)
   Djungarian hamster
   South African hamster
   Pbb/Ld mouse
Hybrids
   Wellesley hybrid mouse (C3Hf × I)
Unknown heredity with major environmental component
   Sand rat
   Spiny mouse
   Tuco-tuco
Unknown
   BB rat
   Celebese ape (*Macaca nigra*)
   Keeshond dog

Source: Mordes and Rossini (1981).

nary similarities exist between the various available models in the morphological characteristics of the disease regardless of the cause of the diabetes (Tables 7-4, 7-5). In other words, the nephropathy of spontaneous diabetes in man is very similar to those of other species with spontaneous diabetes, alloxan or streptozotocin, pancreatectomy, or growth hormone diabetes. Despite these remarkable similarities, no single diabetic animal model develops diabetic nephropathy identical to that in humans. For example, the rodent does not develop glomerular arteriolar hyalinosis or increased linear extracellular basement membrane localization of plasma proteins.

2. EXAMPLE OF ANIMAL MODELS OF DIABETES WITH CARDIOVASCULAR SEQUELLAE

Unfortunately, the utilization of animal models for the evaluation of the cardiovascular implications of diabetes have been hampered by difficulty in the availability of the most appropriate model (i.e., nonhuman higher primates) and by the knowledge that the most widely available models (i.e., rodents) develop cardiovascular disease to a much more limited

Example of Animal Models of Diabetes with Cardiovascular Sequellae

Table 7-4. Structural Changes of the Kidney in Diabetes

	Humans	Other Primates	Rodents	Dog
Renal enlargement	Yes	?	Yes	?
Glomerular enlargement	Yes	Yes	Yes	Yes
Increase in mesangial area (diffuse glomerulosclerosis)	Yes	Yes	Yes	Yes
Increased mesangial matrix area	Yes	Yes	Yes	Yes
Increased mesangial cell area	Yes	Yes	Yes	Yes
Kimmelsteil-Wilson lesions (nodular glomerulosclerosis)	Yes	Yes	No	Yes
Glomerular basement membrane thickening	Yes	Yes	Yes	Yes
Hyaline (exudative) glomerular lesions	Yes	Yes	Yes	Yes
Glomerular arteriolar hyalinosis	Yes	Yes	No	Yes
Increased extracellular basement membrane plasma protein localization	Yes	?	No	?

Source: Brown et al. (1982).

extent than humans. This is unfortunate since the cardiovascular implications of diabetes in humans are of considerable consequence since arteriosclerotic vascular disease usually takes place at an earlier age and progresses more rapidly in diabetics than in nondiabetic subjects often resulting in myocardial infarcts, strokes, and gangrene.

The difficulty with rodents is clearly illustrated in studies of rats that

Table 7-5. Animal Models of Diabetes in which Glomerular Basement Membrane and Mesangial Thickening Have Been Demonstrated

Animal	Model System
Rat	Alloxan
	Streptozotocin
	Pancreatectomy
	High sucrose diet (genetic)
Mouse	KK (genetic)
	C57/BLKsJ-*db/db* (genetic)
Dog	Alloxan
	Growth hormone
Monkey	Alloxan

Source: Mauer et al. (1981).

have been made diabetic by alloxan or streptozotocin. In these cases, no significant macrovascular disease has been reported unless dietary manipulations are employed. Such difficulties have led to the use of alloxan-treated rabbits since they develop a stable nonketotic form of diabetes with the growth rate being close to normal over a prolonged period of time. These attributes make this model of potential value for chronic evaluation.

A task-force studying the current state of knowledge about the predicted efficacy of diabetic animal models for human cardiovascular disease have recently concluded that:

Nonhuman primates have many desirable features for studies on the macrovascular and cardiac complications of the disease as well as risk factor alterations, but their availability, cost, and maintenance present practical disadvantages. The spontaneous rodent models of diabetes currently are not considered very useful for cardiovascular research, but they have not been well characterized with respect to most aspects of their cardiovascular system. Alloxan-diabetic rabbits offer some promise for examining the effects of diabetes on atherogenesis, lipoprotein metabolism, and cardiomyopathy, but additional research is required to validate their usefulness. Insufficient data are available on canine and swine models of diabetes to judge their merits for cardiovascular research.

The Task Force recommends: (1) additional long-term investigations to determine the extent and severity of cardiovascular complications in the well-characterized rodent models and in diabetic rodents with multiple risk factor abnormalities; (2) further studies on the macrovascular disease and lipoprotein abnormalities of the alloxan-diabetic rabbit and the development of rabbit colonies with spontaneous diabetes; (3) increased emphasis on such potentially important but neglected areas of research in diabetic animals as the intramyocardial circulation, adventitial blood vessels, blood pressure, platelet function, blood coagulation, blood rheology, and autonomic nervous function; (4) long-term studies on the influence of control of hyperglycemia and of insulin therapy on cardiovascular complications in diabetic animals; and (5) encouragement of use of diabetic nonhuman primates for cardiovascular research and institution of measures to increase their supply and availability of expanding current colonies, screening newly imported animals for diabetes, and establishing a visiting scientist's program allowing investigators to study diabetic primates at resource centers.

3. EXAMPLE OF DIABETIC ANIMAL MODELS WITH CENTRAL AND PERIPHERAL NERVOUS SYSTEM PROBLEMS

Brown et al. (1982) has reported that at least 10% of all diabetic patients develop symptomatic neuropathy and greater than 80% of diabetics have electrophysiologic and morphologic abnormalities of peripheral nerves. These authors claim that animal models may be useful for assessing the metabolic and electrophysiologic abnormalities associated with diabetic neuropathy. Of particular potential are genetic models such as the Chinese hamster, *ob/ob* mouse, *db/db* mouse, BB-Wistar rat, and SSDR rat as well as the chemically induced or nutritional models. However, only limited similarities with human diabetic neuropathy have been shown in any of the above genetic models. Studies with the chemically (alloxan and streptozotocin) induced models, have resulted in streptozotocin as the chemical of choice, because the use of alloxan has been found to result in a higher mortality and prolonged time for the development of nerve fiber pathology.

As in many other domains of biomedical science this area of diabetes research places extraordinary demands on the predictive capability of animal models. The problem in the case of diabetes-induced central and peripheral nervous system disorders is that there are several types of neuropathy with different underlying mechanisms. In order to know whether a model fits the human disease will demand an extensive detailing of the human disease. At the present time, our understanding of the human disease and animal models are too limited for extensive and thorough evaluation of the animal models.

4. SUMMARY

There are numerous potential animal models that investigators use to study diabetes. Characterization of such models is a rapidly evolving area as indicated by the scope and breadth of a conference proceedings on this topic published in 1982. The problem is that diabetes itself is a complex disease with genetic, aging, dietary, and environmental components exhibiting a broad spectrum of pathophysiologic sequellae. In addition, the physiological manifestations of diabetes must be superimposed on other physiological and/or disease processes in operation. This results in the need to define very narrowly those aspects of diabetes any particular model is desired to simulate, be it renal, circulatory, or nervous. At present, the best predictive models are those concerned with renal toxicity.

REFERENCES

Brown, D. M., Andres, G. S., Hostetter, T. H., Mauer, S. M., Price, R., and Venkatachalam, M. A. (1982). Kidney complications. *Diabetes Suppl. 31* **1**: 71–81.

Brown, M. S., Dyck, P. J., McClearn, G. E., Sima, A. A. F., Powell, A. C., and Porte, D., Jr. (1982). Central and peripheral nervous system complications. *Diabetes* **31** Suppl. 65–70.

Chobanian, A. V., Aquilla, E. R., Clarkson, T. B., Eder, H. A., Howard, C. F., Jr., Regan, T. J., and Williamson, J. R. (1982). Cardiovascular complications. *Diabetes* **31** Suppl. 64–64.

Mauer, S. M., Steffes, M. W., and Brown, D. M. (1981). The kidney in diabetes. *Am. J. Med.* **70:** 603–612.

Mordes, J. P. and Rossini, A. A. (1981). Animal models of diabetes. *Am. J. Med.* **70:** 353–360.

Schiller, C M., Walden, R., and Fowler, B. A. (1981). Interaction between arsenic and alloxan-induced diabetes: effects on rat urinary enzyme levels. *Biochem. Pharmacol.* **30:** 168–170.

Traiger, G. J. and Plaa, G. L. (1971). Differences in the potentiation of carbon tetrachloride in rats by ethanol and isopropanol pretreatment. *Toxicol. Appl. Pharmacol.* **20:** 105–112.

Traiger, G. J. and Plaa, G. L. (1972). Relationship of alcohol metabolism to the potentiation of CCl_4 hepatotoxicity induced by aliphatic alcohols. *J. Pharmacol. Exp. Ther.* **183:** 481–488.

Traiger, G. J. and Plaa, G. L. (1973). Effect of aminotriazole on isopropanol- and acetone-induced potentiation of CCl_4 hepatotoxicity. *Can. J. Physiol. Pharmacol.* **51:** 291–296.

Traiger, G. and Robert, T. (1973). Relationship of alcohol metabolism to the potentiation of CCl_4 hepatotoxicity by 2-butanol. *Pharmacologist* **15:** 260.

C. Genetic Diseases

1. CYSTINURIA, CYSTINOSIS, AND TYROSINEMIA OF THE KIDNEY AND HEAVY METAL TOXICITY

A broad variety of factors including genetic diseases, nutritional deficiencies, and environmental pollutants may adversely affect the functioning of the kidney so that in certain cases abnormally high urinary levels of water, phosphate, sodium, potassium, glucose, amino acids, and other organic acids may occur (Schneider and Seegmiller, 1972).

Several of the more important genetic diseases that may adversely affect resorption of nutrients by the kidney resulting in Fanconi's syndrome (hyperaminoaciduria, glycosuria, and hypophosphatemia) include (1) cystinuria, (2) cystinosis, and (3) tyrosinemia (Schneider and Seegmiller, 1972).

a. Genetic

Cystinuria

Cystinuria is a disorder of amino acid transport affecting the epithelial cells of the renal tubule. The disease is usually noted by the precipitation of cystine in the urethra. Other amino acids including lysine, arginine, ornithine, and cysteine-homocysteine mixed disulfide are also present in excess in the urine. Cystinuria is usually of no clinical importance except under conditions of low dietary protein and when cystine crystals may develop in the urethra. It has been estimated that about 1 per 200–250 individuals excrete excess cystine in the urine and 1 per 20,000–100,000 are homozygous for the trait (Thier and Sogal, 1972).

Screening Tests for Cystinuria. Every patient with urinary calculi or with urinary tract sensitivity indicative of calculi should be examined for the possibility of cystinuria (Thier and Segal, 1972). The cyanide-nitroprusside test has been used as a chemical procedure (Brand et al., 1930). This test usually allows for easy detection of the homozygote; some, but not all, heterozygotes may be identified via this chemical test. More sophisticated and quantitative tests, if necessary, are available. [See Thier and Segal (1972) for a further discussion of identification tests.]

Cystinosis

Cystinosis is noted by a high intracellular content of cystine located in particular cell types. Symptoms of the most severe form, nephropathic cystinosis, result in the above-cited substances present in excess in the urine. The continued loss of excessive quantities of phosphate results in the development of hypophosphatemic rickets resistant to treatment by vitamin D. Growth is often severely retarded and death usually occurs by the age of 10 from uremia. In less severe nephropathic cases (intermediate forms), clinical symptoms are considerably reduced and such people live to the second or third decade of life. In the least severe form (benign cystinosis), the principal clinical difference between the benign and the nephropathic form of the disease is the absence in these patients of renal dysfunction. However, they do exhibit crystalline deposits in the cornea, bone marrow, and leukocytes, but these do not result in disability and the patient lives to adult life. Both the nephropathic and benign types of cystinosis have an autosomal recessive form of inheritance. The frequency of the disease in the population has not been assessed (Schneider and Seegmiller, 1972).

Screening Tests for Cystinosis. According to Schneider et al. (1967), fibroblasts derived from cystinotic patients and maintained for several generations have greatly increased quantities of free cystine. In fact, the levels of cystine from nephropathic cystinotic patients may have greater than 100 times the normal level of free cystine, while patients with benign cystinosis have approximately 50 times the normal free cystine content (Schneider et al., 1967, 1968). Schneider and Seegmiller (1972) suggest that these results imply that there is a quantitative, and therefore, an identifiable biochemical difference between the clinical expression of these two cystinotic variants, as well as the normal segment of the population. Although there is occasional overlap between the control and heterozygote (for the nephropathic conditions), the differences between these groups remain highly significant for leukocyte ($P < 0.001$) and fibroblast ($P < 0.01$) values. The authors suggest that since the cystine levels of heterozygote tissues are substantially less than half those present in tissues of homozygote patients, it means that the cystine level is not directly proportional to gene dosage in this recessively inherited disease. Finally, individuals with benign cystinosis are usually identified only by "accident" during a normal ophthalmologic examination with a slit lamp (due to cystine crystals in the cornea) (Schneider and Seegmiller, 1972).

Tyrosinemia

In individuals with tyrosinemia, *p*-hydroxyphenyllactic acid accumulates and is thought to be the cause of tubular damage in such individuals (Gentz et al., 1965). It is of considerable interest that these same compounds accumulate in individuals with scurvy since ascorbic acid is needed by the enzyme (*p*-hydroxyphenyl-pyruvic acid oxidase), which is deficient in tyrosinemia. The clinical description is highly variable depending on whether the disease is chronic or acute. However, very often, afflicted individuals exhibit congenital cirrhosis of the liver, renal tubular defects with aminoaciduria, and vitamin D resistant rickets (hypophosphatemic condition). Although the frequency of tyrosinemia in the general population is not known, it has been estimated that the frequency of heterozygote carriers in the Chicoutimi region of northeastern Quebec is one carrier for every 20–31 people (Laberge and Dallaire, 1967; La Du and Gjessing, 1972).

Screening Tests for Tyrosinemia. In the chronic stage, tyrosinemia is clearly diagnosed by liver cirrhosis, methioninemia, tyrosyluria with *p*-hydroxyphenyllactic acid as the major metabolite, generalized amino aciduria of a specific form, glucosuria, proteinuria, and hyperphosphaturia with rickets. The acute stage is harder to identify since fructosemia appears similar to transitory hypertyrosinemia combined with liver disease. Thus, in order to definitely distinguish tyrosinemia, it is necessary to perform fructose, galactose, ascorbic acid, and phenylalanine loading tests. No biochemical test has been developed to identify heterozygous carriers of the disease (La Du and Gjessing, 1972; Bodegard et al., 1969; Lindemann et al., 1970).

b. **Nutritional**

Jonxis and Wadman (1950) reported an increased output of some free and bound amino acids in the urine of a child with scurvy. Other clinical studies with two children with scurvy also revealed an increased output of amino acids. However, following ascorbic acid administration for at least 3 weeks, normal levels of amino acids in the urine were approached (Jonxis and Huisman, 1954). Dinning (1953) has reported that vitamin E deficiency in rabbits also acts to significantly increase the levels of amino acids in the urine.

c. Environmental Pollutants

Hyperaminoaciduria has been reported in experimental plumbism with clinical lead intoxication (Wilson et al., 1953; Bickel and Souchon, 1955; Maggioni et al., 1960) and in asymptomatic industrial workers exposed to lead (Clarkson and Kench, 1956). Chisolm (1962) further showed lead intoxication may also produce glucosuria and hypophosphatemia as well as hyperaminoaciduria. According to Chisolm (1962), lead-induced aminoaciduria is similar to that found in vitamin D deficiency rickets, cystinosis, hyperparathyroidism with renal calcinosis, and congenital renal aminoaciduria. These conditions result in a generalized renal aminoaciduria in which the reabsorption of all amino acids by the proximal renal tubular cells is uniformly impaired. Chisolm (1962) suggested that "some renal tubular mechanism which is common to and essential for the reabsorption of all amino acids is affected" in these diseases.

Clarkson and Kench (1956) have demonstrated that aminoaciduria may occur in asymptomatic industrial workers exposed to cadmium, mercury, and uranium in addition to lead. Exposure to cadmium and uranium resulted in more extreme changes than did lead and mercury in the urinary amino-nitrogen levels and the amino acid patterns. Other chemical agents including maleic acid (Harrison and Harrison, 1954), and copper in Wilson's disease (Bearn et al., 1957; Stein et al., 1954) have also produced the Fanconi syndrome.

Renal tubular impairment has also been caused by intoxication from a variety of aromatic compounds. Otten and Vis (1968) have noted that ingestion of methyl-3-chromone induced the Fanconi syndrome. They also pointed out the structural similarity of methyl-3-chromone to tetracycline. Benitz and Diermeier (1964) had previously indicated that anhydro-4-epitetracycline, which is the degradation product of tetracycline, is toxic to the renal tubules.

If the suggestion by Chisolm (1962) that the same renal tubular mechanism that is common to and essential for the absorption of all the amino acids is similarly affected by the various genetic diseases, nutritional deficiencies, and pollutants such as lead and cadmium, is selected as a working model, then individuals affected by one or more of the related genetic diseases and/or nutritional deficiencies should be considered at high risk to the effects of lead, cadmium, mercury, and uranium. Since these individuals are already predisposed to the development of the Fanconi syndrome, the concomitant exposure of such pollutants would provide an additional stress at a critical biological weak point. It is not possible at this time to determine whether the exposure of these substances (separately or together) in these instances (e.g., genetic disease or nutritional deficiency) would result in toxic effects of an additive or synergistic nature.

REFERENCES

Bearn, A. G., Yu, T. F., and Guttman, A. B. (1957). Renal function in Wilson's Disease. *J. Clin. Invest.* **36:** 1107.

Benitz, K. F. and Diermeier, H. F. (1964). Renal toxicity of tetracycline degradation products. *Proc. Soc. Exp. Biol. Med.* **115:** 930.

Bickel, H. and Souchon, F. (1955). Die papierchromatographie in der Kinderheilkunde. *Arch. Kinderheilkd. Suppl.* **31:** 101.

Bodegard, G., Gentz, J., Lindblad, B., Lindstedt, S., and Zetterstom, R. (1969). Hereditary tyrosinemias. III. On the differential diagnosis and the lack of effect of early dietary treatment. *Acta Paediatr. Scand.* **58:** 37.

Brand, E., Harris, M. M., and Biloon, S. (1930). Cystinuria: excretion of a free cystine. *J. Biol. Chem.* **86:** 315.

Chisolm, J. J. (1962). Aminoaciduria as a manifestation of renal tubular injury in lead intoxication and a comparison with patterns of aminoaciduria seen in other diseases. *J. Pediatr.* **60:** 1.

Clarkson, T. W. and Kench, J. E. (1956). Urinary excretion of amino acids by men absorbing heavy metals. *Biochem. J.* **62:** 361.

Dinning, J. S. (1953). Some effects of vitamin E on amino acid metabolism. *Fed. Proc.* **12:** 412.

Gentz, J., Jagenburg, R., and Zetterstrom, R. (1965). Tyrosinemia. An inborn error of tyrosine metabolism with cirrhosis of the liver and multiple renal tubular defects. (Le Toni-Debré-Fanconi Syndrome). *J. Pediatr.* **66:** 670.

Harrison, H. E. and Harrison, H. C. (1954). Experimental production of renal glycosuria, phosphaturia and aminoaciduria by injection of maleic acid. *Science* **120:** 606.

Jonxis, J. H. P. and Huisman, T. H. J. (1954). Amino aciduria and ascorbic acid deficiency. *Pediatr.* **14:** 238.

Jonxis, J. H. P. and Wadman, S. K. (1950). De vitscheiding van aminozuren in de urine bij een patient met scorbuut. *Maandschr. Kindergeneeskd.* **81:** 251.

Laberge, C. and Dallaire, L. (1967). Genetic aspects of tyrosinemia in the Chicoutimi region. *Can. Med. Assoc.* **97:** 1099.

La Du, B. N. and Gjessing, L. R. (1972). Tyrosinosis and Tyrosinemia. In: *The Metabolic Basis of Inherited Disease.* J. B. Stanbury, J. B. Wyngaarden, and D. S. Fredrickson (eds.). McGraw-Hill, New York, p. 296.

Lindemann, R., Gjessing, L. R., Merton, B., Christie Loken, A., and Haluorsen, S. (1970). Amino acid metabolism in fructosemia. *Acta Paediatr. Scand.* **59:** 141.

Maggioni, G., Bottini, E., and Biagi, G. (1960). Contributo alla conoscenza dell' aminoaciduria qualitativa e quantitativa dell' intossicazione da piomba del bambino. *Bull. Soc. Ital. Biol. Sper.* **36:** 193.

Otten, J. and Vis, H. L. (1968). Acute reversible renal tubules dysfunction following intoxication with methyl-3-chrome. *J. Pediatr.* **73:** 422.

References

Schneider, J. A., Rosenbloom, F. M., Bradley, K. H., and Seegmiller, J. E. (1967). Increased free-cystine content of fibroblasts cultured from patients with cystinosis. *Biochem. Biophys. Res. Commun.* **29:** 527.

Schneider, J. A. and Seegmiller, J. E. (1972). Cystinosis and the Fanconi Syndrome. In: *The Metabolic Basis of Inherited Disease.* J. B. Stanbury, J. B. Wyngaarden, and D. S. Fredrickson (eds.). McGraw-Hill, New York, p. 1581.

Schneider, J. A., Wong, V., Bradley, K. H., and Seegmiller, J. E. (1968). Biochemical comparisons of the adult and childhood forms of cystinosis. *N. Engl. J. Med.* **279:** 1253.

Selkurt, E. E. (1971). Renal function. In: *Physiology.* E. E. Selkurt (ed.). Little, Brown, Boston, p. 487.

Stein, W. H., Bearn, A. E., and Moore, S. (1954). The amino acid content of the blood and urine in Wilson's disease. *J. Clin. Dis.* **33:** 410.

Thier, S. O. and Segal, S. (1972). Cystinuria. In: *The Metabolic Basis of Inherited Disease.* J. B. Stanbury, J. B. Wyngaarden, and D. S. Fredrickson eds.). McGraw-Hill, New York, p. 1504.

Wilson, V. K., Thompson, M. L., and Dent, C. E. (1953). Amino-aciduria in lead poisoning. *Lancet* **2:** 66.

8 Immunological Associations/Deficiencies

A. HLA Associations

Within the past decade, striking advances in the development of identifying genetic markers that are associated with the occurrence of a wide range of human diseases have been made. Just as each individual is thought to have his own unique fingerprints, it is now known that each individual also has a biochemical fingerprint as determined by the presence of specific proteins on the surface of cell membranes. This array of cellular surface proteins has been best studied with leukocytes and is called the human leukocyte–antigen (HLA) system (McDevitt and Bodmer, 1974). While the recognition of the presence of interindividual differences in cellular antigens has been of critical importance in transplantation surgery for the matching of appropriate donors according to tissue types, biomedical studies have now revealed that various HLAs display striking associations with many human diseases (Svejgaard, 1975; Dausset and Svejgaard, 1977; Braun, 1979; Schultz et al., 1981).

The identification of the different antigens has also involved an assessment of their inheritance. It is now known that they are dependent on a set of very closely linked genes on a short segment of chromosome number 6. Limited human population studies have begun to assess the frequency of the various antigens in the population (Mourant et al., 1978). It should be noted that the first HLA was discovered in 1958; by 1967 only five additional antigens were isolated and four more by 1970. However, by 1980 the number of HLAs had approached 100. Since each person is thought to inherit a total of 10 HLA antigens, the number of possible combinations is immense being in the hundreds of millions (Harsanyi and Hutton, 1981).

The classic and most striking example of an HLA association is that between ankylosing spondylitis (AS) and HLA-B27. Among Europeans approximately 90% of patients with AS display this antigen while it is present in only 8–9% of the general population. Statistical associations have revealed that someone possessing the HLA-B27 antigen has between 150 and 200 times the normal risk of developing AS. A number of other arthropathies including Reiter's syndrome, yersinia, salmonella arthritis, juvenile arthritis, and acute anterior uveitis also display strong positive associations with the B-27 antigen while psoriatic arthropathy shows a moderately strong association. In addition to arthropathies, several diseases of the endocrine system have been shown to have moderate associations (relative risk 2–10) with HLA-B8. These diseases include individuals with dermatitis herpetiformis, chronic autoimmune hepatitis, myasthenia gravis, celiac disease, juvenile diabetes mellitus, Graves' disease, idiopathic Addison's disease, systemic lupus erythematosus, and Behat's syndrome (Emery, 1976; Svejgaard et al., 1975; McDevitt and Bodmer, 1974). Other ailments including some allergies, cardiovascular disease, immune system diseases, dermatological disorders,* renal disease, ophthalmologic disorders, gastroenterological diseases, and certain malignancies have been associated with the presence of one or more HLAs (Mourant et al., 1978; Ryder et al., 1981).

Despite some striking statistical associations of certain diseases with specific HLAs, any mechanistic relationship is yet to be uncovered, thereby at present precluding the possibility of knowing whether the relationship is causal or associational. Nevertheless, the recognition of the statistical relationship of HLAs with a wide range of human diseases some of which are known to also be related to occupation such as bladder cancer, asbestosis,† farmer's lung, and others suggest that inherent genetic factors are affecting the occurrence of the disease within the population (Harsanyi and Hutton, 1981).

---

*Psoriasis vulgaris has been associated with antigens B13, and B17 in 11 studies involving 836 patients with a relative risk of 4.8 for both antigens. Other antigens (e.g., B37 and Cw6) have also been associated with an increased risk of this disease but the supporting data are not as extensive. Dermatitis herpetiformis has been associated with both Dw3 and DRw3 with respective relative risks of 15.4 and 56.4 as a result of very limited numbers of patients studied (61 and 29, respectively) (Ryder et al., 1981).

†Several studies have demonstrated a statistically significant higher frequency of HLA-B27 antigen in asbestosis patients than in controls (Merchant et al., 1975; Matej and Lange, 1976). However, other research has not supported these findings (Evans, 1977).

The methodology for determining HLA types is considered simple and is frequently conducted in numerous medical centers in the United States. The typical cost is now less than $100.00 for a complete analysis (Harsanyi and Hutton, 1981).

REFERENCES

Braun, W. E. (1979). *HLA and Diseases: A Comprehensive Review.* CRC, Cleveland.

Dausset, J. and Svejgaard, A. (1977). HLA and disease. Predisposition to disease and clinical implications. First International Symposium on HLA and Disease. Inserm., Paris. (Cited in Mourant et al., 1978.)

Emery, A. E. H. (1976). *Methodology in Medical Genetics.* Churchill Livingstone, New York, pp. 103–106.

Evans, C. C., Levinsohn, H. C., and Evans, J. M. (1977). *Br. Med. J.* **1:** 603.

Harsanyi, Z. and Hutton, R. (1981). *Genetic Prophecy: Beyond the Double Helix.* Rawson, Wade, New York.

McDevitt, H. O. and Bodmer, w. F. (1974). HLA immune response genes and disease. *Lancet* **1:** 1269–1275.

Matej, H. and Lange, A. (1976). First International Symposium on HLA and Disease. Inserm., Paris, p. 256. (Cited in Mourant et al., 1978.)

Merchant, J. A., Klouda, P. T., Soutar, C. A., Parkes, W. R., Lawler, S. D., and Turner-Warwick, M. (1975). *Br. Med. J.* **1:** 189.

Mourant, A. E., Kopec, A. C., and Domaniewska-Sobezak, K. (1978). *Blood Groups and Diseases.* Oxford University, New York.

Ryder, L. P., Platz, P., and Svejgaard, A. (1981). Histocompatibility antigens and susceptibility to disease—genetic considerations. In: *Current Trends in Histocompatibility,* Vol. 2. R. A. Reisfeld and S. Ferrone (eds.). Plenum, New York, pp. 279–301.

Schultz, J. S., Good, A. E., Sing, C. F., and Kapur, J. J. (1981). HLA profile and Reiter's Syndrome. *Clin. Genet.* **19:** 159–167.

Svejgaard, A., Hauge, M., Jersild, C., Platz, P., Ryder, L. P., Nielsen, L., Staub, L., and Thomsen, M. (1975). *The HLA system. An introductory survey.* Karger, Basel.

Svejgaard, A., Jersild, C., Nielsen, L., Staub, L., and Bodmer, W. F. (1974). HLA antigens and disease: Statistical and genetical considerations. *Tissue Antigens* **4:** 95–105.

B. Susceptibility to Infectious Diseases

Differential susceptibility to infectious agents such as fungi, protozoans, bacteria, and viruses in animal species including humans is well known. The causes of this differential susceptibility may be found, in part, within host factors such as genetic makeup (Nathanson and Martin, 1979) or dietary status (Gopalan and Srikantia, 1973). As in the case with much that is known of differential susceptibility to chemical toxins, the knowledge of genetic determinants of susceptibility to infectious agents is derived from studying animals, especially rodents. The following review will assess: (1) the usefulness of animal models in providing an insight into the role that genetic factors may play in affecting susceptibility to infectious agents, especially viruses; (2) evidence of human genetic factors affecting susceptibility to infectious agents; and (3) the environmental health implications of this information with particular reference to drinking water quality and treatment as well as water reclamation concerns.

1. ANIMAL MODELS

According to Brinton and Nathanson (1981) mice are the best species with which to assess the genetic basis of susceptibility to viruses. This is because there are a sizeable number of inbred strains in each testing conducted. In addition, new congeneric strains may be readily developed. In Tables 8-1–8-3, these authors have summarized the current state of knowledge concerning the relationship of host genetic factors of experimentally induced viral disease in laboratory mice for RNA viruses (Table 8-1), RNA tumor viruses (Table 8-2), and DNA viruses (Table 8-3).

Based on their extensive assessment of this literature, Brinton and Nathanson (1981) have offered several generalizations:

1. Sometimes a single genetic locus may be sufficient to explain the difference between susceptible and resistant mouse strains. In other situations, two or more loci may be involved.
2. The dominant gene may cause either resistance or susceptibility; on occasion, codominance occurs.
3. Each locus seems to affect susceptibility to only a very specific group of viruses.
4. Only a limited number of the genes influencing susceptibility to viruses map to the H-2 locus.

Based on the extensive work with animal models that has unequivocally demonstrated genetic predispositions to an extraordinarily wide range of viral agents (Tables 8-1–8-3), it logically follows that humans should behave in a qualitatively similar manner. Much research in this area has focused on the development of methods for histocompatibility testing in humans and the delineation of the various HLA and their association with various human diseases. The book by Braun (1979) on HLA associations with various diseases should be consulted by the interested reader.

2. HUMORAL IMMUNE DYSFUNCTION

Brinton and Nathanson (1981) have stated that poliomyelitis may represent an example where the course of a viral disease may be affected by

Table 8-1. Host Genetic Control of Experimental Viral Disease in Laboratory Mice (RNA Viruses)[a]

Virus	Disease	Susceptible Prototype Strain	Resistant Prototype Strain	Most Inbred Strains	Number of genes or Designation[b]	Dominant Trait	Maps to H-2
Toga							
Flavi	Encephalitis	C3H	PRI	Susceptible	1	Resistant	No
LDV	Polioencephalitis	C58	C3H	Resistant	1+	Resistant	No
Orthomyxo							
Influenza A	Pneumonia	A/J	A2G	Susceptible	1	Resistant	No
Paramyxo							
Measles	Encephalitis	C3H	S/JL	Susceptible	1+	Resistant	No
Corona							
MHV2	Hepatitis	PRI	C3H	Resistant	1	Susceptible	No
MHV3	Persistence	C3H/S	C3H	Resistant	2	Undetermined	No
JHM	Demyelination	C3H	S/JL	Susceptible	Rhv-1	Resistant	No
					Rhv-2	Susceptible	No
Picorna							
EMC/M	Diabetes	SWR/J	C57BL/6	Resistant	1	Resistant	No

Source: Brinton and Nathanson (1981).

[a] Abbreviations: LDV, lactic dehydrogenase elevating virus; MHV, mouse hepatitis; JHM, JHM variant of MHV; EMC/M, encephalomyocardites/M variant.
[b] One or more.

Table 8-2. Genetic Control of Experimental Viral Disease in Laboratory Mice (RNA Tumor Viruses: Selected Examples)[a]

Murine Leukemia Virus	Disease or Outcome	Susceptible Prototype Strain	Resistant Prototype Strain	Most Inbred Strains	Designation of Gene	Dominant Trait	Maps to H-2 (Chromosome)
Ecotropic	Replication	NIH	BALB/c	E	Fv-1	Resistant	No (4)
FLV	Splenic foci	Many	C57BL/6	Susceptible	Fv-2	Susceptible	No (9)
Endogenous	Virus expression	AKR	NIH	Resistant	Akv-1[b]	Susceptible	No (7)
Endogenous	Virus expression	AKR	NIH	Resistant	Akv-2[b]	Susceptible	No (16)
GLV	Leukemia	H-2[k]	H-2[b]	Resistant	Rgv-1	Resistant	Yes (17)
GLV	Leukemia	H-2[k]	H-2[b]	Resistant	Rgv-2	Resistant	No (?24)[d]
FLV	Recovery from splenomegaly	H-2[k]	H-2D[b]	E	Rfv-1	R/S	Yes (17)[c]
FLV	Recovery from splenomegaly	H-2[k]	H-2K[b]	E	Rfv-2	R/S	Yes (17)[c]
FLV	Recovery from splenomegaly	BALB/c	C57BL/6	Susceptible	Rfv-3	Resistant	(?)[d]

Source: Brinton and Nathanson (1981).

[a] Abbreviations: R/S, intermediate phenotype; E, both genotypes equally represented in inbred strains; H-2, located on mouse chromosome 17; FLV, Friend leukemia virus; GLV, Gross leukemia virus.
[b] Akv-1 and Akv-2 are transcripts of virus genetic information.
[c] Probably map outside I (immune) region.
[d] ?, undetermined.

Table 8-3. Host Genetic Control of Experimental Viral Disease in Laboratory Mice (DNA Viruses)

Virus	Disease	Susceptible Prototype Strain	Resistant Prototype Strain	Number of Genes Involved[a]	Dominant Trait[a]	Maps to H-2[a]
Herpes						
Herpes simplex virus-1	Encephalitis	A/J	C57BL/6	2+	Resistant	No
Cytomegalovirus	Encephalitis	BALB/c	C3H/He	1+	Susceptible	Yes
Papova						
Polyoma	Tumors	AKR	C57BL/6	1+	Susceptible	?
	Runting	AKR	C57BL/6	1+	Resistant	No
Pox						
Ectromelia	Mousepox	CBA	C57BL	1+	?	?

Source: Brinton and Nathanson (1981).

[a] +, one or more; ?, undetermined.

genetic factors. They based this notion on an earlier paper (Nathanson and Martin, 1979) that reported the ratio of paralytic cases to total infections often varies from about 1 in 50 up to 1 in 500, thereby suggesting that the risk of infection may be affected by host factors. In support of their position is evidence from the U.S. Center for Disease Control (1977) indicating that about 10% of all paralytic poliomyelitis in the United States since 1969 has affected children with immunodeficiencies. According to Wyatt (1973, 1974, 1975), from 1961 to 1971 there were 10 vaccine-associated cases of poliomyelitis in children with immune disorders of a total of 110 cases in the United States. This amounts to 9%. These 10 cases comprise about 25% of hypogammaglobulinemic infants born in that decade. Thus, the risk of contracting vaccine-associated cases of paralytic polio by those patients with hypogammaglobulinemia was about 10,000 times greater than that of normal children.

In addition to poliomyelitis, it has been hypothesized that genetic factors may play an important role in affecting susceptibility to hepatitis B. This idea stemmed from observations indicating striking variations in the overall frequency of infection and in the relative frequency of hepatitis B surface antigens and anti-hepatitis B in studies of various populations (Blumberg et al., 1969).

3. GRANULOCYTOPENIA

This condition is very common in persons with acute leukemia, after bone marrow transplantation, after myelosuppressive drug therapy for malignancy, or because of aplastic anemia. The incidence and severity of infection is inversely related to the absolute granulocyte count. The most common pathogens to which persons affected with this condition may be susceptible to include various gram-negative bacilli such as *Pseudomonas aeruginosa* and *Klebsiella pneumoniae*, yeasts such as *Candida* and *Torulopsis*, and filamentous fungi such as *Aspergillus*. However, since the number of affected people is generally quite low and under close medical supervision, this population subgroup would not normally be considered as a factor in environmental regulatory decision making (Schimpff, 1979).

4. CELLULAR IMMUNE DYSFUNCTION

The cellular immune system is thought to have primary responsibility for protecting the body against "(a) certain bacteria (*Listeria monocylogenes*, nontyphosa salmonella species, mycobacterium tuberculosis and the atypical mycobacteria, and *Nocardia asteroides*), (b) viruses (varicella zoster, cytomegalovirus, and herpes simplex), (c) fungi (*Cryptococcus neofor-*

mans) and (d) protozoa (*Pneumocystis carinii, Toxoplasma gondii*) and perhaps *Strongyloides stercoralis*" (Schimpff, 1979). According to Schimpff (1979), patients with depressed cellular immunity have much more devastating infections from these agents than persons of the general population.

5. GENETIC SUSCEPTIBILITY TO FUNGAL INFECTIONS

Tinea imbricata is a superficial skin fungal disease caused by *Trichophyton concentricum* found mostly in tropical areas in Asia, Indonesia, and the South Pacific Islands. Several studies which indicated that the prevalence of this disease was considerably higher in some races than others even though they resided in the same country and habitat suggested that racial characteristics may affect susceptibility to this disease (Dey and Marplestone, 1942; Polunin, 1952; Reid, 1976). Familial studies by Schofield et al. (1963) and Serjeantson and Lawrence (1977) have supported the hypothesis that genetic factors affected susceptibility to this disease. In fact, the report of Serjeantson and Lawrence (1977) supports a single autosomal locus. A more recent study by Ravine et al. (1980) involving segregation analysis on 228 family pedigrees has also supported the proposed autosomal recessive inheritance pattern hypothesis. While this disease may be of little relevance to Western industrialized societies, the concept of genetic susceptibility to fungal agents is relevant and worthy of further research.

6. SUMMARY

There is strong support for the hypothesis that genetic factors affect susceptibility to infectious diseases by animal model studies as well as in humans with well-characterized immunodeficiency states. Unfortunately, there are a genuine lack of data on the more subtle yet important genetic influences operating in the general population that may account for differences in susceptibility to infectious agents.

7. PRESENCE OF INFECTIOUS AGENTS IN WASTES AND DRINKING WATER

The presence of viral agents in waste water and drinking water is of considerable public health concern. The NAS has addressed this issue and has summarized the relationship between the occurrence of human disease and waterborne viral agents. Among the human diseases associated with waterborne viral agents are poliomyelitis, hepatitis, gastroen-

Schimpff, S. (1979). Infections in the compromised host. In: *Principles and Practice of Infectious Diseases*. G. L. Mandell, R. Gordon-Douglas, and J. E. Bennett (eds.). Wiley, New York, pp. 2257–2262.

Schofield, F. D., Parkinson, A. D., and Jeffrey, D. (1963). Observations on the epidemiology, effects and treatment of *Tinea imbricata*. *Trans. R. Soc. Trop. Med. Hyg.* **57**: 214–217.

Serjeatson, S. and Lawrence, C. (1977). Autosomal recessive inheritance of susceptibility to *Tinea imbricata*. *Lancet* **1**: 13–15.

Wyatt, H. V. (1973). Poliomyelitis in hypogammaglobulinemias. *J. Infect. Dis.* **128**: 802–806.

Wyatt, H. V. (1974). The immunity gap and vaccination against poliomyelitis. *Lancet* **1**: 784–785.

Wyatt, H. V. (1975). Risk of live poliovirus in immunodeficient children. *J. Pediatr.* **87**(1): 152–153.

Wyatt, H. V. (1975a). Is poliomyelitis a genetically determined disease? I. A genetic model. *Med. Hypoth.* **1**(1): 35–42.

Wyatt, H. V. (1975b). Is poliomyelitis a genetically determined disease? II. A critical examination of the epidemiological data. *Med. Hypoth.* **1**(2): 23–32.

C. Immunodeficiency Diseases

Considerable interest has been directed in recent years to the association of immunological function and the occurrence of cancer. The increasing incidence of cancer in the aged has been hypothesized to be caused by the progressive weakening of immune defenses. Figure 8-1, which circumstantially supports this theory, illustrates the inverse association of cancer incidence with the size of the thymus, the plasma levels of the hormone thymosin, and the age of the members of the population. This hormone is known to play a significant role in maintaining the proper functioning of the immunosurveillance system as reflected in the activity of cell-mediated immunity. However, the most extensively documented data associating immune defects to carcinogenesis result from the study of individuals receiving kidney transplants and large doses of immunosuppressive agents to prevent rejection. Fraumeni and Hoover (1977) have reviewed kidney transplant registry data of about 9000 transplant recipients with respect to the occurrence of cancer as compared to the general public. They reported that lymphomas occurred at a rate 25 times greater than expected. They noted that the increase was confined to non-Hodgkin's lymphoma with a 45-times excess, being mainly reticulum cell sarcoma. Large increased risks were also noted for the development of liver and biliary tract cancer. In contrast to the striking increase in lymphoma and liver cancer incidence, the risk of other cancers was increased by only a factor of approximately 2; no across-the board effect was observed, however, as would probably have been ex-

Immunodeficiency Diseases

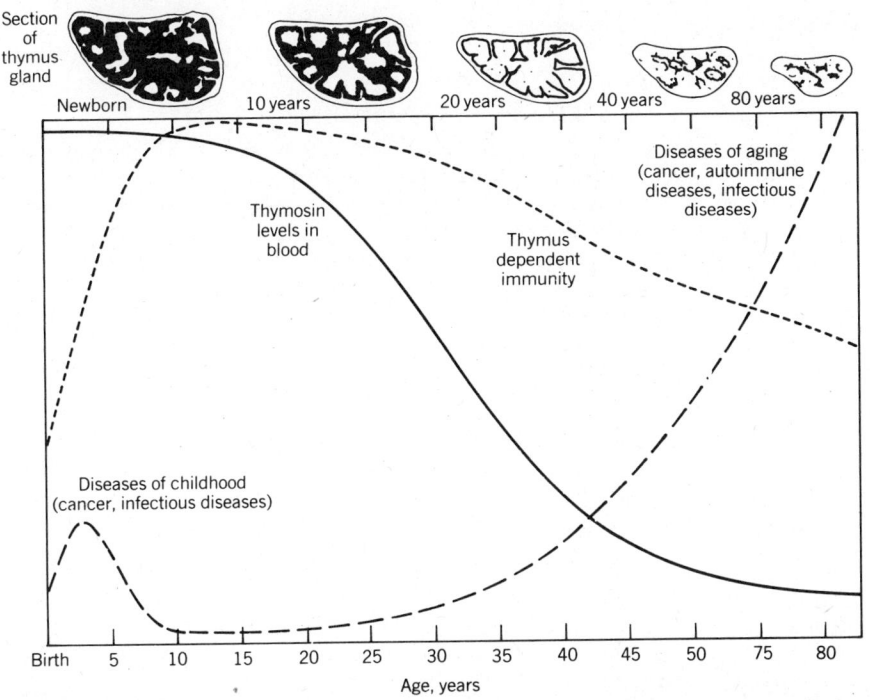

Figure 8-1. Thymus-dependent senescence of immunity in man. *Source:* Goldstein et al. (1975).

pected by the immunosurveillance theory. Cancers such as those of the breast and large bowel, however, did not show any excess occurrence.

These data collectively support the notion that disruptions in the immune surveillance system are a predisposition factor with respect to a variety of human cancers, especially lymphoma. It may be reasoned that if drug-induced immunosuppression enhanced the risk of developing cancer, then individuals born with immunodeficiency disease should also display an increased risk of developing malignancy.

Studies of naturally occurring immunodeficient diseases (NOIDs) reveal that they are a highly variable group. Fourteen such NOIDs have been classified in the Immunodeficiency Cancer Registry. According to Kersey et al. (1974), the cancer mortality rate in immunodeficient children was estimated to be 100 times greater than for the general pediatric population, with the majority of malignancies being lymphoid tumors (Filipovich et al., 1980). However, the overall risk of the various specific subgroups of those with immunodeficient diseases to cancer occurrence is not well defined as a result of a very limited number of entries into the Immunodeficiency Cancer Registry for each specific NOID. Further research along these lines is in order.

REFERENCES

Filipovich, A. H., Spector, B. D., and Kersey, J. (1980). Immunodeficiency in humans as a risk factor in the development of malignancy. *Rev. Med.* **9:** 252–259.

Fraumeni, J. F. and Hoover, R. (1977). Immunosurveillance and cancer: Epidemiologic observations. *Natl. Cancer Inst. Monogr.* **47:** 121–126.

Goldstein, A. L., Wara, D. W., Amman, A. J., Sakai, H., Harris, W. S., Thurman, G. B., Hooper, J. A., Cohen, G. H., Goldman, H. S., Constanzi, J. J., and McDaniel, M. D. (1975). First clinical trial with thymosin: Reconstruction of T-cells in patients with cellular immunodeficiency diseases. *Transplant. Proc.* **7**(1): 681.

Kersey, J. H., Spector, B. D., and Good, R. A. (1974). Cancer in children with primary immunodeficiency diseases. *J. Pediatr.* **84:** 263–264.

D. Susceptibility to Isocyanates

Stokinger and Scheel (1973) have reported that industrial exposure to organic isocyanates may result in an asthmalike syndrome that develops on subsequent exposure to even very small quantities of isocyanates. This is referred to as an immediate type of hypersensitivity comparable to the allergic reactions associated with hay fever. In contrast, other workers display a "delayed" type of response that is characterized by a low-grade fever, and headache without activation of respiratory tract secretions. The delayed response requires a much greater exposure to activate it than the immediate reaction. The issue that emerges is how can one predict which persons will be hypersensitive? Since the responses in this case are severe ones, animal models have been used in serum screening tests to predict human responses. In essence, the tests for detecting the allergic responses are founded on the assumption that the allergic responses to the isocyanates are mediated through reaginic, fixed tissue antibodies, while the toxic responses result in precipitating (circulating) antibodies. The tests for reaginic antibodies involve the injection of human serum into the skin of a monkey, subsequently followed by an intravenous injection of isocyanate antigen. As for the toxic responses study, the guinea pig model is used since it gives a positive percutaneous anaphylactic response following inhalation of isocyanate.

Stokinger and Scheel (1973) indicated that the data collected from the guinea pig study allow the determination of whether the person can produce circulating precipitating antibodies (i.e., a positive guinea pig test) as a sign of toxicity. The monkey test indicates whether the individual can produce fixed reaginic antibodies (i.e., a positive monkey test), a sign of a hypersensitive response.

Based on the testing of the sera of greater than 1000 individuals, Stokinger and Scheel (1973) reported that 0.5% of the worker population displayed a delayed hypersensitivity, while 1.5% exhibited the immediate hypersensitivity to organic isocyanates. Of the remaining 98%, clinical signs were seen in 38% when initially exposed. However, these became asymptomatic on continued exposure. The final 60% made precipitating antibodies (gammaglobulin G) and developed immunity without symptoms. While these data appear very interesting, the information was presented only in summary format with no described methodology or validation procedures. Nevertheless, these authors state that they knew of two U.S. factories manufacturing isocyanates which employed the above-mentioned animal model procedure as part of a preemployment screening program. They stated that this program has been very effective in lowering the number of clinical reactors. Unfortunately, the names of the companies were not given nor a description of the programs and their methodologies. Without such information, it is not possible to objectively assess the reported success as noted by Stokinger and Scheel (1973).

It is this type of research that the scientific and regulating communities need in order to judge the efficacy of screening procedures. To keep the information fundamentally restricted not only retards the growth of scientific knowledge but also retards information transfer and possibly increases the likelihood that workers can have any new knowledge to work for the benefit of their health. Finally, this type of restriction also promotes fuel for those with an untrusting eye on the activities of many industries.

REFERENCE

Stokinger, H. E. and Scheel, L. D. (1973). Hypersusceptibility and genetic problems in occupational medicine—a consensus report. *J. Occup. Med.* **15:** 564–573.

9

Skin Disorders

A. Nonallergic Contact Dermatitis

Primary irritants are chemicals that damage skin by direct cytotoxic activity. This is different from the action of contact allergens that incite inflammation by indirect immunologic reactions affecting responses in the whole individual. Primary irritants are generally recognized as the dominant cause of dermatitis as the result of contact with an offending chemical (Kligman and Wooding, 1967). The likelihood of damaging the skin continues to be a growing concern with respect to safety evaluation of new compounds with which one may come in contact at work, at home, or nearly any place in urbanized areas as a result of exposure to medicines, cosmetics, cleaners, clothing, and chemical manufacturing processes. That industrial chemicals may be important causes of skin irritation has been long recognized by industrial hygienists and toxicologists. The threshold limit values of the American Conference of Governmental Industrial Hygienists provide for a skin notation when effects on the skin contribute to the hazard involved in the use of industrial substances (McCreesh and Steinberg, 1977). Of such a magnitude is the prevalence of occupational skin disease and disorders that they accounted for more than 40% of all reported occupational diseases each year from 1972 to 1976 (Wang, 1978). Further along these same lines, Lubowe (1972) has pointed out the dermatological effects of indus-

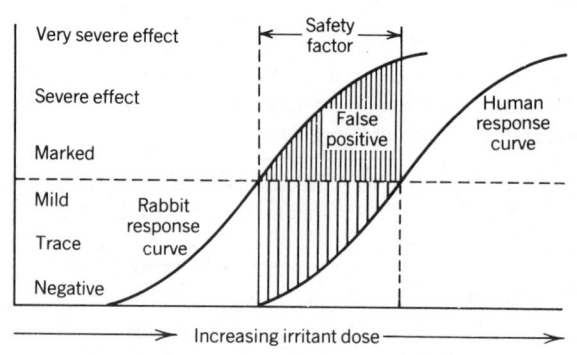

Figure 9-1. These hypothetical dose response curves show how irritant effects on rabbit and human skin are related. Rabbit skin responses to a given irritant exposure are more likely to simulate responses of very sensitive persons than the average human response. This difference can be exploited for determination of risks in use of a new substance but it precludes direct estimation of human responses. *Source:* These curves are based on data of Phillips, Steinberg, Maibach and Akers (1972).

moderate irritant on human skin. However, substances only producing a small degree of irritation on rabbits are often hard to differentiate with respect to human predictability.

According to Mershon and Callahan (1975), the enhanced sensitivity of the rabbit to dermal irritants, while not offering a precise simulation of human responses, may be considered as "ideal for the pharmaceutical manufacturer who needed a margin of safety for the protection of the more sensitive members (i.e., high-risk groups) of the human population." Figure 9-1, which is based on the data of Phillips et al. (1972), offers a graphic representation of the relationship of rabbit and human responses to dermal irritants.

The use of animal models such as the rabbit, therefore, has a very long history of predicting dermal irritants in the pharmaceutical, military, and industrial areas. However, there are limitations to its usefulness as has been amply illustrated by Nixon et al. (1975). The earlier adoption of the rabbit skin irritancy test into federal regulatory processes has tended to deemphasize other models useful for investigation. Nevertheless, research with a wide variety of nonrabbit models reveals it is the least likely to yield a false negative. However, because of its enhanced sensitivity, most investigators feel that a positive irritancy test with rabbit skin should not necessarily lead to a regulatory denial of the substance (Steinberg et al., 1975). The use of the rabbit test should represent the initial dermatologic screening process that eliminates those more potent irritants. However, it is evident that human testing of potential irritants is a necessity in order to avoid considerable future dermatological problems.

REFERENCES

Draize, J. H., Woodard, G., and Calvery, H. O. (1944). Methods for the study of irritation and toxicity of substances applied topically to the skin and mucous membranes. *J. Pharmacol. Exp. Ther.* **82:** 377–390.

Kligman, A. M. (1979). Etiologic agents and mechanisms: Irritants. Presented at the Occupational Industrial and Plant Dermatology Meeting, March 26–28, 1979. San Francisco, California.

Kligman, A. M. and Wooding, W. M. (1967). A method for the measurement and evaluation of irritants on human skin. *J. Invest. Dermatol.* **49**(1): 78–94.

Lubowe, I. J. (1972). Dermatitis urbis. *Soap Perfum. Cosmet.* **45:** 35–38.

McCreesh, A. H. and Steinberg, M. (1977). Skin irritation testing in animals. In: *Dermatotoxicology and Pharmacology*. E. N. Marzulli and H. I. Maibach (eds.). Hemisphere, Washington, D.C., pp. 193–210.

Magnussen, G. and Hellgren, L. (1962). Skin irritation and adhesive characteristics of some different adhesive tapes. *Acta. Derm. Venereol.* **42:** 463.

Mershon, M. M. and Callahan, J. F. (1975). Exploiting the differences. In: *Animal Models in Dermatology: Relevance to Human Dermatopharmacology and Dermatotoxicology*. H. Maibach (ed.). Churchill Livingstone, New York, pp. 36–52.

Nixon, G. A., Tyson, L. A., and Wertz, W. C. (1975). Interspecies comparisons of skin irritancy. *Toxicol. Appl. Pharmacol.* **31:** 481–490.

Phillips, L., Steinberg, M., Maibach, H. F., and Akers, W. A. (1972). A comparison of rabbit and human skin response to certain irritants. *Toxicol. Appl. Pharmacol.* **21:** 369–382.

Schwartz, L., Tutipan, L., and Birmingham, D. (1957). *Occupational Diseases of the Skin.* Lea and Febiger, Philadelphia, p. 31.

Shmunes, E. (1980). The importance of pre-employment examination in the prevention and control of occupational skin disease. *J. Occup. Med.* **22**(6): 407–409.

Steinberg, M., Akers, W. A., Weeks, M., McCreesh, A. H., and Maibach, H. I. (1975). A comparison of test techniques based on rabbit and human skin responses to irritants with recommendations, regarding the evaluation of mildly or moderately irritant compounds. In: *Animal Models in Dermatology: Relevance to Human Dermatopharmacology and Dermatotoxicology*. H. Maibach (ed.). Churchill Livingstone, New York, pp. 1–11.

Wang, C. L. (1978). The problem of skin disease in industry. Office of Occupational Safety and Health Statistics. U.S. Department of Labor.

B. Allergic Contact Dermatitis

Allergic contact dermatitis is known to be a serious concern for persons employed in a wide variety of occupations, especially those involving contact with such potentially potent sensitizers as nickel and chromium. For example, the extent of chromium use is so widespread that it is commonly found in industries where there is frequent contact with chemicals, leather, metal, paint, cement, paper pulp, timber, building material, and a variety of household products including detergents and glue (Marzulli and Maibach, 1977).

Assessing the occurrence of contact allergy has long been employed diagnostically via patch tests. In such diagnostic tests, a preparation is applied to the clinical patient's skin under an occlusive patch for 48 hr and the skin is evaluated for the occurrence of erythema, edema, or more severe skin alterations occurring 24, 48, and 72 hr after removal of the patch. Allergic substances are thereby identified by reproducing skin disease on a minor scale with offending chemicals (Marzulli and Maibach, 1977).

A rapidly growing data base using standard screening procedures is beginning to reveal the frequency of skin reactions of clinical patients to a wide variety of sensitizing agents. For example, of 16 known sensitizing materials, nickle sulfate was the most frequent sensitizer among U.S. and Canadian dermatologic patients with a reaction rate of 11%. *P*-phenylenediamine, potassium dichromate, and thimersal had 8% reaction rates, ethylene diamine had a 7% reaction rate, while the fungicide Thiram had a 4% reaction rate (Marzulli and Maibach, 1977).

While these findings are of significance to the overall understanding of allergic dermatitis, they are studies on the frequency of reactions in already diagnosed clinical patients. The present issue does not deal with retrospective assessments as would occur diagnostically but with predictive test procedures. Systematic predictive test procedures for skin sensitization were initially developed in the 1940s and early 1950s (Draize et al., 1944; Rostenberg and Sulzberger, 1937; Schwartz, 1941, 1951, 1960; Shelanski, 1951). At present, such tests usually require multiple occlusive patches for induction of sensitization (followed by a 2-hr rest period and then challenge with a patch at a new skin site) (Marzulli and Maibach, 1973). There are a variety of modifications of this procedure some of which (e.g., the maximization test) are designed to elicit a response if it is physiologically possible. Thus, it is usually the case that the higher percentage of positive responses occurs when the maximization tests are used.

The real issue for predictive testing is not so much whether the tests provide a reasonably accurate account of the sensitizing properties of a product but whether they can predict what may happen under conditions of actual use. For example, despite the very widespread use of the potent contact allergen formaldehyde in cosmetic products, the prediction rate is about the same as the incidence observed in diagnostic tests. Why formaldehyde has not sensitized a greater number of patients who use cosmetics is thought to be because it is principally confined to use at low concentrations such as a preservative in shampoos (Marzulli and Maibach, 1977).

Predictive tests for contact allergy in occupational settings should reflect the types of exposure to be encountered. Tests such as the maximization procedure whose main function appears to be in providing an indication of the uppermost limits of sensitization are probably going to result in a high percentage of practical false positives.

The use of predictive contact allergy patch testing has a long history. For example, the U.S. Army's Industrial Hygiene Laboratory applied 28,201 test patches to 2036 volunteer human test subjects during 1948–1949 in an attempt to predict the sensitization capability of various items of military supplies and equipment (Holland et al., 1950). Numerous other examples of the use of predictive patch testing have been reported especially with respect to cosmetics.

The use of patch testing in occupational settings usually concerns diagnostic testing rather than predictive testing. It would appear that the adoption of a systematic predictive testing scheme would certainly help to identify persons with biological predispositions for developing job related sensitizations. Various standardized methods have been developed for predictive patch testing and are available. The results of such testing are of course dependent on a number of factors including the vehicle, the concentration, amount of material used, and test sites, among others (Hjorth, 1977). The possibility of both false positives and negatives is real and must be carefully evaluated. In addition, predictive patch testing has been widely reported to cause side effects in some individuals such as itching, possible sensitization to the test substance, and in some extreme cases fainting and fever have occurred. Despite the strict requirements of the testing procedures and the possible confounding variables, predictive patch testing offers an outstanding possibility to identify individuals at high risk to developing allergic contact dermatitis. Nevertheless, the issue may be legitimately raised as to whether it is necessary to subject all prospective employees to such tests. It is hoped that initial screening via animal models such as the guinea pig (Calabrese, 1983) should provide initial predictive value so that the substances in question should not pose a significant risk to the general population of workers.

REFERENCES

Calabrese, E. J. (1983). *Principles of Animal Extrapolation*. Wiley, New York.

Draize, J. H., Woodard, G., and Calvery, H. D. (1944). Methods for the study of irritation and toxicity of substances applied topically to the skin and mucous membranes. *J. Pharmacol. Exp. Ther.* **83**: 377–390.

Hjorth, N. (1977). Diagnostic patch testing. In: *Dermatotoxicology and Pharmacology*. F. N. Marzulli and H. I. Maibach (eds.). Hemisphere, Washington, D.C., pp. 341–351.

Holland, B. D., Cox, W. C., and Dehne, E. J. (1950). "Prophetic" patch test. *Arch. Dermatol. Syphilol.* **61**: 611–618.

Marzulli, F. N. and Maibach, H. I. (1973). Antimicrobials: Experimental contact sensitization in man. *J. Soc. Cosmet. Chem.* **24**: 399–421.

Marzulli, F. N. and Maibach, H. I. (1977). Contact allergy; predictive testing in humans. In: *Dermatotoxicology and Pharmacology*. F. N. Marzulli and H. I. Maibach (eds.). Hemisphere, Washington, D.C., pp. 353–372.

Rostenberg, A., Jr. and Sulzberger, M. B. (1937). Some results of patch tests. *Arch. Dermatol. Syphilol.* **35**: 433–455.

Schwartz, L. (1941). Dermatitis from new synthetic resin fabric finishes. *J. Invest. Dermatol.* **4**: 459–470.

Schwartz, L. (1951). The skin testing of new cosmetics. *J. Soc. Cosmet. Chem.* **2**: 321–324.

Schwartz, L. (1960). Twenty-two years experience in the performance of 200,000 prophetic-patch tests. *South. Med. J.* **53**: 478–483.

Shelanski, H. A. (1951). Experience with and consideration of the human patch test method. *J. Soc. Cosmet. Chem.* **2**: 324–331.

C. Genetic Susceptibility to Ultraviolet-Induced Skin Cancer

Considerable controversy has focused on the effects of ultraviolet radiation on human health as a result of studies that have indicated that supersonic jets (e.g., Concorde) and spray propellants (e.g., chlorofluoromethanes) may be affecting a reduction in the stratospheric ozone layer. The ozone layer absorbs potentially harmful wavelengths of ultraviolet radiation (250–320 nm), thereby preventing such radiation from reaching and adversely affecting human life on earth. It is generally accepted that the formation of the ozone layer during our geological past was a necessity for the emergence and evolution of terrestrial life. However, despite the efficiency of the shielding activity of the ozone layer, a biologically important amount of ultraviolet radiation does ultimately reach the earth's surface causing both beneficial and harmful effects on humans.

1. BENEFICIAL ASPECTS OF ULTRAVIOLET RADIATION

Ultraviolet radiations (280–320 nm) are known to penetrate the skin and have been reported to effect the conversion of provitamin D to vitamin D. Since vitamin D is present in large amounts in only a few foods (e.g., bony fishes), the formation of vitamin D via the action of ultraviolet radiation has been suggested as playing an important role in acquiring sufficient amounts of vitamin D for proper development. Without adequate quantities of vitamin D in the diet, proper development of skeletal tissue would be prevented and children would become rachitic. Conversely, the presence of excessive vitamin D can result in a toxic condition known as hypervitaminosis D, which is characterized by calcification of soft tissues (e.g., kidneys). To effectively regulate the amount of ultraviolet radiation that penetrates the skin, humans have evolved a melanin pigment system which can effectively screen out ultraviolet radiation or permit penetration depending on the degree of pigmentation. Black skin is considerably more efficient in preventing penetration of ultraviolet radiation than lighter skin. According to this theory, skin color, including the phenomenon of reversible tanning, is an evolutionary adaptation to regulate the level of vitamin D in the body by controlling the ultraviolet mediated synthesis of vitamin D (see Loomis, 1967, for a detailed discussion of the skin color–vitamin D hypothesis). Calabrese (1977) has suggested that elevated levels of "breathable" ozone in our urban areas may act to actually increase our daily dietary requirement of vitamin D by absorbing ultraviolet radiation that would usually reach the body and lead to the formation of vitamin D.

2. CARCINOGENIC ACTIVITY OF ULTRAVIOLET RADIATION

In addition to being involved with the synthesis of vitamin D, considerable evidence indicates that ultraviolet radiation possesses carcinogenic activity. The evidence associating ultraviolet radiation with skin cancer has a relatively long history, stretching back to several reports in the 1890s (Unna, 1894; Dubreuilh, 1896; Shield, 1899), and continuing up to the present (Urbach, 1975). During this time, the evidence of the ultraviolet–skin cancer association has been derived from complementary areas of research, including animal studies and human epidemiological and clinical observations. A short summary of the evidence implicating midultraviolet (250–320 nm) as a significant factor in the development of most human skin cancer is as follows (see Urbach, 1975):

1. Skin cancer appears on parts of the body most exposed to sunlight especially the head, neck, arms, and hands (Anchev et al., 1966; Silverstone and Gordon, 1966; and McGovern, 1966).
2. Squamous cell and basal cell carcinomas occur primarily on skin locations most often exposed to solar radiation (Urbach, 1966; Allison and Wong, 1967).
3. Anchev et al. (1966) and Blum (1959) reported a significantly greater incidence of skin cancer in outdoor workers (e.g., farmers).
4. For similar skin-types the rate of skin cancer increased dramatically as the latitude decreased, especially within the middle latitude (Urbach, 1975). Figure 9-2, which strongly supports the work of Urbach (1975), represents the annual skin cancer incidence rates in Honolulu county and selected mainland areas. Scotto et al. (1974) have indicated that the Third National Cancer Survey revealed that by age 50, 1% of all male Caucasians in the area around Dallas, Texas developed skin cancer malignancy. By age 80, the figure was 3% per year. These rates were twice those of individuals from Minneapolis-St. Paul.
5. Susceptible racial groups, especially those of Celtic origin, are more susceptible to the development of skin cancer than other whites. The most susceptible individuals have skin with little pig-

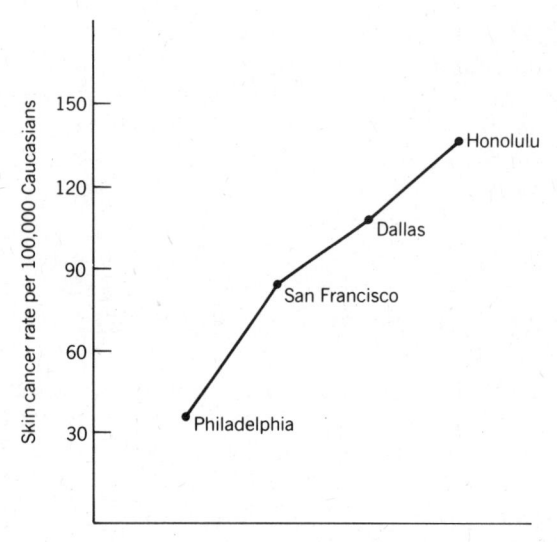

Figure 9-2. Skin cancer annual incidence rates per 100,000 white population, Honolulu County, 1955 and 1956; Selected mainland areas, 1947, 1948. *Source:* Allison, S. D. and Wong, K. L. (1967). Skin cancer: some ethnic differences. In: *Environments of Man,* Addison-Wesley, Reading, Massachusetts, p. 69.

ment, scattered freckles, light colored hair, blue or gray eyes, and a facility to sunburn (O'Beirn et al., 1970; Urbach et al., 1972). Blacks and Orientals are generally much less susceptible than Caucasians (Oettle, 1966; Allison and Wong, 1967). In fact, skin cancer is 45 times more common among Caucasians as compared to non-Caucasians in Honolulu, Hawaii. Figure 9-3 shows the skin cancer incidence rates by race in Honolulu County.

6. Finally, animal studies, most often using mouse skin, have repeatedly demonstrated that skin cancer can be produced by repeated exposure to wavelengths between 250 and 320 nm with 280–320 nm the most carcinogenic (Epstein, 1966; Blum, 1948).

3. HEALTH IMPLICATIONS OF A REDUCTION IN THE OZONE LAYER

Based on the previously summarized data concerning the carcinogenic effects of ultraviolet radiation, much public and scientific commotion has resulted from the possible depletion of the ozone layer.

It has been estimated that supersonic transports that cruise in the stratosphere would reduce the ozone layer by 0.07% for a small fleet (NAS, 1975) and by 50% for a fleet of 500 (Johnson, 1973). Depletion of the ozone layer by chlorofluorocarbons has been projected to be 16% by

Individuals at High Risk to Ultraviolet Radiation

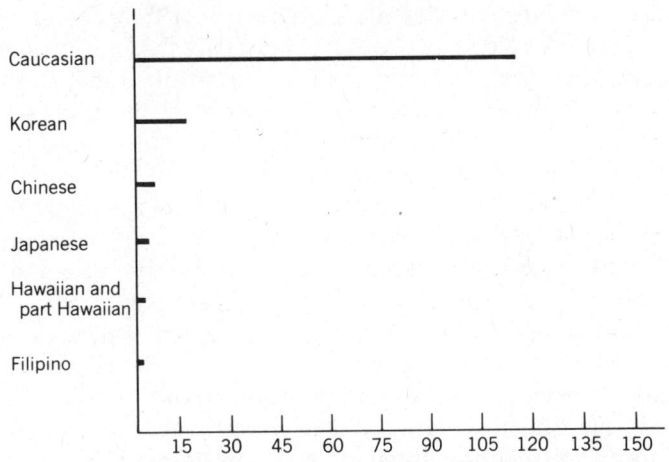

Figure 9-3. Skin cancer incidence rates per 100,000 population by race, Honolulu County, 1955 and 1956. *Source:* Allison, S. D. and Wong, K. L. (1967). Skin cancer: Some ethnic differences. In: *Environments of Man,* Addison-Wesley, Reading, Massachusetts, p. 69.

the year 2000 if usage of these propellant chemicals were to increase at 10% per year (Wofsy et al., 1975). A National Research Council report has concluded that the continued release of halocarbons at the 1973 rate will ultimately produce between a 2 and 20% reduction in stratospheric ozone with the probable reduction about 7% (Maugh, 1976). Schultze (1974) has calculated that if the ozone concentration in the stratosphere were diminished by 10%, the total ultraviolet radiation dose would be increased by 19% in the middle latitudes (40°N to 60°N) and by 22% in the equatorial zones. Fears et al. (1976) have developed a mathematical model that has allowed these investigators to predict the relative increases in skin cancer incidence and mortality associated with changes in ultraviolet levels (erythema dose). Specifically, they have predicted an increase by 15–25% in the melanoma incidence of males depending on the degrees north latitude. Similar effects are predicted for females.

4. INDIVIDUALS AT HIGH RISK TO ULTRAVIOLET RADIATION

One of the important recommendations of the National Research Council report was to undertake research to identify population groups with a drastically higher than normal susceptibility to malignant melanoma. Previously cited research has noted the marked enhancement of ultraviolet radiation induced skin cancer in Caucasians, especially those with fair complexions. However, there are other groups, although less numerous, who should also be considered at increased risk to skin cancer from the ultraviolet radiation. These groups include: people with birth marks such as moles (Oettle, 1966; McGovern, 1966); the elderly, especially those who develop Hutchinson's melanotic freckle, a lesion present in skin severely affected by the sun (Anchev et al., 1966; Oettle, 1966); oculocutaneous albinism including both tyrosinase positive and negative phenotypes and other reduced melanin conditions (discussed below); and individuals with xeroderma pigmentosum.

Albinism is a genetically transmitted disorder of the melanin pigment (melanocyte) system that is recognized by dramatic reduction of melanin in the skin, hair, and eyes. Albinism occurs not only in man, but also in other mammals, as well as in birds, reptiles, amphibians, and fish. Melanocytes are specialized cells where melanin synthesis occurs. In man, mature melanocytes are usually found in definite regions: skin (hair bulbs, dermis, and dermoepidermal junction), mucous membranes, nervous system (piarachnoid), and eye (uveal tract and retinal pigment epithelium). The melanin pigment is synthesized in the cytoplasm on melanosomes by the enzymatic conversion of its precursor (tyrosine) by tyrosinase. With the exception of those located in the hair bulbs and retinal pigment epithelium, all melanocytes seem to have the ability to develop into malignant melanomas. The metabolic disorder may affect the entire melanocyte system (oculocutaneous albinism) or the melanocytes at a specific site (ocular albinism) (Fitzpatrick and Quevedo, 1972). Oculocutaneous albinism is known to occur in the different races of man with the affected having significant reductions of melanin in the hair, skin, and eyes. There are two phenotypic expressions of oculocutaneous albinism that are distinguished by either the presence or absence of tyrosinase and are thus referred to as tyrosinase positive or tyrosinase negative (Fitzpatrick and Quevedo, 1972).

Associated with the reduction of the melanin pigment is an enhanced susceptibility to the toxic effects of solar radiation. Individuals with albinatic skin are known to have a high frequency of solar keratosis and basal cell and squamous cell carcinomas of the skin of the exposed area (Curban, 1951; Shapiro et al., 1953). Malignant melanomas are also known to occur in these individuals (Young, 1958; Garrington et al., 1967).

The incidence of tyrosinase-negative oculocutaneous albinism within the Black and Caucasian population is approximately 1:34,000–36,000 persons (Witkop, 1972). Tyrosinase-positive oculocutaneous albinism has an incidence of approximately 1:14,000 in Blacks and 1:60,000 in Caucasians (Witkop, 1972). Oculocutaneous albinism is present with a high incidence among various American Indians (Tule Cuna Indians, 1:143; Hopi, 1:227; Jemez, 1:140; Zuni, 1:247; Navajo, 1:3750) (Witkop, 1972). Furthermore, several investigators have reported that the total incidence of oculocutaneous albinism in Ireland is approximately 1:10,000–15,000 with an estimated mutation frequency for albinism of 3.3×10^{-5} to 7.0×10^{-5} per gene per generation (Froggatt, 1957; Witkop, 1972). Interestingly, it has been noted that Ireland has the third highest skin cancer rate in the world, despite the fact that the sun shines only 30% of the possible sunshine hours (MacDonald, 1966).

According to Herndon and Freeman (1976), patients with vitiligo (skin disease manifested by smooth milk-white spots on various parts of the body), phenylketonuria (PKU) or localized loss of pigmentation because of injury or inflammatory skin disease also represent groups at high risk to skin cancer. With regard to PKU, the frequency of carriers in the general population is 1:80, while the frequency of the afflicted PKU individual is 1:25,000 (Knox, 1972). The major biochemical changes in PKU are related to the accumulation in the tissues of that part of the dietary L-phenylalanine which would normally be converted to tyrosine. No harmful effects have been reported in the heterozygote as yet, despite the fact that the serum of heterozygotes may have elevated phenylalanine levels. To investigate whether carriers of PKU are more susceptible to ultraviolet induced skin cancer is especially relevant when one considers that there are greater than 2.5 million PKU carriers within the United States.

In order to survive the daily exposure to appreciable amounts of ultraviolet radiation, plants and animals, including man, have evolved DNA repair processes to protect affected cells and to assist in the recovery from the damaging radiation effects. Urbach (1975) has reported that three major kinds of repair processes have been described (see Chapter 11):

1. The damaged molecule may be repaired to its active condition *in situ*. This is achieved either by an enzyme catalyzed repair or by "decay" of the damage to some inconsequential (or harmless) form.
2. The damaged part can be excised and replaced with undamaged components that restore normal function.
3. The damage may not be specifically repaired, but the cell may either bypass the defect or make some other functional adaptation.

A genetic defect in DNA repair processes has been reported by Cleaver (1968) in cultured cells from ultraviolet radiation sensitive skin cancer-prone patients with the condition xeroderma pigmentosum (XP). Usually, XP fibroblasts are extremely sensitive to ultraviolet light and perform reduced amounts of repair replication during repair of damage to DNA. Subsequent findings by Cleaver and Carter (1973) reported the existence of genetic variants of XP with regard to their different abilities to repair ultraviolet damage to the cells. This phenomenon is thought to be of potential significance for the hypothesis that genetic background may be an important variable in one's susceptibility to developing skin cancer. It is of interest to note that a variety of cell types including liver, kidney, and brain have excision repair mechanisms that may assist against the clastogenic effects of systemic mutagens.

According to Urbach (1975), in mouse skin and in most cancer patients, the DNA repair process does not appear to be impaired. Consequently, he concluded that the lack of DNA repair cannot be considered the basis of most skin cancers. Commenting on the research of Epstein et al. (1971) and Zajdela and Latarjet (1973), he suggested that the development of skin cancer by ultraviolet radiation is actually initiated by repair of DNA, permitting the cell to survive, yet facilitating the occurrence of errors in DNA replication resulting in an enhanced likelihood of malignant change. This, of course, implies a defect in the repair system also. However, the defect would be more subtle and difficult to detect. It would seem that the discovery by Cleaver and Carter (1973) of the existence of genetic variants with respect to xeroderma pigmentosum and their variable DNA repair capabilities, clearly underlines the potential differential susceptibility of the general population if genetic polymorphisms for repair processes do exist and are widespread. At this point, further research is needed to clarify the role of DNA repair mechanisms in carcinogenesis before any sound generalizations can be made.

D. Noncarcinogenic Skin Irritation by Ultraviolet Radiation

1. LUPUS ERYTHEMATOSUS

It is well-known that sunlight adversely affects the skin of those afflicted with both systemic and cutaneous lupus. Dubois (1974) has reported that approximately 25–50% of patients with systemic lupus erythematosus (SLE) will have adverse responses to typically tolerated levels of ultraviolet radiation, while nearly twice that percentage of patients with discoid lupus erythematosis will have adverse effects. Herndon and Freeman (1976) have reported that irritation of existing skin lesions, fever, and progression of systemic activity may occur following excessive exposure; however, sunlight is not considered an integral component of the pathogenesis of lupus erythematosus. They further indicate that numerous SLE patients easily tan, and may even sustain sunburn without relapse. However, individuals with lupus erythematosis are usually well advised to avoid overexposure.

2. PHOTOALLERGIC DRUG REACTIONS

Herndon and Freeman (1976) have reported that ultraviolet radiation interacts with various substances to form complete photoantigens that may cause anaphylactic responses in sensitive individuals. Examples of such photoallergic substances include halogenated antiseptic compounds used in soaps and cosmetics. Freeman et al. (1970) have reported that during typical usage of these photoallergic substances sensitive individuals, even those of dark skin racial background, became susceptible to painful erythema 4 times more quickly (5 minutes compared to 20 minutes) than controls in Texas noonday sun. The frequency of these hypersensitive individuals within the general population remains to be determined.

REFERENCES

Allison, S. D. and Wong, K. L. (1967). Skin cancer: Some ethnic differences. In: *Environments of Man*. J. B. Bresler (ed.). Addison-Wesley, Reading, Massachusetts, pp. 67–71.

Anchev, N., Popov, I., and Ikonopisov, R. L. (1966). Epidemiology of malignant melanoma in Bulgaria. In: *Structure and Control of the Melanocyte*. G. Della Porta and O. Muhlbock (eds.). Springer-Verlag, New York, pp. 286–291.

Blum, H. F. (1948). Sunlight as a causal factor in cancer of the skin of man. *J. Natl. Cancer Inst.* **9**: 247–258.

Blum, H. F. (1959). *Carcinogenesis by Ultraviolet Light*. Princeton University, Princeton, New Jersey.

Calabrese, E. J. (1977). Environmental quality indices predicted by evolutionary theory. *Med. Hypoth.* **3**(6): 241–244.

Cleaver, J. E. (1968). Defective repair replication of DNA in xeroderma pigmentosum. *Nature* **218**: 652–656.

Cleaver, J. E. and Carter, P. M. (1973). Xeroderma pigmentosum: Influence of temperature on DNA repair. *J. Invest. Dermatol.* **60**: 29–32.

Curban, G. V. (1951). Multiple cutaneous carcinomatosis in 2 albino brothers. *Rev. Paul. Med.* **39**: 440.

Dubois, E. L. (1974). *Lupus Erythematosus*. 2nd ed. University Press, Los Angeles, p. 798.

Dubreuilh, W. (1896). Des Hyperkeratoses Circonscriptes. *Ann. Dermatol. Syphiligr.* (Ser. 3) **7**: 1158–1204.

Epstein, J. H. (1966). Ultraviolet light carcinogenesis. In: *Advances in Biology of Skin. Vol. VII: Carcinogenesis*. W. Montagna and R. L. Dobson (eds.). Appleton-Century-Crofts, New York.

Epstein, W. L., Fukuyama, K., and Epstein, J. H. (1971). UV light, DNA repair and skin carcinogenesis in man. *Fed. Proc.* **30**: 1766–1771.

Fears, T. R., Scotto, J., and Schniederman, M. A. (1976). Skin cancer, melanoma, and sunlight. *Am. J. Public Health* **66**(5): 461–464.

Fitzpatrick, T. B. and Quevedo, W. C. (1972). Albinism. In: *The Metabolic Basis of Inherited Disease*. J. B. Stanbury, J. B. Wyngaarden, and D. S. Fredrickson (eds.). McGraw-Hill, New York, pp. 326–337.

Freeman, G., Hudson, A., Carnes, R., and Knox, J. M. (1970). Salicylanilide photosensitivity. *J. Invest. Dermatol.* **54**: 145–149.

Froggatt, P. (1957). Albinism: A statistical, genetical and clinical appraisal based upon a complete ascertainment of the condition in Northern Ireland. Thesis, Trinity College, Dublin, Ireland.

Garrington, G. E., Scofield, H. H., Cornyn, J., and Lacy, G. R. (1967). Intraoral malignant melanoma in a human albino. *Oral Surg.* **24:** 224.

Herndon, J. H., Jr. and Freeman, R. G. (1976). Human disease associated with exposure to light. *Ann. Rev. Med.* **27:** 77–87.

Johnson, H. S. (1973). Report LBL-2217. Lawrence Berkeley Laboratory, Berkeley, California.

Knox, W. E. (1972). Phenylketonuria. In: *The Metabolic Basis of Inherited Disease.* J. B. Stanbury, J. B. Wyngaarden, and D. S. Fredrickson (eds.). McGraw-Hill, New York, pp. 326–337.

Loomis, W. F. (1967). Skin-pigment regulation of vitamin D biosynthesis in man. *Science* **157:** 501–506.

MacDonald, E. J. (1966). Discussion comment. In: *Structure and Control of the Melanocyte.* G. Della Porta and O. Muhlbock (eds.). Springer-Verlag, New York, p. 317.

McGovern, V. J. (1966). Melanoblastoma in Australia. In: *Structure and Control of the Melanocyte.* G. Della Porta and O. Muhlbock (eds.). Springer-Verlag, New York, pp. 312–315.

Maugh, T. H. (1976). The ozone layer: The threat from aerosol cans is real. *Science* **194:** 170–172.

National Academy of Sciences. (1975). Estimates of increase in skin cancer due to increases in ultraviolet radiation caused by reduced stratospheric ozone. Appendix C. In: *Environmental Impact of Stratospheric Flight*, pp. 117–221.

O'Beirn, S. F., Judge, P., Urbach, F., MacCon, C. F., and Martin, F. (1970). The prevalence of skin cancer in County Galway Ireland. In: *Proceedings of the 6th National Cancer Conference.* Lippincott, Philadelphia, pp. 489–500.

Oettle, A. G. (1966). Epidemiology of melanomas in South Africa. In: *Structure and Control of the Melanocyte.* G. Della Porta and O. Muhlbock (eds.). Springer-Verlag, New York, pp. 292–307.

Schultze, R. (1974). Increase of carcinogenic ultraviolet radiation due to reduction in ozone concentration in the atmosphere. *Proceedings of the International Conference on Structure, Composition and General Circulation of the Upper and Lower Atmospheres and Possible Anthropogenic Perturbations.* Atmospheric Environment Service, Ontario, Canada.

Scotto, J., Kopf, A. W., and Urbach, F. (1974). Non-melanoma skin cancer among caucasians in four areas of the United States. *Cancer* **34:** 1333–1338.

Shapiro, M. P., Keen, P., Cohen, L., and Murray, J. F. (1953). Skin cancer in the South African Bantu. *Br. J. Cancer* **7:** 45.

Shield, A. M. (1899). A remarkable case of multiple growths of the skin caused by exposure to the sun. *Lancet* **1:** 22–23.

Silverstone, H. and Gordon, D. (1966). Regional studies in skin cancer. 2nd report: Wet tropical and sub-tropical coast of Queensland. *Med. J. Aust.* **2:** 733–740.

Unna, P. G. (1894). Die Histopathologie der Hautkrankheiten. Hirschwald, Berlin.

Urbach, F. (1966). Ultraviolet radiation and skin cancer in man. In: *Advances in Biology of Skin. Vol. VII: Carcinogenesis.* W. Montagna and R. L. Dobson (eds.). Pergamon, Oxford.

Urbach, F. (1975). Ultraviolet radiation: Interaction with biological molecules. In: *Cancer 1: Etiology: Chemical and Physical Carcinogenesis.* F. F. Becker (ed.). Plenum, New York, pp. 441–451.

Urbach, F., Rose, D. B., and Bonnem, M. (1972). Genetic and environmental interactions in skin carcinogenesis. In: *Environment and Cancer.* Williams and Wilkins, Baltimore, pp. 355–371.

References

Witkop, C. J., Jr. (1972). Albinism. In: *Advances in Human Genetics, Vol. 2.* H. Harris and K. Hirschhorn (eds.). Plenum, New York.

Wofsy, S. C., McElroy, M. B., and Sze, N. D. (1975). Freon consumption: Implications for atmospheric ozone. *Science* **187:** 535–536.

Young, T. E. (1958). Malignant melanoma in an albino. *Arch. Pathol.* **64:** 186.

Zajdela, F. and Latarjet, R. (1973). The inhibiting effect of caffeine on the induction of skin cancer by UV in the mouse. *C. R. Acad. Sci. Paris* (Aug.) Vol. 29. (Cited in Urbach, 1975.)

10 Ocular Disorders

A. Genetic Susceptibility to Glaucoma

Glucocorticoids are widely used anti-inflammatory agents. A side effect of the action is their ability to produce an increase in intraocular pressure when applied topically to the eye (Francois, 1954; Armaly, 1963; Becker and Mills, 1963), or when given systemically (Bernstein and Schwartz, 1962). Of particular concern is that this hypertensive effect was comparable to that occurring spontaneously in open angle glaucoma in the sense that it was symptomless and could achieve clinically relevant levels that lead to glaucomatous destruction of visual function (Goldmann, 1962; Armaly, 1964).

The glucocorticoid induced increase in intraocular pressure was found to occur within families, thereby suggesting a genetic foundation (Armaly, 1963; Becker and Hahn, 1964). While there is considerable similarity within families for the intraocular responsiveness to glucocorticoids, there is marked heterogeneity of response within randomly selected samples. Based on such a heterogenetic response, Armaly (1968) reported that at least three levels of response existed: a low, intermediate, and high degree of responsiveness. This led to the notion that the response was controlled by a pair of alleles, P^L for the low pressure and P^H for high pressure, and that the three levels of response are phenotypic expressions of the three possible combinations: the low level $P^L P^L$, the intermediate $P^L P^H$, and the high level $P^H P^H$.

Armaly (1966) tested this genetic hypothesis for the occurrence of glucocorticoid sensitivity in family units. The findings were extremely consistent with the proposed genetic association.

The next step in his exploratory progression was to test the association of the responsiveness to increases in ocular pressure with the occurrence of open angle glaucoma. A comparison of individuals with glaucoma with those of a randomly selected population sample clearly revealed that persons with glaucoma displayed a greater frequency of P^H (i.e., $P^L P^H$, $P^H P^H$) and a great reduction in $P^L P^L$. More specifically, 66% of those in the random sample had an inferred $P^L P^L$ genotype and only 29% and 5% with the intermediate ($P^L P^H$) and high ($P^H P^H$) phenotypes, respectively. In contrast, > 90% of those with open-angle hypertensive glaucoma, low tension glaucoma, and normal eyes in recessed angle glaucoma displayed moderate or high phenotype responses to the glucocorticoid.

The findings of Armaly (1968) clearly illustrate the role of genetic factors in affecting the differential response of individuals to glucocorticoid antiflammatory agents. The data also strongly implicate the genetic factors as predisposing influences in the occurrence of glaucoma.

REFERENCES

Armaly, M. F. (1963). Effect of corticosteroids on intraocular pressure and fluid dynamics: I. The effect of dexamethasone in the normal eye. *Arch. Ophthalmol.* **70**: 482.

Armaly, M. F. (1963a). Effect of corticosteroids on intraocular pressure and fluid dynamics: II. The effect of dexamethasone in the glaucomatous eye. *Arch. Ophthalmol.* **70**: 492.

Armaly, M. F. (1964). Effect of corticosteroids on intraocular pressure and fluid dynamics: III. Changes in visual function and pupil size during topical dexamethasone application. *Arch. Ophthalmol.* **71**: 636.

Armaly, M. F. (1965). Statistical attributes of the steroid hypertensive response in the clinically normal eye: I. The demonstration of three levels of response. *Invest. Ophthalmol.* **4**(2): 187.

Armaly, M. F. (1966). Dexamethasone ocular hypertension in the clinically normal eye: II. The untreated eye, outflow facility, and concentration. *Arch. Ophthalmol.* **75**: 776.

Armaly, M. F. (1966a). The heritable nature of dexamethasone-induced ocular hypertension. *Arch. Ophthalmol.* **75**: 32.

Armaly, M. F. (1967). Dexamethasone ocular hypertension and eosinopenia and glucose tolerance test. *Arch. Ophthalmol.* **78**: 193.

Armaly, M. F. (1967a). Inheritance of dexamethasone hypertension and glaucoma. *Arch. Ophthalmol.* **77**: 747.

Armaly, M. F. (1967b). The genetic determination of ocular pressure in the normal eye. *Arch. Ophthalmol.* **78**: 187.

Armaly, M. F. (1968). Genetic factors related to glaucoma. *Ann. N.Y. Acad. Sci.* **151**: 861–874.

Becker, B. and Hahn, K. A. (1964). Topical corticosteroids and hereditary in primary open-angle glaucoma. *Am. J. Ophthalmol.* **57**: 543.

Becker, B. and Mills, D. W. (1963). Corticosteroids and introcular pressure. *Arch. Ophthalmol.* **70**: 500.

Becker, B. and Shaffer, R. (1965). *Diagnosis and Therapy of the Glaucomas*, 2nd ed. Mosby, St. Louis.

Bernstein, H. N. and Schwartz, B. (1962). Effects of long-term systemic steroids on ocular pressure and tonographic values. *Arch. Ophthalmol.* **68:** 742.

Francois, J. (1954). Cortisone et tension oculaire. *Ann. Ocul. (Paris)* **187:** 805.

Goldmann, H. (1962). Cortisone glaucoma. *Arch. Ophthalmol.* **68:** 621.

B. Leber's Optic Atrophy

In 1871, Leber reported the occurrence of a rare genetic disorder in which those affected experience visual failure either subacutely or insidiously. The condition often finally results in more or less severe bilateral central scatomata and pallor of the optic disks.

The condition is usually first recognized in males in late teenage years or early 20s although the range of onset is quite wide. In western cultures only about 15% of those with Leber's optic atrophy are women (Wilson, 1965). Based on the hereditary nature of the disease and clinical phenomena, Wilson (1965) suggested that the disease is caused by an inborn metabolic error whose clinical expression is significantly affected by exogenous (environmental) factors; smoking is considered one such factor. The association between smoking and Leber's optic atrophy was suggested by Wilson (1965a) because of the recognized presence of several neurotoxic agents in cigarette smoke including cyanide as well as the knowledge that repeated exposure of cyanide to experimental animals has produced neurological effects similar to those of Leber's optic atrophy.

Normally, cyanide is highly reactive and is rapidly metabolized to less toxic compounds. Consequently, a significant proportion of cyanide is quickly converted to thiocyanate and excreted. With this considered, Wilson (1965a) experimentally tested the hypothesis that smoking may adversely affect those with Leber's optic atrophy. The results demonstrated that patients with the disease had significantly lower levels of thiocyanate in the plasma and urine than did the controls when both groups were smoking. The author suggested that such patients may have an inborn metabolic error with respect to cyanide metabolism and may be unable to detoxify sufficiently cyanide to thiocyanate.

Such evidence certainly implies that individuals with Leber's optic atrophy may be at increased risk to cyanide poisoning. However, the results are certainly of a preliminary nature and considerably more research is needed before any reliable generalizations can be made. Presently, research on the gene frequency of this trait is necessary.

The toxicity of cyanide is also thought to be markedly influenced by the nutritional status of the individual. As previously mentioned, cyanide is thought to be detoxified, in part, by its metabolic conversion to thiocyanate. This detoxification pathway involves substrates derived from cysteine. It is significant to note that epidemiological studies involving 320 Nigerian patients (14–24/1000 adults) with "tropical neuropathy" quite similar to Leber's disease, have revealed that the geographical distribution of the "tropical neuropathy" is closely associated with the consumption of cassava, a tuber whose outer integument has high levels of the cyanogenetic glycoside, linamarin. Furthermore, the diet of the 320 Nigerians was found to be generally quite low in cysteine and other sulphur-containing amino acids. It was suggested that the combination of low dietary levels of these amino acids and their increased utilization for the detoxification of cyanide may be the cause of the severely depleted levels of plasma cysteine and methionine in these patients. However, the mode of action by which toxic symptoms are affected is not precisely known, although it may include direct intraneuronal enzymic inhibition by cyanide or by the inactivation of vitamin B_{12} (Anonymous, 1969).

REFERENCES

Anonymous (1969). Chronic cyanide neurotoxicity. *Lancet* **2** (Nov. 3): 942–943.

Leber, T. (1871). Leber hereditare und congenitalangelegte schnerbenleiden. *Albrecht von Graefes Arch. Ophthalmol.* **17** (2 Abt.): 249–291.

Wilson, J. (1965). Skeletal abnormalities in Leber's hereditary optic atrophy. *Ann. Phys. Med.* **8**: 92–95.

Wilson, J. (1965a). Leber's hereditary optic atrophy: A possible defect of cyanide metabolism. *Clin. Sci.* **29**: 505–515.

11 DNA Repair and Chromosome Instability Diseases

A. Xeroderma Pigmentosum (XP)

The principal clinical feature of XP is a heightened sensitivity of the skin to solar radiation as seen in pigmentation changes, elevated erythema, and multipleneoplasms (Robbins et al., 1974; Kraemer, 1977). Basal and squamous cell carcinomas are the most frequent types while malignant melanomas also occur. Tumor incidence increases rapidly with age and by early teens many patients may have as many as 100 or more cancers. Deaths commonly occur by the fourth decade and are attributable to metastatic carcinoma and wasting (Pathak and Epstein, 1971).

There are two forms of XP that are typically distinguished by clinicians. First, there is the classical XP with skin complications only. Second, there is a neurological form of XP whereby a broad spectrum of central nervous system deficiencies occur along with skin lesions (Robbins et al., 1974; Kraemer, 1977). These two forms of XP are known to occur with a similar prevalence and appear to be genetically different.

XP cells are usually more sensitive than normal cells to killing (Cleaver, 1970a; Arlett et al., 1975) (Table 11-1), the occurrence of chromosomal aberrations (Sasaki, 1973), and ultraviolet induced mutagenesis (Maher et al., 1976). It

Table 11-1. Sensitivity of Cells from Various Human Diseases[a] (Fibroblast Survival Curves Except Where Noted Otherwise)

Disease	Ratio of 10% Survival Doses (Normal:Disease)	Agent
XP A	7 (15–30)[b]	Ultraviolet, many chemical carcinogens
C	6 (3–9)[b]	Ultraviolet, many chemical carcinogens
D	7–12 (28–20)[b]	Ultraviolet, many chemical carcinogens
V	1–1.4 (1.4–1.7)[b]	Ultraviolet, many chemical carcinogens
Ataxia telangiectasia	1.0	Ultraviolet
	3.2	X rays
	2.3	MNNG
Fanconi's anemia	1.4 (4.2)[b]	Ultraviolet
	1.0 (1.0)[b]	X rays
	5–15	Mitomycin C
Cockayne's Syndrome	4.2	Ultraviolet
	1.0	X rays
Retinoblastoma (D-deletion)	1.6	X rays
Progeria	1.6–2.3[c]	X rays
Bloom's Syndrome	2.0	Ultraviolet
	>1.0	EMS
Down's Syndrome	~2[d]	X rays

Source: Cleaver (1980).

[a] Ratios derived from published curves.
[b] Adenovirus survival.
[c] Adenovirus V antigen.
[d] Chromosome aberrations (Sasaki, 1975).

is interesting to note that the relative sensitivity to killing is associated with the extent of central nervous system symptoms and not with cancer incidence (Andrews et al., 1978).

The above two clinical classifications of XP have been additionally refined and subdivided on the basis of biochemical observations in heterokaryons made by cell hybridization. At least eight mutually complementary groups are now known via *in vitro* cell hybridization (Cleaver and Bootsma, 1975). These various complementation groups display considerable variation in sensitivity to ultraviolet induced damage.

All XP strains are now known to be defective in at least some aspect of the basic repair mechanisms affecting ultraviolet photoproducts. Cells derived from all neurological and most classical XP patients display a deficit in excision repair (Cleaver, 1974; Cleaver and Bootsma, 1975; Paterson et al., 1973). As noted above, those XP strains that are deficient in excision repair have been further subdivided on the basis of complementation studies. However, despite the large number of such complementation groups, cell excision repair deficient XP strains seemed to be blocked at the incision (Paterson, 1979), thereby implying that fewer single-strand nicks are detected in XPs as compared to normal cells.

Not only are XP cells more susceptible than normal cells to ultraviolet radiation, they also display a similar enhanced susceptibility to a number of chemical mutagens. More specifically, XP cells are very sensitive to the reactive forms of polycyclic aromatic hydrocarbons (e.g., *K*-region epoxides of BP), aromatic amides [i.e., 4-nitroquinoline-*N*-oxide (4-NQO)], and aflatoxin, substances implicated as causative agents of human cancers. In contrast, XP cells appear to respond normally to ionizing radiation, monofunctional alkylating agents [e.g., *N*-methyl-*N'*-nitro-*N*-nitrosoguanidine (MNNG)], and the DNA-cross-linking agent, mitomycin C (MMC) (Maher et al., 1975; Maher and McCormick, 1976; Arlett, 1977; Fujiwara et al., 1977; Setlow, 1978) (Table 11-2).

XP is an autosomal recessive disease with a frequency in the general population of about 1/250,000, and it occurs in all racial groups (Robbins et al., 1974), but with an apparently unusually high incidence in Arabs in North America (Miller, 1977). The heterozygote is calculated to occur at a rate of 2–4/1000 (Swift, 1976).

Heterozygotes of the XP gene have been studied to discuss if and to what extent they may also display a reduced ability to repair damaged DNA. Stich et al. (1973) obtained heterozygote XP cells from skin biopsies of parents of two XP patients. Thus, the parents were *assumed* (not proved) to be carriers. The DNA repair ability of the XP carrier cells was compared with that of XP homozygous fibroblasts and control cells. Stich et al. (1973) reported that the carrier cells did not show enhanced susceptibility to ultraviolet, 4-NQO, *N*-acetoxy-AAF, and aflatoxin B_1 (Table 11-3). Further studies reported by Cleaver (1980) also indicated that XP heterozygotes are difficult to differentiate both clinically and biochemi-

Table 11-2. Response of XP Cells to Chemical Carcinogens and Mutagens

Normal Repair Capacity	Reduced Repair Capacity
MNNG	4NQO and carcinogenic derivatives
MMS	3-Mc-4NPO
EMS	MCA-epoxide (K-region)
NMU	BA-epoxide (K-region)
HN$_2$	N-acetoxy-AAF
BHP	N-hydroxy-AAF
ICR-191	N-acetoxy-AAS
	N-hydroxy-AAS
	N-acetoxy-4-AABP
	N-acetoxy-2-AAP
	N-hydroxy-2-AAP
	N-myristoyloxy-2-AAF
	N-acetoxy-3-AAF
	DMN, activated
	Aflatoxin B$_1$, activated
X ray	Ultraviolet

Source: Stich et al. (1973).

Table 11-3. Comparative Levels of DNA Repair Synthesis in XP Cells Exposed to Ultraviolet Light, 4NQO, N-Acetoxy-AAF, and Aflatoxin B$_1$ (Activated by a Postmitochondrial Liver Fraction)

Patient	Sex	Ultraviolet 1000 erg/mm^2	4NQO $2 \times 10^{-6} M$ 90 min	N-acetoxy-AAF $5 \times 10^{-5} M$ 5 hr	Aflatoxin B$_1$
XPH$_1$	F	9.8	10.3	10.2	
XPH$_2$	M	12.1	13.1	16.0	
XP$_E$	F	21.5	20.2	22.9	19.7
XP$_V$	F	36.0	33.6	29.1	
XP$_K$	M	56.6	55.7	60.8	61.1
Heterozygotes					
XP$_E$	Mother	102	121	116	
XP$_V$	Mother	101	108.5	100	
XP$_K$	Father	123.5	94.2		
XP$_K$	Mother	116.1	95.2		
Control	M + F	100%	100%	100%	100%

Source: Stich et al. (1973).

cally from normals although Ritter (unpublished abstract) reported slower excision repair at high doses.

The homozygous XP disease has been successfully diagnosed prenatally by identifying an abnormal response of cultured amniotic cells to ultraviolet irradiation (Ramsay et al., 1974). According to Lynch et al. (1977), this early detection is essential for proper prophylactic care since many XP patients diagnosed early can be placed on a photoprohibition scheme from birth and live a reasonably long and fruitful life.

As for the heterozygotes, no accepted methodology exists for their biochemical identification. It is only as an obligate heterozygote (i.e., parent of a homozygous recessive). As for the enhanced susceptibility of heterozygotes to environmental mutagens, Stich et al. (1973) were unable to report any enhanced susceptibility to three well-known mutagens in XP heterozygotes (Table 11-3). Nevertheless, Swift is reported by Marx (1978) to state that carriers of the XP gene while not being at increased risk of dying from cancer, appear to have a higher incidence of skin cancer, although no evidence was presented to support this statement.

REFERENCES

Andrews, A. D., Barrett, S. F., and Robbins, J. H. (1978). Xeroderma pigmentosum neurological abnormalities correlate with colony-forming ability after ultraviolet radiation. *Proc. Natl. Acad. Sci.* **75:** 1984–1988.

Arlett, C. F. (1977). Lethal response to DNA-damaging agents in a variety of human fibroblast cell strains. *Mutat. Res.* **46:** 106.

Arlett, C. F., Harcourt, S. A., and Broughton, B. C. (1975). The influence of caffeine on cell survival in exicision-proficient and excision-deficient xeroderma pigmentosum and normal human cell strains following ultraviolet irradiation. *Mutat. Res.* **33:** 341–346.

Cleaver, J. E. (1970a). DNA repair and radiation sensitivity in human (xeroderma pigmentosum) cells. *Int. J. Rad. Biol.* **18:** 557–565.

Cleaver, J. E. (1974). Repair processes for photochemical damage in mammalian cells. *Adv. Rad. Biol.* **4:** 1–75.

Cleaver, J. E. (1980). DNA damage, repair systems and human hypersensitive diseases. *J. Environ. Pathol. Toxicol.* **3:** 53–68.

Cleaver, J. E. and Bootsma, D. (1975). Xeroderma pigmentosum: Biochemical and genetic characteristics. *Ann. Rev. Genet.* **9:** 19–38.

Fujiwara, Y., Tatsumi, M., and Sasaki, M. S. (1977). Cross-link repair in human cells and its possible defects in Fanconi's anemia cells. *J. Mol. Biol.* **113:** 635–650.

Kraemer, K. H. (1977). Progressive degenerative diseases associated with defective DNA repair: Xeroderma pigmentosum and ataxia telangiectasia. In: *Cellular Senescence and Somatic Cell Genetics: DNA Repair Processes.* W. W. Nichols and D. G. Murphy (eds.). Symposia Specialists, Miami, pp. 37–71.

Lynch, H. T., Lynch, J., and Lynch, P. (1977). Management and control of familial cancer. In: *Genetics of Human Cancer.* J. J. Mulvihill, R. W. Miller, and S. F. Fraumeni, Jr. (eds.). Raven, New York, pp. 235–256.

Maher, V. M. and McCormick, J. J. (1976). Effect of DNA repair on the cytotoxicity and mutagenicity of UV irradiation and of chemical carcinogens in normal and xeroderma pigmentosum cells. In: *Biology of Radiation Carcinogenesis.* J. M. Yuhas, R. W. Tennant, and J. D. Regan (eds.). Raven, New York, pp. 129–145.

Maher, V. M., Ouellette, L. M., Mittlestat, M., and McCormick, J. J. (1975). Synergistic effect of caffeine on the cytotoxicity of ultraviolet irradiation and of hydrocarbon epoxides in strains of xeroderma pigmentosum. *Nature* **258:** 760–763.

Maher, V. M., Ouellette, L. M., Curren, R. D., and McCormick, J. J. (1976). Frequency of ultraviolet light-induced mutations is higher in xeroderma pigmentosum variant cells than in normal cells. *Nature* **261:** 593–595.

Marx, J. L. (1978). DNA repair: New clues to carcinogenesis. *Science* **200:** 518–521.

Miller, R. W. (1977). Ethnic differences in cancer occurrence: Genetic and environmental influences with particular reference to neuroblastoma. In: *Genetics of Human Cancer.* J. J. Mulvihill, R. W. Miller, and J. F. Fraumeni, Jr. (eds.). Raven, New York, pp. 1–14.

Paterson, M. C. (1979). Environmental carcinogenesis and imperfect repair of damaged DNA in *Homo sapiens:* Causal relation revealed by rare hereditary disorders. In: *Carcinogens: Identification and Mechanisms of Action.* A. C. Griffin and C. R. Shaw (eds.). Raven, New York, pp. 251–276.

Paterson, M. C., Lohman, P. H. M., and Sluyter, M. L. (1973). Use of a UV endonuclease from *Micrococcus luters* to monitor the progress of DNA repair in UV-irradiated human cells. *Mutat. Res.* **19:** 245–256.

Pathak, M. A. and Epstein, J. H. (1971). Normal and abnormal reactions of man to light. In: *Dermatology in General Medicine.* T. B. Fitzgerald, K. A. Arndt, W. H. Clark, Jr., A. Z. Eisen, E. J. VanScott, and J. H. Vaughn (eds.). McGraw-Hill, New York, pp. 977–1036.

Ramsay, C. A., Coltart, T. M., Blunt, S., Pawsey, S. A., and Giannelli, F. (1974). Prenatal diagnosis of xeroderma pigmentosum. Report of the first successful case. *Lancet* **2:** 1109–1112.

Ritter (unpublished abstract) (See Cleaver, 1980.)

Robbins, J. H., Kraemer, K. H., Lutzner, M. A., Festoff, B. W., and Coon, H. G. (1974). Xeroderma pigmentosum: An inherited disease with sun sensitivity, multiple cutaneous neoplasma and abnormal DNA repair. *Ann. Intern. Med.* **80:** 221–248.

Sasaki, M. S. (1973). DNA repair capacity and susceptibility to chromosome breakage in xeroderma pigmentosum cells. *Mutat. Res.* **20:** 291–293.

Setlow, R. B. (1978). Repair deficient human disorders and cancer. *Nature* **271:** 713–717.

Stich, H. F., Kieser, D., Laishes, B. A., and San, R. H. C. (1973). The use of DNA repair in the identification of carcinogens, precarcinogens, and target tissue. *Proceedings of the 10th Canadian Cancer Research Conference,* University of Toronto, Toronto, pp. 83–110.

Swift, M. (1976). Malignant disease in heterozygous cancers. In: *Birth Defects: Original Article Series.* **12**(1): 133–136.

B. Ataxia Telangiectasia

Ataxia telangiectasia (AT) is a progressive multifaceted disorder displaying simultaneous neurological, oculocutaneous, and immunological complications (Sedgwick and Boder, 1972; Kraemer, 1977). Diagnosis is typically made in early childhood from the joint appearance of cerebellar ataxia and telangiectasia of the bulbar conjunctivae and skin. Such individuals are predisposed to experiencing repeated respiratory infections that are thought to be related to immune deficiencies (Friedman et al.,

1977; Ammann and Hong, 1971; Ribon and Perera, 1975). The immunodeficiencies are thought to be caused by an absent or rudimentary thymus. Persons with AT have a 10% chance of contracting cancers at an early age, 85% of which are lymphoma and lymphatic leukemias (Paterson, 1979). This cancer risk is over 100 times greater than age-matched controls. Patients typically do not live past early adulthood, usually dying of respiratory ailments and/or lymphoproliferative neoplasms. In addition, some AT patients respond adversely and on occasion fatally to standard radiotherapy (Cunliffe et al., 1975).

Several studies have noted that an AT gene may have some clinical effects in persons who are heterozygous for the AT gene. Among those effects found in some AT heterozygotes are defective immunity (e.g., autoimmunity) (Friedman et al., 1977), oculocutaneous telangiectasias (Sedgwick and Boder, 1972), and enhanced cancer risk (Swift et al., 1976).

Diploid fibroblasts of AT homozygotes display increased sensitivity to inactivation by ionizing radiation. In addition, the degree of hypersensitivity is considerably less variable than that for ultraviolet induced killing of XP cells. In studies of X- and gamma-ray [low linear energy transfer (LET) radiation] induced killing, 11 strains of AT cellular strains displayed a very uniform enhanced response, being 3.2 times more sensitive than normal cells (Taylor et al., 1975). Other researchers demonstrated that AT strains are also more sensitive to high LET radiation; however, the degree of enhanced sensitivity is only approximately 1.4 to neutron radiation (Myers et al., cited by Paterson, 1979), leading Paterson (1979) to conclude that the genetic defect in AT strains is less crucial to the repair of damage caused by high rather than low LET radiation. Several researchers have demonstrated that AT strains are more sensitive to MNNG and actinomycin D (Scudiero, 1978; Hoar and Sargent, 1976), while some, but not all, are more sensitive to MMC and probably MMS (Hoar and Sargent, 1976). Of particular interest is that both near and far ultraviolet irradiation and N-hydroxy-AAF affected AT cells in a similar fashion as normal cells (Arlett, 1977). These responses are opposite those of XP strains that are sensitive to such agents.

The capacity of AT cells to carry out specific DNA repair processes has been investigated. AT strains have been found to rejoin both single- and double-strand breaks caused by ionizing radiation and several radiomimetic substances as well as cell strains from normal people (Vincent et al., 1975; Taylor et al., 1975; Lehmann and Stevens, 1977). In contrast, a majority of strains display decreased efficiencies of repair replication following anoxic irradiation (Paterson et al., 1977). Other studies have indicated that the AT strains, while being proficient in excision repair of ultraviolet photoproducts, are defective in excision repair of gamma radio products (Paterson et al., 1976). These authors assert that their findings are consistent with the hypothesis that the incision step is carried out by different enzymes, gammaendonuclease, or ultraviolet endonuclease depending on the damage.

The fact that variation in response to mutagenic agents exists among AT strains supports the hypothesis of a genetic heterogeneity in this disease in a similar fashion to XP. The discovery of biochemical complementation of different DNA repair defects as noted above and the occurrence of AT strains efficient in repair replication also support the heterogeneity concept (Paterson et al., 1977).

The occurrence of the AT homozygote in the population is relatively rare being approximately 25/1,000,000. However, if such figures are applied to the Hardy-Weinberg equilibrium equation, then it is estimated that about 1% of the general population are AT heterozygotes. The key public health issue then rests with a determination of the potential health risks of being an AT heterozygote. A number of studies have demonstrated that obligate AT heterozygotes (i.e., the parent of an AT homozygote child) are moderately sensitive to ionizing radiation. This has been demonstrated for both transformed lymphoplastoid cell lines (Lavin et al., 1978) and diploid fibroblasts (Paterson et al., 1978). Epidemiological studies by Swift et al. (1976) have also disclosed that blood relatives of AT patients have a predisposition to developing cancer. For AT heterozygotes younger than age 45, the risk of dying from a malignant neoplasm was calculated to be greater than 5 times the risk for the general population. AT heterozygotes were theorized as comprising more than 5% of all persons dying from a cancer before age 45. The types of cancer increased in the AT families studied were ovarian, gastric, and biliary system carcinomas and leukemia and lymphoma. In addition, there is evidence to support some predisposition of AT heterozygotes to basal cell, breast, pancreatic, cervical, and colonic carcinomas.

The results of Swift et al. (1976) clearly suggest the need to develop a laboratory test to detect AT heterozygotes. In 1978, Chen et al. published an article with a very enticing title, "Identification of Ataxia Telangiectasia Heterozygotes, a Cancer Prone Population." They described a laboratory methodology for the identification of a series of obligate heterozygotes for AT based on the sensitivity of lymphoblastoid cell lines to ionizing radiation. In their study, differential colony survival following gamma irradiation occurred with respect to the normal, obligate heterozygote and AT homozygote lymphoblastoid cells with the heterozygote having an intermediate response. These researchers concluded therefore that cell survival via cloning in agar offers a means of distinguishing AT heterozygotes from controls and homozygotes. They noted that the assay is somewhat slow and tedious but can be markedly speeded via the measurement of viability of the irradiated lymphoblastoid cells in a suspension by trypan blue exclusion.

The proposed identification method of Chen et al. (1978) is a valuable starting point in the attempt to identify heterozygotes. Certainly, the epidemiologic study of Swift et al. (1976) would have benefited to a considerable degree had there been a methodology for heterozygote validation rather than just the assumption that a parent of a homozygote is a heterozygote. While this is a very reasonable assumption, it is possible that the assumed parent was not the real biological parent, or that the child may have experienced a mutation for the trait independent of one of the parents. At this stage, we must recognize the very limited population sample used by Chen et al. (1978) and the rather widespread range of values present within each presumed genetic group. As Chen et al. (1978) pointed out, extensive family studies are needed for further validation since various complementation groups exist and all heterozygotes may not fit into the same pattern.

REFERENCES

Ammann, A. J. and Hong, R. (1971). Autoimmune phenomena in ataxia telangiectasia. *J. Pediatr.* **78:** 821.

Arlett, C. F. (1977). Lethal response to DNA damaging agents in a variety of human fibroblast cell strains. *Mutat. Res.* **46:** 106.

Chen, P. C., Lavin, M. F., Kidson, M. F., and Moss, D. (1978). Identification of ataxia telangiectasia heterozygotes, a cancer prone population. *Nature* **274:** 484–486.

Cleaver, J. E. (1980). DNA damage, repair systems and human hypersensitive diseases. *J. Environ. Pathol. Toxicol.* **3:** 53–68.

Cunliffe, P. N., Mann, J. R., Cameron, A. H., Roberts, K. D., and Ward, H. W. C. (1975). Radiosensitivity in ataxia-telangiectasia. *Br. J. Radiol.* **48:** 374–376.

Friedman, J. M., Fialkow, P. J., Davis, S. D., Ochs, H. D., and Wedgwood, R. J. (1977). Autoimmunity in the relatives of patients with immunodeficiency diseases. *Clin. Exp. Immunol.* **28:** 375–388.

Hoar, D. I. and Sargent, P. (1976). Chemical mutagen hypersensitivity in ataxia telangiectasia. *Nature* **261:** 590–592.

Kraemer, K. H. (1977). Progressive degenerative diseases associated with defective DNA repair: Xeroderma pigmentosum and ataxia telangiectasia. In: *Cellular Senescence and Somatic Cell Genetics: DNA Repair Processes.* W. W. Nichols and D. G. Murphy (eds.). Symposia Specialists, Miami, pp. 37–71.

Lavin, M. F., Chen, P. C., and Kidsen, C. (1978). Ataxia telangiectasia characterization of heterozygotes. *J. Supramol. Struct. (Suppl.)* **2:** 75.

Lehmann, A. R. and Stevens, S. (1977). The production and repair of double strand breaks in cells from normal humans and from patients with ataxia telangiectasia. *Biochem. Biophys. Acta* **474:** 49–60.

Levin, S. and Perlov, S. (1971). Ataxia-telangiectasia in Israel with observations on its relationship to malignant disease. *Isr. J. Med. Sci.* **12:** 1535–1541.

Lynch, H. T., Lynch, J., and Lynch, P. (1977). Management and control of familial cancer. In: *Genetics of Human Cancer.* J. J. Mulvihill, R. W. Miller, and J. F. Fraumeni, Jr. (eds.). Raven, New York, pp. 235–256.

Paterson, M. C. (1979). Environmental carcinogenesis and imperfect repair of damaged DNA in *Homo sapiens:* Causal relation revealed by rare hereditary disorders. In: *Carcinogens: Identification and Mechanisms of Action.* A. C. Griffin and C. R. Shaw (eds.). Raven, New York, pp. 251–276.

Paterson, M. C., Anderson, A. K., Smith, B. P., and Smith, P. J. (1979). Enhanced radiosensitivity of cultured fibroblasts from ataxia telangiectasia heterozygotes manifested by defective colony-forming ability and reduced DNA repair replication after hypoxic γ-irradiation. *Cancer Res.* **39:** 3725–3734.

Paterson, M. C., Lohman, P. H. M., and Sluyter, M. L. (1973). Use of a UV endonuclease from *Micrococcus luteus* to monitor the progress of DNA repair in UV-irradiated human cells. *Mutat. Res.* **19:** 245–256.

Paterson, M. C., Smith, B. P., Lohman, P. H. M., Anderson, A. K., and Fishman, L. (1976). Defective excision repair of gamma-ray damaged DNA in human (ataxia telangiectasia) fibroblasts. *Nature* **260:** 444–447.

Paterson, M. C., Smith, B. P., Knight, P. A., and Anderson, A. K. (1977). Ataxia telangiectasia: An inherited human disease involving radiosensitivity, malignancy and defective DNA repair. In: *Research in Photobiology*. Plenum, New York, pp. 207–218.

Paterson, M. C., Smith, B. P., Smith, P. J., and Anderson, A. K. (1978). Radioresponse of fibroblasts from ataxia telangiectasia heterozygotes. *Radiat. Res. (Abs.)* **74:** 83–84.

Purtillo, D., Paquin, L., and Gindhart, T. (1978). Genetics of neoplasia-impact of ecogenetics on oncogenesis. *Am. J. Pathol.* **91**(3): 609–688.

Ribon, A. and Perera, S. (1975). Immunodeficiency with ataxia telangiectasia. *Ann. Allergy* **35:** 104–108.

Scudiero, D. A. (1978). Repair-deficiency in M-methyl-N'-nitro-N-nitrosoguanidine treated ataxia telangiectasia (AT) fibroblasts. *J. Supramol. Struct. (Suppl.)* **2:** 83.

Sedgwick, R. P. and Boder, E. (1972). Ataxia-telangiectasia. In: *Handbook of Clinical Neurology. Vol. 14*. P. J. Vinken and G. W. Bruyn (eds.). North-Holland, Amsterdam, pp. 267–339.

Swift, M., Shulman, L., Perry, M., and Chase, C. (1976). Malignant neoplasms in the families of patients with ataxia-telangiectasia. *Cancer Res.* **36:** 209–215.

Taylor, A. M. R., Harnden, D. G., Arlett, C. F., Harcourt, S. A., Lehmann, A. R., Stevens, S., and Bridges, B. A. (1975). Ataxia telangiectasia: A human mutation with abnormal radiation sensitivity. *Nature* **258:** 427–429.

Vincent, R. A., Jr., Sheridan, R. B., III, and Huang, P. C. (1975). DNA strand breakage repair in ataxia telangiectasia fibroblast-like cells. *Mutat. Res.* **33:** 357–366.

C. Fanconi's Anemia

Fanconi's Anemia (FA) is a complex syndrome that displays hematological disturbances involving cell components of the bone marrow, diverse anatomic malformations, cutaneous lesions, and growth retardation. As is the case with XP and AT, those with FA usually die at a young age. Those dying in childhood usually succumb to problems of bone marrow failure resulting in massive infections or excess bleeding, while those surviving to adulthood have considerably greater risks of developing a wide variety of cancers including acute leukemia, squamous cell carcinoma or mucocutaneous junctions surrounding the oral and anal cavities, and hepatic adenoma (Mulvihill, 1975).

As a result of the enhanced risk of persons with FA to developing cancer, it has been hypothesized that such cells may exhibit an enhanced sensitivity to chemical and physical carcinogenic agents. In general, FA

Fanconi's Anemia

cells are very sensitive to bifunctional alkylating agents (e.g., MMC, nitrogen mustard), as well as psoralen-plus-black light but only slightly more sensitive than normal cells to far ultraviolet or gamma radiation, 4-NQO, and MMS (Paterson, 1979).

A reason for the obvious differential susceptibility of FA cells to known chemical and/or physical carcinogens has been proposed. The substances to which FA cells are hypersensitive are known to induce interstrand DNA cross-links. FA strains are thought to be defective in initiating the repair of these lesions. Experimental evidence that DNA repair defects exist has been reported by Poon et al. (1974). More specifically, damage caused by ultraviolet irradiation remains unrepaired due to a malformed exonuclease. In addition, Remsen and Cerutti (1976) reported that DNA damage following gamma irradiation was inefficiently repaired presumably due to a diminished ability to excise thymine glycols.

Most of the research on FA has focused on the clinical disorders of the homozygote patients. The frequency of the homozygote in the population is thought to be only 3 per 1,000,000. However, the heterozygote carrier is estimated at 1 per 300. A striking report published in 1971 by Swift suggested that close relatives of FA patients are at increased risk (i.e., about threefold) of malignancy and this was thought to reflect the risk of the heterozygote carriers. Based on these initial findings, it was estimated that the FA heterozygote may comprise 1% of all persons dying from cancer (Swift, 1971). However, a recent study by Swift's groups has subsequently not supported the previous striking findings (Caldwell et al., 1979). Nonetheless, limited efforts have been made to assess the potentially enhanced susceptibility of heterozygotes to known mutagenic agents as well as to develop a screening technique for rapid and accurate identification of FA heterozygotes. In studies evaluating the effects of five mutagens on FA homozygotes, FA heterozygotes and normal cells via the micronuclear methods, the results indicated that the normals and heterozygotes did not respond in a statistically significantly different manner (Heddle et al., 1978). As for the homozygote cells, they were 10 and 1.5 times more sensitive to MMC and radiation, respectively. These findings are clearly too preliminary to offer any generalizations about the sensitivity of FA heterozygotes to chemical and/or physical mutagens relative to controls. Future tests should employ a wide range of DNA damage estimates for a wide diversity of substances including those tested in the initial study.

Reports have indicated that several reliable screening methods exist for the designation of the FA heterozygote. The initial report was published in 1978 by Auerbach and Wolman who were able to distinguish FA heterozygous cell strains from normals in chromosomal breakage studies after exposing the cells to the dysfunctional alkylating mutagenic agent diepoxybutane (DEB). These authors noted that several earlier studies (Latt et al., 1975; Schuler et al., 1969) failed to distinguish FA heterozygous lymphocyte cells following exposure to alkylating agents as measured by increased chromosomal aberrations or increased sister chromatid exchange (SCE). Auerbach and Wolman (1978) hypothesized that their system of using fibroblasts may provide a more sensitive way of detecting susceptibility to induced chromosomal breakage. More recently, Auerbach et al. (1979, 1981) have developed successful prenatal and postnatal diagnoses of persons with FA. They reported a cytogenetic procedure involving the determination of chromosome breakage following DEB treatment for rapid diagnosis of FA homozygotes in blood lymphocytes and in amniotic fluid cells and of FA heterozygotes in blood lymphocytes. They indicated that their DEB system offers an advantage in that the characteristic lesion (complex chromatid exchange figure) produced can be identified following an assessment of only a few cells, thereby eliminating the need for extensive chromosome breakage analysis.

Of major significance is that the ability to identify the FA heterozygote may make it possible to evaluate whether FA heterozygotes truly are at greater risk of developing cancer. Earlier reports, both positive and negative, of Swift and colleagues, were forced to presume the genetic status of their heterozygote participants, thereby introducing some uncertainty in their conclusions. Finally, a recent study by Cohen et al. (1982) has challenged the findings of Auerbach and colleagues being unable to discriminate between heterozygote FAs and normal controls using DEB induced chromosomal damage at any concentration. They concluded on a pessimistic note by stating that "the detection of individual heterozygotes does not seem practical, at this point, by the use of the putative discriminant clastogens." Why the results of Auerbach and Cohen may differ is not known, but it may be found in the recognition that FA is a highly heterogeneous disease and may actually comprise several genetic disorders with similar manifestations.

REFERENCES

Auerbach, A. D., Adler, B., and Chaganti, R. S. K. (1981). Prenatal and postnatal diagnosis and carrier detection of Fanconi Anemia by a cytogenetic method. *Pediatrics* **67**(1): 128–135.

Auerbach, A. D., Warburton, D., Bloom, A. O., and Chaganti, R. S. K. (1979). Prenatal detection of the Fanconi Anemia gene by cytogenetic methods. *Am. J. Hum. Genet.* **31**: 77–81.

Auerbach, A. D. and Wolman, S. R. (1976). Susceptibility of Fanconi's anaemia fibroblasts to chromosome damage by carcinogens. *Nature* **261**: 484.

Auerbach, A. D. and Wolman, S. R. (1978). Carcinogen-induced chromosome breakage in Fanconi's anaemia heterozygous cells. *Nature* **271**: 69.

Caldwell, R., Chase, C., and Swift, M. (1979). Cancer in Fanconi anemia families. *Am. J. Hum. Genet.* **31**: 132A.

Cohen, M. M., Simpson, S. J., Honig, G. R., Maurer, H. S., Nicklas, J. W., and Martin, A. O. (1982). The identification of Fanconi Anemia genotypes by clastogenic stress. *Am. J. Hum. Genet.* **34:** 794–810.

Heddle, S. A., Lue, C. B., Saunders, E. F., and Benz, R. D. (1978). Sensitivity to five mutagens in Fanconi's Anemia as measured by the micronucleus method. *Cancer Res.* **38:** 2983–2988.

Latt, S. A., Stetten, G., Juergens, L. A., Buchanan, G. R., and Gerald, P. S. (1975). Induction by alkylating agents of sister chromatid exchanges and chromatid breaks in Fanconi's anemia. *Proc. Natl. Acad. Sci. U.S.A.* **72:** 4066–4070.

Mulvihill, J. J. (1975). Congenital and genetic diseases. In: *Persons at High Risk of Cancer.* J. F. Fraumeni, Jr. (ed.). Academic, New York, pp. 3–37.

Paterson, M. C. (1979). Environmental carcinogenesis and imperfect repair of damaged DNA in *Homo sapiens:* Causal relation revealed by rare hereditary disorders. In: *Carcinogens: Identification and Mechanisms of Action.* A. C. Griffin and C. R. Shaw (eds.). Raven, New York, pp. 251–276.

Poon, P. K., O'Brien, R. L., and Parker, J. W. (1974). Defective DNA repair in Fanconi's anaemia. *Nature* **250:** 223–225.

Remsen, J. F. and Cerutti, P. A. (1976). Deficiency of gamma-ray excision repair in skin fibroblasts from patients with Fanconi's anemia. *Proc. Natl. Acad. Sci. U.S.A.* **73:** 2419–2423.

Schuler, D., Kiss, A., and Fabian, F. (1969). *Humangenetik* **7:** 314–332.

Swift, M. (1971). Fanconi's anaemia in the genetics of neoplasia. *Nature* **230:** 103–107.

D. Bloom's Syndrome and Other Disorders

1. BLOOM'S SYNDROME

Bloom's syndrome is an autosomal recessive disease clinically characterized by decreased growth, enhanced sensitivity to sunlight, immune deficiency, and enhanced likelihood of developing cancer, especially acute leukemia. Unlike the repair diseases such as XP, fibroblasts from Bloom's syndrome patients perform excision repairs of ultraviolet damage normally (Cleaver, 1970). In fact, Bloom's syndrome is not considered a DNA repair disease but a "chromosome breakage syndrome" since lymphocytes and bone marrow cells from such patients display unusually higher frequencies of chromosome breaks and/or SCE (German et al., 1974; Chaganti et al., 1974; Shiraishi et al., 1976). While these enhanced frequencies of nuclear irregularities occur spontaneously in homozygous patients, several studies have revealed that individuals heterozygous for Bloom's syndrome display normal frequencies of chromosomal aberrations (Kuhn and Therman, 1979) and SCEs (Chaganti et al., 1974).

The cells of persons with Bloom's syndrome are considerably more hypersensitive to certain mutagens than normal cells. These cells have displayed enhanced sensitivity to ethyl methanesulphonate (EMS) and MMS as determined via SCE assays. In contrast, the response to MMC did not differ from controls (Krepinsky et al., 1979, 1979a). In addition, no overall enhanced sensitivity of Bloom's syndrome cells was shown with respect to exposure to 254-nm ultraviolet radiation although 1 of 3 Bloom syndrome fibroblast strains studied was quite sensitive (Krepinsky et al., 1980). More recent studies by Krepinsky et al. (1982) reported a striking hypersensitivity of these cells to ethylating agents N-ethyl-N-nitrosourea and N-ethyl-N'-nitro-N-nitrosoquanidine at doses that caused little or no adverse effects on the normal controls. Based on these and their previous findings, Krepinsky et al. (1982) concluded that cells with Bloom's syndrome have a general hypersensitivity to ethylating agents.

2. OTHER POSSIBLE REPAIR-DEFICIENT DISORDERS

There are only three known DNA repair diseases consisting of XP, AT, and FA. However, according to Paterson (1979), several clinical syndromes exist in which repair defects are suspected. These include Bloom's syndrome (see above), Cockayne's syndrome, Rothmund-Thomson syndrome, Werner's syndrome, basal cell nevus syndrome, D-deletion-type retinoblastoma, and progeria. He stated that while the first four diseases are inherited via an autosomal recessive mode, the next two are inherited as autosomal dominants with there being insufficient information to determine the "genetic etiology" of progeria. The first five diseases listed share a common enhanced sensitivity to solar radiation. Little is collectively known about the possible enhanced sensitivity of these disease syndromes to widespread environmental mutagens and carcinogens (see Table 11-1). This is not unexpected since even present medical understanding of the diseases themselves is quite limited.

REFERENCES

Ahmed, F. E. and Setlow, R. B. (1978). Excision repair in ataxia telangiectasia, Fanconi's anemia, Cockayne syndrome, and Bloom's syndrome after treatment with ultraviolet radiation and N-acetyl aminofluorene. *Biochim. Biophys. Acta* **521**: 805–817.

Chaganti, R. S. K., Schonberg, S., and German, J. (1974). A many-fold increase in sister chromatid exchanges in Bloom's syndrome lymphocytes. *Proc. Natl. Acad. Sci. U.S.A.* **71**: 4508–4512.

Cleaver, J. E. (1970). DNA damage and repair in light-sensitive human skin disease. *J. Invest. Dermatol.* **54**: 181–195.

Cleaver, J. E. (1980). DNA damage, repair systems and human hypersensitiveness. *J. Environ. Pathol. Toxicol.* **3**: 53–68.

German, J., Crippa, L. P., and Bloom, D. (1974). Bloom's syndrome. III. Analysis of the chromosome aberration characteristic of this disorder. *Chromosoma* **48**: 361–366.

Krepinsky, A. B., Heddle, J. A., and German, J. (1979). Sensitivity of Bloom's syndrome lymphocytes to ethyl methanesulfonate. *Hum. Genet.* **50**: 151–156.

Krepinsky, A. B., Heddle, J. A., Rainbow, A. J., and Kwok, E. (1979a). Sensitivity of Bloom's syndrome cells to specific mutagens. *Environ. Mutagen* **1:** 188–189.

Krepinsky, A. B., Rainbow, A. J., and Heddle, J. A. (1980). Studies on the ultraviolet light sensitivity of Bloom's syndrome fibroblasts. *Mutat. Res.* **69:** 357–368.

Krepinsky, A. B., Gingerich, J., and Heddle, J. (1982). Further evidence for the hypersensitivity of Bloom's syndrome cells to ethylating agents. In: *Progress in Mutation Research, Vol. 3.* K. C. Bora, G. R. Douglas, and E. R. Nestmann (eds.). Elsevier Biomedical, New York, pp. 175–178.

Kuhn, E. M. and Therman, E. (1979). No increased chromosome breakage in three Bloom's syndrome heterozygotes. *J. Med. Genet.* **16:** 219–222.

Paterson, M. C. (1979). Environmental carcinogenesis and imperfect repair of damaged DNA in *Homo sapiens:* Causal relation revealed by rare hereditary disorders. In: *Carcinogens: Identification and Mechanisms of Action.* A. C. Griffin and C. R. Shaw (eds.). Raven, New York, pp. 251–276.

Remsen, J. F. (1980). Repair of damage by N-acetoxy-2-acetylaminofluorene in Bloom's syndrome. *Mutat. Res.* **72:** 151–154.

Shiraishi, Y., Freeman, A. I., and Sandberg, A. A. (1976). Increased sister chromatid exchange in bone marrow and blood cells from Bloom's syndrome. *Cytogenet. Cell Genet.* **17:** 162–173.

E. Down's Syndrome

Numerous investigations of persons with Down's syndrome have revealed that they experience a greater susceptibility to develop leukemia at all ages than normal people (Schunk and Lehman, 1954; Merrit and Harris, 1956; Krivit and Good, 1957; Stewart et al., 1958; Wald et al., 1961; Holland et al., 1962; Warkany et al., 1963; Miller, 1963, 1968; Miller and Fraumeni, 1968; Jackson et al., 1968; Turner, 1969; Fraumeni and Miller, 1967; Miller, 1968). These findings led O'Brien et al. (1971) to evaluate the possibility that individuals with Down's syndrome may be at enhanced risk to developing pollutant-induced chromosomal abnormalities. They decided to utilize the environmental carcinogen 7,12-dimethylbenz(a)anthracene (DMBA) since it is a potent leukemogen in animals.

This study involved an evaluation of the effects of DMBA on the chromosomes of seven patients with Down's syndrome, five with trisomy 21, and two with translocation trisomies aged 20–29 and 10 healthy volunteers aged 17–39 years. The authors reported that significantly greater numbers of DMBA-induced chromosomal abnormalities occurred in lymphocytes of Down's syndrome patients as compared to those of the controls. The DMBA-induced increase in chromosomal aberrations was dose dependent and consisted principally of chromatid gaps (Figure 11-1). These findings were consistent with earlier research (Todaro et al., 1966; Todaro and Martin, 1967; Aaronson and Todaro, 1968) that cells of Down's syndrome patients are more susceptible to tumor-virus transfor-

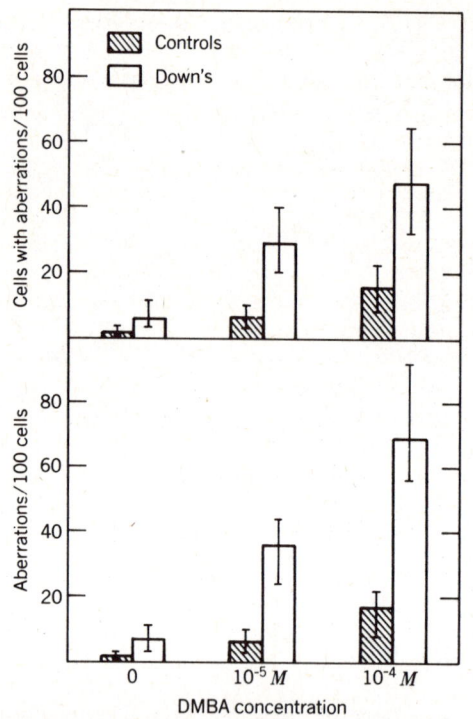

Figure 11-1. DMBA-induced aberrations in chromosomes from patients with Down's syndrome and normal controls. The vertical lines represent the entire range of values found for the ten controls and seven patients with Down's syndrome. *Source:* O'Brien et al. (1971).

mation and to ionizing radiation induced chromosomal damage (Sasaki and Tonomura, 1969).

In an effort to determine the demonstrated susceptibility to DMBA-induced chromosomal abnormalities, O'Brien et al. (1971) assessed the DNA repair capacity of Down's syndrome patients. They found that the lymphocytes of patients with Down's syndrome were not deficient in their ability to repair DNA damage caused by DMBA, ultraviolet light, or 4-NQO. Further research is needed to clarify the biochemical basis of the demonstrated enhanced susceptibility of Down's syndrome patients to environmental mutagens and/or carcinogens.

FREQUENCY IN THE POPULATION

Community surveys of Caucasians of European origin have revealed a frequency for Down's syndrome between 1 in 636 and slightly less than 1 in 1000 births (Penrose, 1961). The actual risk is highly dependent on maternal age where the incidence among women under 30 is about 1 in 2000 births, with the risk increasing significantly thereafter (i.e., aged

30–34—about 1 in 1000 births; aged 35–39—1 in 300 births; aged 40–44—1 in 100 births; and after age 45—1 in 50 births). Based on studies by Kashgarians and Rendtorff (1969) and Penrose (1961), there appears to be a comparable Down's syndrome risk among the races.

REFERENCES

Aaronson, S. A. and Todaro, G. J. (1968). SV40 T antigen induction and transformation in human fibroblast cell strains. *Virology* **36:** 254–261.

Fraumeni, J. E. and Miller, W. W. (1967). Epidemiology of human leukemia: Recent observations. *J. Natl. Cancer Inst.* **38:** 593–605.

Holland, W. W., Doll, R., and Carter, C. O. (1962). The mortality from leukemia and other cancers among patients with Down's syndrome (mongols) and among their parents. *Br. J. Cancer* **16:** 177–186.

Jackson, E. W., Turner, J. H., Klauber, M. R., and Norris, F. D. (1968). Down's syndrome: Variation of leukemia occurrence in institutionalized populations. *J. Chron. Dis.* **21:** 247–253.

Kashgarians, M. and Rendtorff, R. C. (1969). Incidence of Down's syndrome in American Negroes. *J. Pediatr.* **74:** 468.

Krivit, W. and Good, R. A. (1957). Simultaneous occurrence of mongolism and leukemia. *J. Dis. Child.* **94:** 289–293.

Merrit, D. H. and Harris, J. W. (1956). Mongolism and acute leukemia. *J. Dis. Child.* **92:** 41–44.

Miller, R. W. (1963). Down's syndrome (mongolism), other congenital malformations and cancers among the sibs of leukemic children. *N. Engl. J. Med.* **268:** 393–401.

Miller, R. W. (1968). Relation between cancer and congenital defects: An epidemiologic evaluation. *J. Natl. Cancer Inst.* **40:** 1079–1085.

Miller, R. W. and Fraumeni, J. D. (1968). Down's syndrome and neonatal leukaemia. *Lancet* **2:** 404.

O'Brien, R. L., Poon, P., Kline, E., and Parker, J. W. (1971). Susceptibility of chromosomes from patients with Down's syndrome to 7,12-dimethylbenz(a)anthracene induced aberrations *in vitro*. *Int. J. Cancer* **8:** 202–210.

Penrose, L. S. (1961). Mongolism. *Br. Med. Bull.* **17**(5): 184.

Sasaki, M. S. and Tonomura, A. (1969). Chromosomal radiosensitivity in Down's syndrome. *Jpn. J. Hum. Genet.* **14:** 81–92.

Schunk, G. J. and Lehman, W. L. (1954). Mongolism and congenital leukemia. *J. Am. Med. Assoc.* **155:** 250–251.

Stewart, A., Webb, J., and Hewitt, D. (1958). A survey of childhood malignancies. *Br. Med. J.* **1:** 1495–1508.

Todaro, G. J., Green, H., and Swift, M. R. (1966). Susceptibility of human diploid fibroblast strains to transformation by SV40 virus. *Science* **153:** 1252–1254.

Todaro, G. J. and Martin, G. M. (1967). Increased susceptibility of Down's syndrome fibroblast strains to transformation by SV40. *Proc. Soc. Exp. Biol.* **124:** 1232–1236.

Turner, J. H. (1962). Thesis. Sc. D. Hyg., University of Pittsburgh. In: *Epidemiology of Mongolism*. Cited by A. M. Lilienfeld. Johns Hopkins Press, Baltimore (1969).

Wald, N., Borges, W. H., Li, C. C., Turner, J. H., and Harnois, M. C. (1961). Leukaemia associated with mongolism. *Lancet* **1:** 1228.

Warkany, J., Schubert, W. K., and Thompson, J. N. (1963). Chromosome analysis in mongolism (Langdon-Down syndrome) associated with leukemia. *N. Engl. J. Med.* **268:** 1–4.

12 Other Genetic Factors

A. Susceptibility to Environmentally Induced Goiter

One of the most common high school biology demonstrations involving genetics is surveying the class for their ability to taste the compound phenylthiourea (PTC). This capacity to taste PTC is inherited as an autosomal dominant trait.

As far back as 1942, Richter and Clisby reported that PTC administration can induce goiter formation in rats. It is interesting to note that the same bimodality in the ability to taste PTC is similarly seen with respect to substances chemically related to PTC via the presence of an N—C=S group in common as is seen in several antithyroid drugs including methyl- and propylthiouracil. It has long been known that nontasters have a higher than expected (by about 10%) incidence of nodular goiter (Harris and Trotter, 1949; Kitchin et al., 1959). Nontasters are also thought to be more susceptible to athyreotic cretinism (Fraser, 1961; Shepard and Gartler, 1960), and also to adenomatous goiter. However, tasters develop toxic diffuse goiter more often than nontasters (Vesell, 1973).

It is interesting to note that the chemical group S—C=N which is goitrogenic, is found in various foods such as turnip, cabbage, brussels sprouts, kale, and rape (Greer, 1957; Clements and Wishart, 1956).

The frequency of the nontasters in the

population has been reported to be 31.5% of Europeans (Saldanha and Becak, 1959), 10.6% of Chinese, and 2.7% of Africans (Barnicot, 1950).

The significance of these findings remains somewhat obscure since the relationships established in earlier studies are associational and not causal. In addition, Price-Evans et al. (1962) could not find any differences between tasters and nontasters to metabolize methylthiouracil and thiopentose. Omenn and Motulsky (1978) raised the tentative speculation that the differential frequency of goiter may not be due to a difference in inherent susceptibility but to the possibility that nontasters in the population currently studied eat more goitrogenic substances. In any case, the resolution to this issue is not available in the current literature. Given the large numbers of persons with the nontasting phenotype, this area is worth pursuing.

REFERENCES

Barnicot, N. A. (1950). Taste deficiency for phenylthiourea in African Negroes and Chinese. *Ann. Eugen.* **15:** 248–254.

Clements, F. W. and Wishart, J. W. (1956). A thyroid-blocking agent in the etiology of endemic goiter. *Metab. Clin. Exp.* **5:** 623–629.

Fraser, G. R. (1961). Cretinism and taste sensitivity to PTC. *Lancet* **1:** 964–965.

Greer, M. A. (1957). Goitrogenic substances in food. *Am. J. Clin. Nutrit.* **5:** 440–444.

Harris, H. and Trotter, W. H. (1949). Taste sensitivity to PTC in goitre and diabetes. *Lancet* **2:** 1038–1039.

Kitchin, F. D., Howel-Evans, W., Clarke, C. A., McConnell, R. B., and Sheppard, P. M. (1959). PTC taste response and thyroid disease. *Br. Med. J.* **1:** 1067–1074.

Omenn, G. S. and Motulsky, A. G. (1978). "Eco-genetics": Genetic variation in susceptibility to environmental agents. In: *Genetic Issues in Public Health and Medicine.* Thomas, Springfield, Illinois, pp. 83–111.

Price-Evans, D. A. (1962). *Pharmogenetique. Med. Hyg.* **20:** 905–909.

Richter, C. P. and Clisby, K. H. (1942). Toxic effects of the bitter-tasting phenylthiocarbamide. *Arch. Pathol.* **33:** 46–57.

Saldanha, P. H. and Becak, W. (1959). Taste thresholds for phenylthiourea among Ashkenazic Jews. *Science* **129:** 150–151.

Shepard, T. H. and Gartler, S. M. (1960). Increased incidence of nontasters of PTC among congenital athyreotic cretins. *Science* **131:** 929.

Vesell, E. S. (1973). Advances in pharmacogenetics. *Prog. Med. Genet.* **9:** 291–367.

B. Susceptibility to Frostbite

According to Abe et al. (1970) there is considerable variation to susceptibility to frostbite within the population. They noted that frostbite is thought to be a result of the exaggeration of the normal arteriole contraction and capillary dilatation following exposure to low temperature (see Yoshimura and Ihda, 1951, 1952; Kellerman, 1955). Several researchers have suggested a genetic basis for the differential susceptibility to frostbite. For example, Harris (1947) has concluded that susceptibility to chilblains is genetically transmitted as an irregular dominant based on familial studies. Furthermore, Curtius and Feiereis (1960) examined 97 monozygotic and 38 dizygotic twin pairs in which one or both members exhibited lability of the autonomic nervous system and found a rate of agreement of 75.3% in monozygotic and 23.8% in dizygotic twin pairs with respect to a predisposition to frostbite. Subsequent studies by Abe et al. (1970) specifically evaluated the susceptibility to frostbite in about 800 3-year-old children in Japan via parental interviews and mailed questionnaires to grandparents. The incidence of frostbite was highest if both parents displayed symptoms during childhood and was lowest if both parents had never been affected. While the findings were statistically significant, it must be emphasized that there was no clinical diagnosis of frostbite and the reliability of their recall-interview methodology was never assessed. In addition, no mention was made of variables such as the relationship of frostbite to socioeconomic status, how extensively the children were dressed, etc. The authors did mention that resistance to frostbite may be increased by the intake of large amounts of protein and sodium chloride (up to 45 g/day), a factor also not considered in the questionnaire. Given the uncertainties mentioned above, the hypothesis that susceptibility to frostbite is genetically determined remains to be proven.

REFERENCES

Abe, K., Amatomi, M., and Kajyama, S. (1970). Genetic and developmental aspects of susceptibility to motion sickness and frost-bite. *Hum. Hered.* **20:** 507–516.

Curtius, F. and Feiereis, H. (1960). Zwillingsuntersuchungen über die Erbveranlagung zum vegetativ-endokrinea Syndrom der Frau. *Z. Kreislaufforsch.* **49:** 44.

Harris, H. (1947). Genetical factors in perniosis. *Ann. Eugen.* **14:** 32.

Kellerman, G. M. (1955). Observations on the critical temperature for vasomotor reaction in the fingers of chilblain subjects. *Aust. J. Exp. Biol. Med. Sci.* **33:** 215.

Yoshimura, H. and Ihda, T. (1950). Studies on the reactivity of skin vessels to extreme cold. I. A point test on the resistance against frost-bite. *Jpn. J. Physiol.* **1:** 647.

Yoshimura, H. and Ihda, T. (1951/52). Studies on the reactivity of skin vessels to extreme cold. II. Factors governing the individual difference of the reactivity or the resistance against frost-bite. *Jpn. J. Physiol.* **2:** 177.

Yoshimura, H. and Ihda, T. (1951/52). Studies on the reactivity of skin vessels to extreme cold. III. Effects of diets on the reactivity of skin vessels to cold. *Jpn. J. Physiol.* **2:** 310.

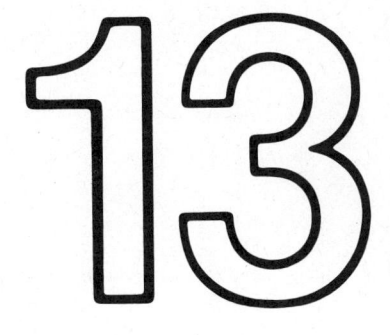

13 Summary and Conclusions

This book evaluated the general issue of whether susceptibility to environmentally related diseases may be enhanced by one's genetic constitution and to what extent any such genetic factors may be identified (as in the form of biochemical markers) in ways that are applicable to widespread screening within the general population or work force.

The findings of this study have revealed that there are excellent theoretical foundations to support the hypothesis that genetic conditions are likely to enhance the risk of affected individuals to developing environmentally induced adverse health effects. However, the extent to which supporting data are available from studies with appropriate predictive animal models and/or human epidemiologic studies is highly variable.

Of the approximately 50 human genetic conditions identified (Table 13-1) as potentially enhancing one's susceptibility, only G-6-PD deficiency has a demonstrable causal history of enhancing one's susceptibility to industrial pollutants. Several clinical investigations have suggested that β-thalassemia may enhance risk to benzene and lead toxicity while no occupational studies have reported on the effects of industrial hematological toxins on persons with sickle-cell trait and NADH reductase deficiency. Even in the case of G-6-PD deficiency there is a near total lack of precise dose-response relationships since any adverse effects have

Table 13-1. Identification and Quantification of Genetic Factors Affecting Susceptibility to Environmental Agents

High-Risk Groups	Estimated Occurrence	Environmental Agent(s) to which the Group is (May be) at Increased Risk
	Red Blood Cell Conditions	
G-6-PD deficiency	American Black males, 16% Mediterranean Jewish males, 11% Greeks, 1–2% Sardinians, 1–8%	Environmental oxidants such as ozone, nitrogen dioxide, and chlorite
Sickle-cell trait	7–13% of American Blacks are heterozygotes	Aromatic amino and nitro compounds; carbon monoxide; cyanide
The thalassemias	β, 4–5% in Americans of Italian and Greek descent α, 2–7% American Blacks; 2–3% American Greeks	Benzene; lead
NADH dehydrogenase deficiency (MetHb reductase deficiency)	Estimated 1% of the population are heterozygotes	MetHb-forming substances
Catalase		
Hypocatalasemia	About 2% of U.S. population based on the Swiss gene frequency	Ozone; radiation
Acatalasemia	1 in 10,000–20,000 of the U.S. population based on Swiss gene frequency	
Low SOD activity	Frequency of genetic variants in population one to 2/10,000; normal population exhibits a unimodal distribution; persons at low end of the distribution may be at increased risk	A wide variety of environmental oxidants; paraquat; radiation; ozone
ALA dehydratase deficiency	Unknown, but thought to be rare	Lead
Hb M	Unknown, but rare	Carbon monoxide

Erythrocyte porphyria	1.5/100,000 in Sweden, Denmark, Ireland, West Australia; 3/1000 in South African Whites; rare in Blacks	Chloroquine; hexachlorobenzene; lead; various drugs, including barbiturates, sulfonamides, others
GSH-Px deficiency	Rare	Environmental oxidants
GSH reductase deficiency	Rare	Environmental oxidants

Liver Metabolism

Defect in glucuronidation Gilbert's syndrome	6% of the normal, healthy adult population	Wide variety of xenobiotics including PCBs
Crigler-Najjar syndrome	Few individuals live to adulthood	
Defect in sulfation	Unknown	Wide variety of xenobiotics; best association is with tyramine containing foods
Acetylation phenotype, slow vs. fast	Slow, 50% Caucasian; 50% Blacks; 10% Japanese	Aromatic amine induced bladder cancer; numerous drugs, e.g., isoniazid and peripheral neuropathies
	Fast, 50% Caucasian; 50% Blacks; 90% Japanese	Numerous drugs, e.g., isoniazid and hepatitis
Gout	0.27–0.3% prevalence in the United States and Europe	Lead
Oxidation center defects	9% of British Caucasians; 8% of Nigerians; 6% Ghanians; 1% Saudi and Egyptians are poor oxidizers	Numerous xenobiotics requiring oxidative metabolism for detoxification
OCT deficiency	Unknown, but thought to be rare	The insect repellant DET
Paraxonase variant	25–30% of population	Parathion
Rhodanese variant	Unknown	Cyanide
Sulfite oxidase deficiency heterozygotes	Unknown	Sulfite, bisulfite, sulfur dioxide
Inadequate carbon disulfide metabolism	Upward of 30–40%	Carbon disulfide

Table 13-1. (Continued)

High-Risk Groups	Estimated Occurrence	Environmental Agent(s) to which the Group is (May Be) at Increased Risk
Alcohol dehydrogenase variant	5% of English; 20% Swiss; 70% Japanese	Metabolize (e.g., ethanol) more quickly than normal
Wilson's disease	Homozygous <1/100,000 while the heterozygote may approach 1/500	Copper, vanadium
Serum Variants		
Albumin variants	Less than 1/1000 in Europeans, much higher frequency in North American and Mexican Indians	Unknown
Pseudocholinesterase variants	Highly sensitive homozygous and heterozygous individuals of European ancestry have a combined frequency of about 1/1250; moderately sensitive genotypic variants of European ancestry have a frequency of 1/15,000	Organophosphate and carbamate insecticides; muscle relaxant drugs
Ig A deficiency	1/400–600	Respiratory irritants
Serum α-antitrypsin deficiency—heterozygote	4–9% of Northern Europeans	Respiratory irritants; smoking
Cardiovascular Conditions		
Hyperlipidemia	5% exhibit polygenic hypercholesterolemia 1–2% in population	Carbon disulfide, carbon monoxide, smoking
Homocystinuria heterozygotes		Carbon disulfide, carbon monoxide, smoking
Respiratory Conditions		
High aryl hydrocarbon hydroxylase inducibility	Approximately 9%	Polycyclic aromatic hydrocarbons

CF heterozygotes	4% of Caucasians	Respiratory irritants

Renal Diseases

Renal nephropathy of diabetes	5% of adult population	Heavy metals; organic solvents
Cystinosis	Most common among Caucasians of European ancestry; unknown frequency	Heavy metals
Cystinuria	1/200–250 individuals, although asymptomatic are heterozygote carriers; 1/20,000–100,000 are homozygous	Heavy metals
Tyrosinemia	Frequency in general population is unknown; frequency of heterozygous carriers in the Chicoutimi region of northern Quebec is a carrier for every 20–31 people	Heavy metals

Immunological Disorders

Immunodeficiency diseases hypoagammaglobulinemia Ig A deficiency (see serum variants)	Very rare	Viruses; bacteria
Immunologic hypersensitivity	2% of some worker populations	Isocyanates

Dermatological Conditions

Albinism		
Tyrosinase positive	1/14,000 Blacks; 1/60,000 Caucasians; very high frequency in American Indians	Ultraviolet radiation
Tyrosinase negative	1/34,000–36,000 Blacks and Caucasians; 1/10,000–15,000 in Ireland	Ultraviolet radiation

PKU	1/80 in U.S. is a carrier; 1/25,000 has the disease	Ultraviolet light
Ocular Diseases		
Glaucoma	5% at high risk; 30% at moderate risk	Glucocorticoid drugs
Leber's optic atrophy	Thought to be rare	Cyanide
Thyroid Disorders		
Nontasters of PTC	30% Europeans; 10% Chinese; 3% African	Goitrogenic substances
DNA Repair and Chromosome Instability Diseases		
XP	1/250,000 homozygous recessive 2–4/1000 heterozygote carriers Higher incidence in North American Arabs	Wide range of mutagens and/or carcinogens
AT	2.5/100,000 homozygous recessive 1% heterozygote carriers	
FA	1/300 heterozygote carrier	
Bloom's syndrome	3/1,000,000 homozygous recessive Unknown frequency	
Down's syndrome	1/636–1/1000 births	

Summary and Conclusions

never adequately addressed the levels of pollutant exposures associated with the hemolytic occurrence. In addition, the role of potential confounding variables such as nutritional and overall health status has never truly been assessed. Thus, the published policies and recommendations to prevent persons with G-6-PD deficiency, sickle-cell trait, and the thalassemias from occupations involving exposure to blood toxins are not based on adequate scientific investigations. One of the greatest historical failings has been the lack of predictive animal models for the hereditary blood disorders. Nevertheless, continued research on the enhanced susceptibility associated with these hereditary blood disorders is of considerable social importance since it affects large numbers of various ethnic subgroups (e.g., 16% of Black males are G-6-PD deficient, 8% of blacks have the sickle-cell trait, and 4% of Italians and Greeks are carriers of β-thalassemia).

There is considerable evidence to suggest that the 4% of the general population who are carriers of the SAT deficiency are at enhanced risk (2 times normal) to developing pollutant-induced emphysema. However, the vast majority of carriers of this condition do not develop emphysema. This is because there are other important genetic variables affecting such susceptibility. Screening for SAT heterozygotes should be combined with other potential genetic risk factors such as high protease levels if it is to identify truly those at high risk to developing emphysema. Research in this area should focus on risk factors in addition to SAT so that it may be possible to have a much greater predictability than now available.

Genetic susceptibility to PAH-induced lung cancer has an excellent foundation based on impressive animal model studies along with limited and controversial human epidemiological studies. The hypothesis that persons with a high ability to induce the synthesis of AHH are at increased risk of developing PAH-induced lung cancer, while of enormous public health importance, remains insufficiently conclusive but should continue to be a high priority for research funding.

The hypothesis that slow acetylators are at increased risk to arylamine-induced bladder cancer has excellent theoretical foundations in animal model studies. Epidemiological evidence, while meager, is consistent with the animal data. Since 50% of the Caucasian and Black populations are slow acetylators, it is of considerable social importance to further evaluate this hypothesis. As in the case with emphysema, identification and quantification of other genetic factors such as deacetylation capability need to be incorporated in subsequent studies. Consideration of only one factor (i.e., acetylator phenotype) is not likely to explain a considerable amount of the variance within a population.

DNA repair diseases represent an extremely important class of human genetic diseases worthy of extraordinary government attention. While the data base in all instances remains very limited, sufficient evidence exists to suggest that one or more of these diseases may be a contributing factor affecting susceptibility to environmentally induced carcinogenesis. However, whether and to what extent these diseases are contributory to environmental illness is a scientific and medical issue at present. Insufficient data exist for any environmental and/or occupational regulatory policy concerning this subgroup of the population.

Nonallergic contact dermatitis is a major problem in occupational health. Those with predispositions to such adverse effects are thought to be individuals with preexisting dermatological diseases. However, adequate epidemiologic studies remain to be conducted to quantify human risk. In addition, the possibility of using predictive testing methodologies in preemployment physicals is hampered by a lack of current methodologies to incorporate typical workplace exposure situations. Allergic contact dermatitis has a much lower prevalence than its nonallergic counterpart. It also suffers from current testing procedures not adequately reflecting workplace exposures.

HLA markers have been found to be associated with a wide variety of human diseases. Some of the associations are so strong as to be of considerable predictive value when assessing risk to specific disease entities. The use of HLA markers in predicting risk to occupationally related diseases remains to be more fully assessed.

Numerous additional genetic variants exist that are thought to have potential with respect to affecting one's susceptibility to toxic substances. For the most part, these additional variants have not been evaluated sufficiently either in appropriate animal models or epidemiologically with respect to susceptibility to environmental and/or occupational diseases.

In conclusion, there is insufficient scientific justification to conduct genetic screening procedures for the purpose of job denial or transfer in order to avoid potential occupationally related diseases. Despite the fact that various industrial practices either in the United States or abroad have at times reported and/or recommended such actions with respect to G-6-PD deficiency, sickle-cell trait, thalassemia, and SAT heterozygotes, there is insufficient evidence at present to justify such programs, policies, and recommendations.

The identification of genetic factors that may predispose the occurrence of job-related disease is a science truly in its infancy. The biological foundations of the concept of genetic screening to identify predispositions to occupational disease are on sound biological grounds. While the future suggests a very widespread and influential role in regulatory processes, the present state of the art can only recommend more study and not the implementation of any job screening programs except as related to the acquisition of data in properly designed studies. Nevertheless, it appears that genetic differences contribute in important ways in explaining the variability of responses to environmental agents. What percentage of the total variability in disease incidence may be explained by genetic factors

Summary and Conclusions

is uncertain. It should be recognized that other biological variables such as age, nutritional status, preexisting diseases, and life-style also contribute in highly important ways in affecting the body's susceptibility to a variety of environmental insults. The study of factors affecting susceptibility to environmental and/or occupational diseases therefore should not stop with a quantification of genetic influences, as important as they may be, but should be comprehensive and incorporate the above-mentioned host variables.

Index

Acatalasemia, and catalase deficiency, 73–76. *See also* Catalase deficiency
Acetylation, 118–124
  arylamine-induced bladder cancer susceptibility, 118–123
  carcinogenic aromatic amine susceptibility, 119–123
  drug susceptibility, 119–120
  fast *vs.* slow, 118–119
  industrial pollutant susceptibility, 329
  screening tests, 123–124
Acetylator phenotype, 119–121
N-Acetyltransferase activity for arylamine substrates, 121–123
N-Acetyltransferase phenotype, 119–121
Agar diffusion test, 174
Air Force Academy, sickle-cell anemia policy, 40
Alcohol dehydrogenase, *see* Liver metabolism, alcohol dehydrogenase variants
Alcoholism, and liver metabolism, 130–136
Aldehyde dehydrogenase, *see* Liver metabolism, alcohol dehydrogenase variants
Alloxan, and diabetes, 237, 240–246
American Conference of Governmental Industrial Hygienists (ACGIH), 7, 11
Aminolevulinic acid:
  dehydratase deficiency, 105
  porphyria, 100–103
Animal models:
  atherosclerosis, 205
  carboxyhemoglobin formation, 106–107
  cardiovascular disease, 200–204
  catalase deficiency, 79–83
  diabetes:
    cardiovascular sequellae, 252–254
    glomerular basement membrane, 252–253
    mesangial thickening, 252–253
    nephropathy, 250–252
    nervous system problems, 254–255
  G-6-PD deficiency, 35–38
  hypertension, 204, 206
  infectious disease susceptibility:
    DNA virus, 267, 269
    RNA tumor virus, 266, 268
    RNA viruses, 266, 267
  isocyanate susceptibility, 276–277
  methemoglobin:
    aniline hydroxylation, 61–64
    causes of interspecies differences, 64–67
    comparative species sensitivity, 63
    differential gastrointestinal tract/bacterial metabolism, 64–66
    differential hydroxylation, 66
    4-dimethylaminophenol injections, 63–64
    formation, 59–70
    influence of conjugation rates, 66–67
    interspecies comparisons, 60–64
    MetHb reductase activity, 60–61, 67–70
    models of choice, 67–70
    substrate available as cause of interspecies differences, 65
  nonallergic contact dermatitis, 281–282
  polycyclic aromatic hydrocarbon-induced lung cancer, 215–218
  predictive selection, 84–89
  serum alpha$_1$ antitrypsin deficiency, 226–228
    emphysema, 226–230
  sickle-cell anemia, 44–47
  sulfite oxidase deficiency, 150–152
  sulfur dioxide toxicity, 148–149

Index

Animal models (*Continued*)
superoxide dismutase:
oxygen toxicity, 93
paraquat toxicity, 95-97
thalassemias, 54-56
Belgrade rat, 54-55
Arteriosclerosis, and homocystinuria, 196-199
Aryl hydrocarbon hydroxylase inducibility:
benzo(a)pyrene-DNA binding, 217-218
benzo(a)pyrene metabolism, 215-216
confounding variables, 216-220
desmethylimipramine metabolism, 217, 220
genetic factors, 210-211
3-methylcholanthrene induction, 210-212
mitogen-stimulated lymphocytes, 213-215
nutritional factors, 212-214, 217-218
polycyclic aromatic hydrocarbon-induced lung cancer, 209-216
research methodology, 210-216
Ataxia telangiectasia:
clinical features, 306-307
DNA repair levels, 307-308
population frequency, 308-309
radiation sensitivity, 306-308
screening tests, 308-309
Atherosclerosis:
animal models, 200-205
hyperlipidemia, 189-190

Benzoylcholine assay with dibucaine, 174
Blacks:
G-6-PD deficiency, 28
sickle-cell anemia, 40-41, 43
Bladder cancer:
acetylation, 120-123
smoking, 164-165
Bloom's syndrome, clinical features, 313-314
Breast milk, PCBs in, 112-113

Cancer:
as immunodeficiency disease, 274-276
smoking-related, 2-3
Carbamate insecticides, and pseudocholinesterase variants, 173-174
Carbon disulfide metabolism, 160-161

Carbon monoxide toxicity, and hemoglobin variants, 105-107
Carboxyhemoglobin, 105-107
Cardiovascular disease, animal models, 200-206. *See also* Atherosclerosis; Homocystinuria; Hyperlipidemia; Hypertension
Catalase deficiency, 73-76
animal models, 78-83
erythrocyte catalase activity, 78-83
GSH-Px levels in, 75-76
irradiation treatment, 81-83
population frequency, 75-76
radiation sensitivity, 74
role in cellular detoxification, 74-75
screening tests, 76
Cats:
comparative species sensitivity to methemoglobin, 63
glucorinidation in, 110-113
Cellular immune dysfunction, 270-271
Cellulose-acetate electrophoresis (CAE), 42-43
Ceruloplasmin levels, and Wilson's disease, 163
Chickens, and atherosclerosis, 202-204
Chloramphenicol toxicity, 112-113
Chlorates, and G-6-PD deficiency, 25-26
Chlorine, and G-6-PD deficiency, 25-26
Chlorites, and G-6-PD deficiency, 25-26
Chromium, allergic contact dermatitis, 283-285
Chromosome 6, human leukocyte-antigen associations, 263-264
Chromosome instability diseases, xeroderma pigmentosa, 301-305. *See also* specific diseases
Common cold, waterborne agents, 272
Copper deficiency, in Wilson's disease, 162-163
Cotinine-nicotine-1'-*N*-oxide ratios, 164-165
Crigler-Najjar syndrome, 113-115
Cyanide toxicity, rhodanese variants, 105
Cyanosis, 57
Cystic fibrosis:
adult mucoviscidosis, 223
screening tests, 223
Cystine levels, in Wilson's disease, 162-163
Cystinosis:
heavy metal toxicity, 257-258

Index

screening tests, 257
Cystinuria:
  heavy metal toxicity, 256-257
  screening tests, 257

Dapsone, acetylation of, 123-124
Debrisoquine metabolism:
  alleles in, 126-128
  EM phenotype ratios, 126-128
  PM phenotype, 126-128
  in Saudi Arabia, 128-130
Deer, and sickle-cell anemia, 44-45
Diabetes:
  animal models:
    cardiovascular sequellae, 252-254
    glomerular basement membrane, 252-253
    mesangial thickening, 252-253
    nephropathy, 250-252
    nervous system problems, 254-255
  arsenic-alloxan-induced, 237-238, 240-246
  clinical features, 235-236
  environmental agents, 244-246
    carbon disulfide, 246
    nitrosamine, 245-246
    PNU rodenticide, 245-246
  fluoride, 237, 239
  hepatoxic effects of organic solvents, 243-244
  juvenile onset diabetes, 236, 239-241
  kidney structure, changes in, 252-253
  prevalence, 236-237
  radiation susceptibility, 239-240
  renal nephropathy, 235-247
  smoking, 240-241
  xenobiotic metabolism, 241-243
N,N-Diethyltoluamide (DET), and ornithine carbamoyl transferase deficiency, 159-160
Differential susceptibility, 1-5
  developmental processes, 3-4
  environmental agents, 2-3
  genetic factors, 5-9
  host factors, 5-6
  nutritional factors, 3-5
  screening tests, 330-331
Diisopropylfluorophosphonate, 173-174
Dithionite-phosphate solubility test, 42-43
DNA repair diseases:
  industrial pollutant susceptibility, 329-330
  possible undiscovered disorders, 314
  ultraviolet-induced skin cancer, 291-292
  see also specific diseases
DNA replication, and sulfur dioxide toxicity, 153-154
Dogs, Dalmatian, and hyperuricemia, 143-144
Down's syndrome:
  clinical features, 315-316
  7,12-dimethylbenz(a)anthracene, 315-316
  pollutant-induced chromosomal abnormalities, 315
  population frequency, 316-317
Drug toxicity:
  pharmacogenetics, 6-8
  racial differences in response, 6

Embden-Meyerhoff pathway, 13-15
  and red blood cell production, 15-16
Emphysema:
  genetic predisposition, 224-226
  serum alpha$_1$ antitrypsin deficiency, 224-226
  specific types, 228-229
Environmental Protection Agency, 69
Enzyme deficiencies:
  genetic disorders, 6-10
  hemolysis, 13-15
  see also specific deficiencies
Erythrocytes:
  catalase activity for, 78-83
  environmental oxidant effects on, 4, 84-89
  enzyme activity, interspecies comparison, 85-89
  G-6-PD activity, variation in animal models, 35-36
  sicklings, occurrence in animals, 45-47
Erythrocyte superoxide dismutase:
  normal ranges, 92
  ontogeny, 91-92
Europeans, alcohol dehydrogenase variants in, 135
Experimental Pharmacogenetics, 6

Fanconi's anemia:
  cell sensitivity, 311
  clinical signs, 310-311
  population frequency, 311-312
  screening tests, 311-312
Fanconi's syndrome:

Fanconi's syndrome (*Continued*)
  and environmental pollutants, 259-260
  and vitamin E, 258-259
Ferrihemoglobin, *see* Methemoglobin
Flex tailed anemia, 55
Fluorescent spot test, and G-6-PD
    deficiency, 29-30
Frostbite, genetic susceptibility to,
    320-321

Gastroenteritis, waterborne agents, 272
Genetic screening, 10-11
Genetics, and pollutant effects, 6-10,
    330-331
Gilbert's Syndrome, 113-115
Glaucoma, and glucocorticoids, 297-298
Glucuronidation, 109-115
  animal models for, 110-113
  chloramphenicol toxicity, 112-113
  Crigler-Najjar Syndrome, 113-115
  developmental immaturity, 110-113
  Gilbert's Syndrome, 113-115
  hyperbilirubinemia identification,
      114-115
  susceptibility, 110-113
Glucose-6-phosphate dehydrogenase:
  animal models, 35-38
  cancer susceptibility, 27-28
  defined, 17-18
  differential susceptibility to, 3-5
  environmental oxidants:
    chlorite and related agents, 24-27
    copper, 21
    fungicide, 27
    lead, 27
    ozone, 21-24
  erythrocyte sensitivity to chemicals, 20
  frequency in population, 29
  glutathione levels, 16-17, 21-22, 30-31
  hemolytic agents, 18-19
    drug and chemical, 19
    normal therapeutic dosage, 19
  industrial oxidants, 30-31
  industrial pollutant susceptibility, 6-7,
      18-21, 323
  nutritional status, 18
  ozone, in vitro experiments, 23-24
  screening tests, 29-30
Goiter, environmental-induced
    susceptibility, 319-320
Gout, 138-142
  classification of, 139
  hyperuricemia in, 138-139
  lead relationship, 140-141
  prevalence and incidence, 138-140
Granulocytopenia, 270

H_2O_2:
  catalase deficiency, 73-75
  induction of MetHb formation, 80
  in sickle-cell anemia, 41-42
Hegesh assay, for methemoglobinemia,
    58
Heme hypoxia, 57
Hemoglobin:
  background information, 16-17
  carboxyhemoglobin, 105-107
    production, 15-16
  sickle, 40-43
    in animal models, 44-47
    phagocytizing capacity, 45
  variants, 105-107
Hemolytic anemia, and G-6-PD
    deficiency, 27
Hemolytic susceptibility, and hereditary
    blood disorders, 13-15
Hepatitis A, water transmission, 272
Hepatitis B, and humoral immune
    dysfunction, 270
Hereditary blood disorders, hemolytic
    susceptibility, 13-15
Hexachlorobenzene, and porphyria,
    101-102
Hexose monophosphate (HMP) Shunt
    Pathway, 13-15
  red blood cell production, 16
  G-6-PD deficiency, 28
Homocystinuria, 196-199
  cystathionine synthase deficiency, 197
  environmental implications, 199
  genetic tendency, 196-197
  heterozygote identification, 197-199
  methionine metabolism, 198-199
  population frequency, 197
  vitamin B_6 deficiency, 197, 199
Human genetic conditions, and
    differential susceptibility, table of,
    323-328
Human leukocyte-antigen association,
    see Immunological associations
Humoral immune dysfunction, 266, 270
Hydroxylation, and liver metabolism,
    125-128
Hyperbilirubinemia, and glucorinidation,
    115
Hyperglycemia, and diabetes, 250, 252

Index

Hyperlipidemia:
  atherosclerosis, 189-190
  cholesterol, 190-191
    HMG CoA reductase suppression, 190-191
    low density lipoprotein, 191-194
  environmental agents, 192-193
  genetic predisposition, 193-194
  hypercholesterolemia genetic predisposition, 191-192
  LDL-receptor defect, 191-194
Hypertension, 204, 206
Hyperuricemia:
  animal models, 142-145
  classification, 139-140
  gout, 138-139
  lead relationship to, 140-141
  urate levels, 142-145
Hypocatalasemia, see Catalase deficiency
Hypochlorite, 25-26

Immunodeficiency diseases:
  ankylosing spondylitis and HLA-B27, 264-265
  cancer as, 274-276
  human leukocyte-antigen markers, 330
  naturally occurring (NOID), 275-276
  thymus-dependent senescence, 275
Immunological associations:
  human leukocyte-antigen associations, 263-265
    HLA-B8, 264-265
    HLA-B27 and ankylosing spondylitis, 264-265
  immunodeficiency diseases, 274-276
  infectious disease susceptibility, 265-266
    animal models, 266-269
    cellular immune dysfunction, 270-271
    DNA virus, 266, 269
    fungal infections, 271
    granulocytopenia, 270
    humoral immune dysfunction, 266, 270
    RNA tumor virus, 266, 268
    RNA virus, 266-267
    water, agents in, 271-273
    waste, agents in, 271-273
  isocyanate susceptibility, 276-277
Immunosuppressive agents, immunodeficiency disease, 274-276

Industrial toxicology, 6-7
Intraocular pressure, see Glaucoma
Iodide-azide test, for carbon disulfide metabolism, 161

Japanese, alcohol dehydrogenase variants in, 135-136
Jews, and G-6-PD deficiency, 27-28

Lead toxicity, and ALA dehydratase deficiency, 105
Leber's optic atrophy, cyanide toxicity, 299-300
Liver metabolism:
  alcohol dehydrogenase variants, 130-136
    blood alcohol concentrations, 135
    drug metabolism, 133-134
    gene frequencies at ADH_2 locus, 135-136
    genetic variants for alcohol metabolism, 130-136
    interracial studies, 131-132
    screening tests, 136
    twin studies, 132-133
  carbon disulfide metabolism, 160-161
    tetraethylthiuram disulfide, 160-161
  conjugation mechanism defects, 109-125
  diabetes, 241-244
  ornithine carbamoyl transferase deficiency, 159-160
  oxidation center data, 125-130
    interspecies differences, 125-128
    toxicological implications, 128-129
    genetic differences in debrisoquine metabolism, 126-128
  paraoxonase variants, 158-159
  smoking-induced bladder cancer, 164-165
  sulfite oxidase deficiency, see Sulfite oxidase deficiency
  Wilson's disease, see Wilson's disease
  see also Acetylation; Glucuronidation; Gout; Sulfation
Lung cancer, see Polycyclic aromatic hydrocarbon-induced lung cancer
Lupus erythematosus, 292
Lymphocytes, and bisulfite toxicity, 152-153
Lymphomas, as immunodeficiency disease, 274-276

Index

Malaria, and thalassemia, 55–56
Mean corpuscular volume (MCV) screening test, 51–52
Metabolic Basis of Inherited Disease, 6–7
Mental retardation:
  and methemoglobinemia, 57
  and sulfite oxidase deficiency, 149
Methemoglobin:
  animal models, 59–70
    aniline hydroxylation, 61–64
    causes of interspecies differences, 64–67
    comparative species sensitivity, 63
    differential gastrointestinal tract/bacterial metabolism, 65–66
    differential hydroxylation, 66
    4-dimethylaminophenol injections, 63–64
    influence of conjugation rates, 66–67
    interspecies comparisons, 60–64
    models of choice, 67–70
    spontaneous MetHb reductase activity, 60–61, 67–70
  chlorite formation, 26–27
  formation in erythrocytes, 79
  interspecies comparisons:
    substrate availability, 65
    causes, 64–67
  oxidant stressor agents, 59–60
  reductase deficiency, 56–58
  reduction to hemoglobin, 15
Methemoglobinemia:
  causes, 57–58
  population frequency, 58
  screening tests, 58
Migraine headache, tyramine-induced, 117–118
Molybdenum, sulfite oxidase deficiency, 150–152
Monoamine oxidase metabolism, 117–118
Mouse:
  catalase deficiency, 82
  G-6-PD deficiency, 36
  microcytic anemia, 55
  model for methemoglobin, 67–70

NADH:
  methemoglobin cofactor, 60
  reductase deficiency, industrial pollutant susceptibility, 323
  regeneration, 15
NADPH:
  dehydrogenase deficiency, 56–58
  methemoglobin cofactor, 60

  regeneration, 15
NADPH/NADP ratio, G-6-PD deficiency, 28
National Football League, sickle-cell anemia, 40
National Institute for Occupational Safety and Health, 7
Neonatal jaundice, and PCBs, 113–115
Nickel, allergic contact dermatitis, 283–285
Nicotine metabolism, 164–165
Nonallergenic contact dermatitis, industrial pollutant susceptibility, 330
Nonglycolytic Pathway, 13–16
Nutrition and Environmental Health, 5

Ocular disorders:
  genetic susceptibility, 297–300
  glaucoma, 297–298
  Leber's optic atrophy, 299–300
  see also Glaucoma
Office of Technology Assessment (OTA), 11
Ouchterlony's double diffusion test, and immunoglobulin A deficiency, 187
Oxonic acid, and hyperuricemia, 143–144
Ozone intermediates, in G-6-PD deficiency, 23–24

Paper spot test, 174
Paraoxonase variants, biotransformation of, 158–159
Pentose phosphate pathway, and red blood cell production, 16
Pharmacogenetics, history of, 2–7
Phenols, toxicity, 110–111
Phenylketonuria, and ultraviolet-induced skin cancer, 291–292
Phenylthiourea tasting, as autosomal trait, 319–320
Photoallergic drug reactions, 292–293
Poliomyelitis, paralytic, and humoral immune dysfunction, 270
Pollutants and High Risk Groups, 4–5
Polychlorinated biphenyls:
  Crigler-Najjar Syndrome, 113–115
  Gilbert's Syndrome, 113–115
  infants, 111–115
  liver metabolism, 109–113
Polycyclic aromatic hydrocarbon-induced lung cancer:
  antipyrine saliva distribution, 215–216
  aryl hydrocarbon hydroxylase

Index

inducibility, 209-216
benzo(a)pyrene metabolism, 210-212
confounding variables, 216-220
cytochrome P-450, 215-216
diol epoxide binding to DNA, 310
genetic factors, 214-216
research methodology, 215-217
Porphyria, 99-103
  erythropoietic porphyria, 100
  hepatic porphyria, 100-103
    acute intermittent, 100, 102-103
    porphobilinogen (PBG), 100, 103
    variegated porphyria, 102-103
  hexachlorobenzene, 101-102
  lead toxicity, 102
  major types, 100-101
  pollutant interactions, 101-102
  population frequency, 102
  protoporphyria, 101
  screening tests, 102-103
  urinary PBG levels, 103
Porphyrin ring, 99-100
Predictive contact allergy patch testing, and allergic contact dermatitis, 284-285
Premature infants, and chloramphenicol toxicity, 112-113
Psoriasis vulgaris, and HLA associations, 264

Rabbits, and atherosclerosis, 201-204
Rat models:
  G-6-PD deficiency, 37-38
  sickle-cell anemia, 45
Red blood cells, normal functioning, 15-16. *See also* specific blood conditions
Renal disorders, genetic diseases, 256-260. *See also* specific diseases
Renal lithiasis, 138
Respiratory diseases, *see* specific diseases
Rhodanses variants, 105
Rickets:
  cystinosis, 257
  Fanconi's syndrome, 259-260

Serum alpha$_1$ antitrypsin deficiency
  animal models, 226-230
  antitrypsin activity, 227-228
  Cadmium levels, 225-226
  chronic obstructive pulmonary disease, 176-182
  emphysema, 176-182, 224-226
  population frequencies, 225-226
  animal models for, 226-230
  ethnic distribution, 176-177
  industrial pollutant susceptibility, 329
  leukocyte lysosomal protease-antiprotease imbalance, 225-226
  leukocyte chemotactic factors, 179-181
  MZ phenotype, 176-179, 181
  ozone levels, 181
  Pi$_2$ genotype, 176-179
  PIMZ type, 176-179
  protease levels, 178-179, 181-183
  screening tests, double ring antibody precipitation pattern, 182
  smoking, 177-182
Serum variants, 167-169
  albumin variants:
    binding capabilities, 170-171
    bisalbuminemia, 168-169
    campothecitin binding, 168
    polymorphic variants, 169
    population frequencies, 170-171
    toxicological implications, 169-171
  immunoglobulin A deficiency, 185-189
    population frequencies, 185-186
    screening tests, 185-189
    upper respiratory tract infections, 185-187
  protein binding of drugs, 167-169
    interspecies differences, 167-168
  pseudocholinesterase variants, 171-175
    acetylcholine, 172
    acetylcholinesterase, 172
    dibucaine variant, 173
    gene frequencies, 173
    screen of drug reactions, 174
    suxamethonium, 172
  see also Serum alpha$_1$ antitrypsin deficiency
Sheep, and G-6-PD deficiency, 36-37
Sickle-cell anemia, 13-15
  animal models for, 44-47
  environmental pollutant susceptibility, 41
  GSH-Px levels, 41-42
  health hazards of, 40-41, 43
  industrial pollutant susceptibility, 323
  oxidant stress, 41-42
  population frequency, 42
  screening tests, 42-43
  sickle-cell hemoglobin, 39-40
  sickle erythrocytes, 41-42
Skin disorders:
  allergic contact dermatitis, 283-285

Skin disorders (*Continued*)
  industrial pollutant susceptibility, 283–285
    screening tests, 284–285
  nonallergic contact dermatitis, 279–283
    animal models, 281–282
    genetic factors, 280–281
    industrial pollutant susceptibility, 279–283
    research methodologies, 281
    sexual factor, 280
  ultraviolet-induced noncarcinogenic skin irritation, 292–293
  ultraviolet-induced skin cancer, 286–292
  see also Ultraviolet-induced skin cancer
Smoking:
  bladder cancer, 164–165
  diabetes, 240–241
  serum alpha₁ antitrypsin deficiency, 177–182
Solubility testing, for sickle-cell anemia, 42–43
Stokinger, Herbert, 7
Sulfation, tyramine-induced migraine headache, 117–118
Sulfite oxidase deficiency, 153–155
  bisulfite toxicity, 146–149, 151–156
  hypothesis testing, 154–156
  in mammalian tissues, 147–149
  mental retardation, 149
  molybdenum and, 150–152
  mutagenicity, 153–155
  sulfite toxicity, 146–149, 151–152, 154–156
  sulfur dioxide toxicity, 146–156
Superoxide dismutase:
  disease levels, 92
  genetic variants, 92
  normal ranges, 92
  ontogeny of erythrocyte SOD activity, 91–92
  oxygen toxicity, 93
  ozone toxicity, 93–94
  paraquat toxicity, 94–97
  radiation toxicity, 94
  in red blood cells, 91–97
Superoxide radicals, superoxide dismutase erythrocytes, 93–97

Tetraethylpyrophosphonate insecticide, 173–174
Tetraethylthiuram disulfide (TETD) metabolism, 160–161
alpha-Thalassemia, 49–52
  animal models, 55–56
  screening tests, 51–52
beta-Thalassemia, 49–52, 323
  animal models, 55–56
  screening tests, 50–51
Thalassemias, 13–15, 48–52
  animal models for, 54–56
  heterozygote health status, 49
  homozygote health status, 49
  pollutant effects, 49–51
    antioxidant enzyme levels, 50–51
    benzene, 49–50, 52
    environmental oxidant risk, 50–51
    lead, 50, 52
  population frequency, 51
  screening tests, 51–52
TNT, and G-6-PD deficiency, 20–21
Transplants, and immunodeficiency disease, 274
Tyrosinemia, 258

Ultraviolet radiation:
  benefits, 286–287
  carcinogenic activity, 287–288
  ozone layer reduction, 288–289
Ultraviolet-induced skin cancer:
  DNA repair processes, 291
  genetic susceptibility, 286–292
    incidence rates, 287–288
    high-risk groups, 289–292
Uric acid, and hyperuricemia, 142–145
Uricase, and hyperuricemia, 142–145
Urine tests, acetylation of dapsone, 123–124

Vanadium compounds, and Wilson's disease, 163
Vitamin D:
  cystinosis, 257
  Fanconi's syndrome, 259–260
Vitamin E, and superoxide dismutase, 96–97

Water:
  enteric viruses, 272
  infectious agents in, 271–273
  national maximum drinking water standard, 69–70
Watson, James, 11
Wilson's disease, 162–163

Index

screening for, 163
vanadium compounds, 162-163

Xanthine oxidase activity, 150-152
Xeroderma pigmentosum:
  cell hybridization, 303-305
  cell sensitivity, 302-305
  chemical carcinogen and mutagen susceptibility, 303-304
  classical, 301
  clinical features, 301
  DNA repair levels, 323-325
  neurological, 301
  ultraviolet-induced skin cancer, 291-293

INDUSTRIAL POLLUTION CONTROL—Volume I: Agro-Industries
   E. Joe Middlebrooks

BREEDING PLANTS RESISTANT TO INSECTS
   Fowden G. Maxwell and Peter Jennings, Editors

NEW TECHNOLOGY OF PEST CONTROL
   Carl B. Huffaker, Editor

THE SCIENCE OF 2,4,5-T AND ASSOCIATED PHENOXY HERBICIDES
   Rodney W. Bovey and Alvin L. Young

INDUSTRIAL LOCATION AND AIR QUALITY CONTROL: A Planning Approach
   Jean-Michel Guldmann and Daniel Shefer

PLANT DISEASE CONTROL: Resistance and Susceptibility
   Richard C. Staples and Gary H. Toenniessen, Editors

AQUATIC POLLUTION
   Edward A. Laws

MODELING WASTEWATER RENOVATION: Land Treatment
   I. K. Iskandar

AIR AND WATER POLLUTION CONTROL: A Benefit Cost Assessment
   A. Myrick Freeman, III

SYSTEMS ECOLOGY: An Introduction
   Howard T. Odum

INDOOR AIR POLLUTION: Characterization, Prediction, and Control
   Richard A. Wadden and Peter A. Scheff

INTRODUCTION TO INSECT PEST MANAGEMENT, Second Edition
   Robert L. Metcalf and William H. Luckman, Editors

WASTES IN THE OCEAN—Volume 1: Industrial and Sewage Wastes in the Ocean
   Iver W. Duedall, Bostwick H. Ketchum, P. Kilho Park, and Dana R. Kester, Editors

WASTES IN THE OCEAN—Volume 2: Dredged Material Disposal In the Ocean
   Dana R. Kester, Bostwick H. Ketchum, Iver W. Duedall and P. Kilho Park, Editors

WASTES IN THE OCEAN—Volume 3: Radioactive Wastes and the Ocean
   P. Kilho Park, Dana R. Kester, Iver W. Duedall, and Bostwick H. Ketchum, Editors

LEAD AND LEAD POISONING IN ANTIQUITY
   Jerome O. Nriagu